D1415498

# EXPERIMENTAL BIOCHEMISTRY

# A Series of Books in Chemistry

LINUS PAULING
*Editor*

General Chemistry (Second Edition)
*Linus Pauling*

A Laboratory Study of Chemical Principles (Second Edition)
*Harper W. Frantz*

College Chemistry (Third Edition)
*Linus Pauling*

College Chemistry in the Laboratory, No. 1
*Lloyd E. Malm and Harper W. Frantz*

College Chemistry in the Laboratory, No. 2
*Lloyd E. Malm and Harper W. Frantz*

Introductory Quantitative Chemistry
*Axel R. Olson, Charles W. Koch, and George C. Pimentel*

Chemical Thermodynamics: A Course of Study
*Frederick T. Wall*

Principles of Organic Chemistry (Second Edition)
*T. A. Geissman*

The Hydrogen Bond
*George C. Pimentel and Aubrey L. McClellan*

Essentials of Chemistry in the Laboratory
*Harper W. Frantz and Lloyd E. Malm*

Qualitative Elemental Analysis
*Ernest H. Swift and William P. Schaefer*

Fundamental Experiments for College Chemistry
*Harper W. Frantz and Lloyd E. Malm*

Selected Experiments in Organic Chemistry
*George K. Helmkamp and Harry W. Johnson, Jr.*

Experimental Biochemistry
*John M. Clark, Jr., Editor*

# EXPERIMENTAL BIOCHEMISTRY

*John M. Clark, Jr.*, EDITOR

BIOCHEMISTRY DIVISION, DEPARTMENT OF CHEMISTRY, UNIVERSITY OF ILLINOIS

W. H. FREEMAN AND COMPANY
SAN FRANCISCO AND LONDON

*Printed in the United States of America.* (*B2*)

# Preface

This manual, prepared by the staff of the Biochemistry Division at the University of Illinois, is an outgrowth of our view that experimental competence with biological systems and their components at a quantitative level is essential to the understanding and acquisition of new knowledge of the molecular and biological properties of living organisms. The content and emphasis reflect our conviction that experimental biochemistry must be considered in terms of the composition of living matter and its molecular organization at a level at which meaningful problems can be posed and solved. The experiments aim to acquaint the student with such problems and with the more important methods of biochemical research.

Because our courses attract students with varied backgrounds and with wide ranges of primary interests, ranging from X-ray crystallography of biologically derived macromolecules to developmental biology, the introductory sections of the manual emphasize the chemistry of the compounds found in living organisms. A wide variety of experiments has been included to meet the diverse needs of these students. Quantitative separation, characterization, and identification of molecules, both large and small, by chemical, physical, and tracer methods are evident throughout. Analytical procedures emphasize simplicity, speed, and sensitivity so far as these are compatible with accuracy of identification and quantitative measurement. The selection of sources for biological materials is intended both to cover the range of interests of the students and to provide what we feel to be the best experimental material for a given study, or that most widely used in current chemical and biological research.

For nearly ten years, this manual has been used in preliminary form—three mimeographed and two multilithed preprints—hence it is based not only upon earlier teaching experience but upon experience gained during this period of trial use, in the course of which various new laboratory techniques were developed by the staff and those obtained from available manuals and reference works tested empirically. Constant revision—stimulated by the criticisms of students and staff—has, we feel, brought under control numerous critical variables, clarified points of loose experimental definition, and eliminated awkwardness and ambiguity from the writing. In the interest of improving the teaching of biochemistry and making more fruitful the student's laboratory experience, we solicit suggestions and criticisms from students and instructors who use this manual.

We should point out that the manual will accommodate the teaching of an abbreviated course. Dr. J. W. Hastings has used it effectively in teaching *elementary* biochemistry to professional students in Veterinary Medicine and Dietetics. By selecting only those experiments he found suited to the needs of these students, he has adapted the manual to a course involving two laboratory sessions per week instead of three.

The earliest version of this manual was prepared by Dr. W. J. Rutter, Dr. Joe Larner, and myself. Later, many new experiments were added, others redesigned, and the manual assembled in its present form by Drs. Rutter, Geller, and Clark. Among those who contributed to the preparation of the manual but who are no longer with the staff are: Dr. Joe Larner, now a member of the Pharmacology Department, Western Reserve University

Medical School; Dr. David Geller, currently in the Pharmacology Department, Washington University, St. Louis; Dr. L. M. Henderson, now with the Biochemistry Department, University of Minnesota; Dr. E. S. Lennox, Salk Institute, La Jolla, California; Dr. Roberts A. Smith, Chemistry Department, University of California, Los Angeles; Dr. Frank Neuhaus, now of the Chemistry Department, Northwestern University, who contributed much imagination and significantly changed the order of presenting the experiments; and Drs. Robert Barker and Carl S. Vestling, Biochemistry Department, University of Iowa, whose ideas added clarity and experimental precision to the manual. Dr. Vestling, a longtime colleague, deserves special mention, particularly for his careful checking of the manuscript and for his supervision of laboratory instruction during the second semester of the academic year 1962–1963, when he broke in a new teaching staff.

Finally, members of the Division wish to acknowledge their particular indebtedness to John M. Clark, Jr., who, as editor of this manual, made the principal contribution to its organization and format. The responsibility for checking experiments, preparing lists of reagents and equipment, and seeing the manual through proof was his.

*December 30, 1963*

I. C. GUNSALUS, *Head*
*Biochemistry Division*
*Department of Chemistry*
*University of Illinois, Urbana*

# Contents

## PART THREE **PROTEINS AND AMINO ACIDS**



## PART FOUR **NUCLEIC ACIDS**



## PART FIVE **METABOLISM**

## APPENDIXES

# Biochemistry Laboratory Equipment List

Beakers
- 2 50-ml
- 2 100-ml
- 2 150-ml
- 2 250-ml
- 2 400-ml
- 1 800-ml
- 2 1000-ml

Bottles
- 3 500-ml, c.s.t.m.
- 1 16-oz, polyethylene
- 2 4-liter, wide-mouth screw-cap

1 Brush, test tube
1 Burette, 10-ml, and holder
2 Bunsen burners
1 Calcium chloride tube
Centrifuge Tubes
- 2 12-ml glass, graduated
- 4 15-ml, polyethylene
- 2 50-ml, polyethylene for Servall

1 Crucible, 15-ml, porcelain, with cover
1 Crucible tongs
1 Distilling head
Erlenmeyer Flasks
- 4 50-ml
- 4 125-ml
- 2 300-ml
- 2 500-ml
- 1 1000-ml

1 File, triangular
1 Filter flask, 500-ml
1 Filter flask, 1000-ml
Filter Paper
- 1 box, 5.5-cm diam.
- 1 box, 9-cm diam.
- 1 box, 18.5-cm diam.

1 Florence flask, 500-ml, R.B.
1 Funnel rack
Funnels
- 1 65-mm diam. Buechner
- 1 110-mm diam. Buechner
- 1 50 mm-diam. Hirsch
- 2 3-in. diam. glass
- 2 6-in. diam. glass

Graduated Cylinders
- 1 10-ml
- 1 25-ml
- 2 100-ml
- 1 250-ml
- 1 500-ml

Ironware
- 2 Adjustable screw clamps
- 2 Rings
- 2 Universal clamps
- 2 Holders
- 1 Pipestem triangle

Matches (2 boxes)
1 Melting point capillary (1 vial)
3 Microscope slides
1 Mortar and pestle, 110-mm
*p*H papers
- 1 red litmus
- 1 blue litmus

3 dispensers (*p*H paper) with ranges 1–2.5; 3–5.5; 6–8; 8–9.5; 10–12
1 Pipette bulb, 25-ml
Pipettes, graduated
- 4 0.1-ml
- 4 1.0-ml
- 2 2.0-ml
- 2 5.0-ml
- 2 10.0-ml

Pipettes, volumetric
- 2 1.0-ml
- 1 5.0-ml
- 1 10.0-ml
- 1 25.0-ml

1 Porcelain dish, 110-mm dia.
5 Rods, glass, 10-cm
Rubber Tubing
- 1 Pressure
- 3 Standard (6 + 1 mm—at least 3′)

1 Separatory funnel, 60-ml
1 Soap
1 Spatula
1 Sponge
1 Spot plate, porcelain
Test Tubes, Pyrex
- 4 15 × 125-mm
- 24 18 × 150-mm
- 4 25 × 200-mm
- 2 12 × 50-mm

1 Test tube holder
1 Test tube rack
1 Thermometer, 200°C
2 Tripods
Vials, screw-cap
- 5 15 × 45-mm
- 2 21 × 70-mm

Volumetric Flasks
- 4 25-ml
- 4 50-ml
- 1 100-ml

3 Watch glasses (2 100-mm; 1 50-mm)
2 Wire gauzes, asbestos centers

# To the Student

This course aims to interest you in the experimentation which forms an important part of modern biochemistry. This goal necessitates a careful emphasis on the design, control, and completion of good experiments. In particular, you will be expected to familiarize yourself with the theory and manipulative details of each experiment prior to the actual performance. You will be expected to consult the scientific literature of biochemistry frequently. Certain specific references follow each of the experiments. The basic textbooks listed below will provide much valuable reference material:

Fruton, J. S., and S. Simmonds, *General Biochemistry* (2nd ed.), Wiley, New York, 1958.

White, A., P. Handler, E. L. Smith, and D. Stetten, *Principles of Biochemistry* (2nd ed.), McGraw-Hill, New York, 1959.

Clark, W., *Topics in Physical Chemistry* (2nd ed.), Williams and Wilkins, Baltimore, 1952.

Conn, E. E., and P. K. Stumpf, *Outlines of Biochemistry*, Wiley, New York, 1963.

Karlson, P., *Introduction to Modern Biochemistry*, Academic, New York, 1963.

It is imperative that you maintain a complete research notebook listing *all* data and calculations as well as results and conclusions. A good notebook should contain complete data arranged so that *anyone* can tell quickly what was done and how. The following suggestions are pertinent:

1. Use a large-size, bound notebook, preferably one with gridded pages. These permit construction of graphs directly and allow you to attach records of primary accessory data, such as chromatograms.
2. Do *not* record your observations on separate sheets of paper; record all data directly in your notebook. A good scheme is to use one side of the open notebook for raw data and calculations, and the other side for results and interpretation.
3. All graphs and tables must be clear and unambiguous. Be particularly sure to give the units on the abscissa and the ordinate of all graphs.
4. The writeup of an experiment should include:

   (a) a brief statement of the purpose;
   (b) a brief account of the theory;
   (c) a *summary* of the principal manipulative steps;
   (d) the raw data;
   (e) complete calculations (in some instances a sample calculation for one of a series of measurements);
   (f) results;
   (g) conclusions.

One additional suggestion should be made. We highly recommend that students who do not own a set of molecular models purchase a set such as: Molecular Models, Catalog No. 71307, Central Scientific Co., 1700 Irving Park Road, Chicago, Illinois (Price, $6.00). The compelling argument in favor of such a purchase is that many matters which are in reality simple and systematic appear to be complex and arbitrary when studied without ready access to a simple set of molecular models.

# PART ONE **CARBOHYDRATES**

## Introduction

All living cells contain carbohydrates in various forms. These compounds are perhaps best defined as polyhydroxy aldehydes or ketones and their derivatives. The biochemist must, of course, consider the chemical reactivities associated with the functional groups (carbonyl, aliphatic hydroxyl) as the heart of the organic biochemistry of the carbohydrates. We refer you to standard textbooks in organic chemistry and biochemistry and to the biochemical literature for discussion of details. The following sections are intended to provide you with background material that should assist you in your study of the various experiments.

### Structure and Classification

The principal reference carbohydrates are the two trioses, D- and L-glyceraldehyde:

```
 H    O                 H    O
  \  //                  \  //
   C                      C
   |                      |
H—C—OH               HO—C—H
   |                      |
   CH2OH                  CH2OH
```

D-Glyceraldehyde          L-Glyceraldehyde

You will notice that the glyceraldehyde molecule possesses an asymmetric carbon atom. Thus two space modifications are possible, to which the above arbitrary names are assigned. The designations D- and L- refer *only* to the assigned configurations. The projected formulas given above are frequently said to be presented according to the E. Fischer convention—a very useful generalization. This convention places the aldehyde (or reducing) end of the molecule *up* and *behind* and the nonreducing end of the molecule *down* and *behind* as the projection is made. The H and OH groups then appear in *front*. When one form of glyceraldehyde (e.g., that defined as D-glyceraldehyde) is presented by "flattening" it into the plane of the projection, it will appear as shown. The L- form will also appear as shown. We suggest that you assemble these models and make the projections in order to see what a relatively simple matter this is. It will then be clear why these projections must be kept in the plane of projection. (That is, one may not lift a projection up out of the paper and view it from behind. If this is done, the projection will look like that of the other form.)

The point of reference for configuration in the carbohydrates is the next to last carbon. Thus all carbohydrates which can be considered to have been derived from D-glyceraldehyde by lengthening the carbon chain away from this penultimate carbon atom are D-sugars. Consult an appropriate textbook, and note the structural relationships among the D- and L- trioses, tetroses, pentoses, hexoses, and heptoses. You should be able to distinguish pairs of compounds which are *enantiomorphs* from those which are *diastereoisomers*. You should also have a clear understanding of the distinction between configuration (denoted by D- or L-) and the sign of optical rotation [properly designated as (+) or (−), or (dextro) or (levo)].

One of the most illuminating reactions of carbonyl compounds is that of aldehydes or ketones with alcohols:

$$\mathrm{R{-}C(=O){-}H} + \mathrm{HO{-}R'} \underset{}{\overset{\text{reversible equilibrium}}{\rightleftharpoons}} \underset{\text{Hemiacetal}}{\mathrm{R{-}CH(OH){-}OR'}} \xrightarrow[-H_2O]{\mathrm{HO{-}R''}} \underset{\text{Acetal}}{\mathrm{R{-}CH(OR''){-}OR'}}$$

Acid hydrolysis ($H_2O$)

The formation of hemiacetals (or hemiketals) is a readily reversible equilibrium in aqueous systems, hence hemiacetals are often referred to as "potential aldehydes." The conversion of hemiacetals to acetals requires dehydrating (forcing) conditions in the laboratory, and the resulting acetals are characterized by a susceptibility to acid hydrolysis and by stability in aqueous alkali.

Many of the properties of the simpler sugars can be best interpreted if it is agreed that they exist in solution and in the crystal primarily as cyclic hemiacetals. Glucose appears to exist in solution primarily as an equilibrium mixture of the α- and β-glucopyranoses, with much smaller amounts of the α- and β-glucofuranoses present along with traces of aldehydo-glucose:

β-D-Glucopyranose ⇌ D-Aldehydo-glucose ⇌ α-D-Glucopyranose

β-D-Glucofuranose ⇌ D-Aldehydo-glucose ⇌ α-D-Glucofuranose

Sir W. N. Haworth proposed to write *perspective* formulas as shown above in order to indicate the three-dimensional character of the sugars. Such Haworth formulas are drawn from the Fischer projection formulas according to the following rules:

1. Draw four or five consecutive carbon atoms in a clockwise manner, beginning with the aldehyde or keto carbon on the right.
2. Draw the hydroxyls appearing on the right side of the Fischer formulas so that they project downward from their respective carbons. Draw the hydroxyls on the left side so that they project upward.
3. Draw the hemiacetal or hemiketal form of sugar by making ring closure with the desired OH, so as to form a furanose (5-membered) or pyranose (6-membered) ring having the ring closure O in the upper right-hand corner of the molecule.
4. Draw (in the Fischer convention) any remaining carbon atom(s) attached to the carbon of the OH involved in ring closure in a direction opposite to that anticipated (rule 2 above) for the ring closure OH.
5. The positioning of the new asymmetric carbon created by hemiacetal or hemiketal formation is dictated by the D- or L- configuration of the sugar. For D-sugars, draw the OH of this anomeric carbon down for $\alpha$-anomers, up for $\beta$-anomers. For L-sugars, do the converse.

Sketching Haworth formulas of such sugars as $\alpha$-D-ribopyranose, $\beta$-D-fructofuranose, and $\alpha$-D-galactofuranose will illustrate the fine points of Haworth structures. It is useful to compare such sketches with the actual sugar models.

In considering the matter of the Haworth formulas, attention should be directed to the matter of the conformation of pyranoses and related compounds (see Reeves [1]). The Haworth representation is somewhat of an oversimplification, since one pictures the pyranose ring as being essentially flat. In all likelihood one has to deal with the appropriate "chair" or "boat" forms (refer to discussions of the cyclohexane ring). When a hexopyranose model is made, it becomes apparent that there can be interactions between (or among) groups which are *axial* in orientation, and that equatorially oriented groups should interact to a smaller extent. That is, the most stable conformations will be those with the larger number of equatorial bulky substituents (see bottom of page).

From the drawings below (and better from the models) we can see that the C1 chair conformation of $\alpha$-D-glucopyranose should be a much more stable arrangement than the 1C chair conformation. We can also deduce that $\beta$-D-glucopyranose would have zero axial OH groups and would be more stable than $\alpha$-D-glucopyranose. This checks with the known state of the equilibrium between $\alpha$- and $\beta$-D-glucopyranose, which at room temperature shows approximately $\frac{2}{3}\beta$- to $\frac{1}{3}\alpha$-. The phenomenon of mutarotation (change of optical rotation with time toward an equilibrium value, usually not zero) is best considered in this light.

Sugar acetals (and ketals), called glycosides, are formed by the reaction of hemi-

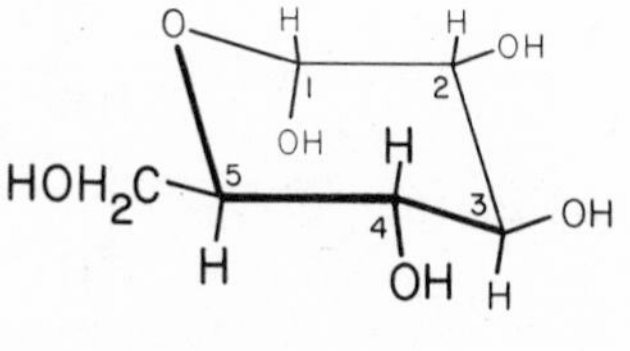

α-D-Glucopyranose
C1 Chair conformation (Reeves)
(1 axial OH group)

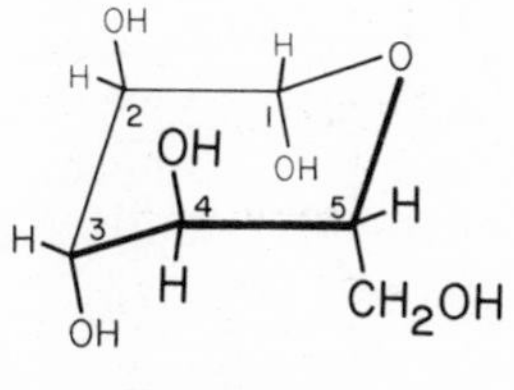

α-D-Glucopyranose
1C Chair conformation
(4 axial OH groups)

acetals (or hemiketals) with alcohols. Many such compounds exist in nature. When the hemiacetal is α-D-glucopyranose and the alcoholic functional group happens to be that on carbon 4 of α-D-glucopyranose, the resulting acetal is maltose, a disaccharide (written here as α-maltose): Oligosaccharides (formed from a "few" (3–8) monosaccharide units) and polysaccharides (formed from many such units) exist widely in nature. The monosaccharides are the sugar hemiacetals, which can be produced by acid (or enzymatic) hydrolysis of oligo- or polysaccharides.

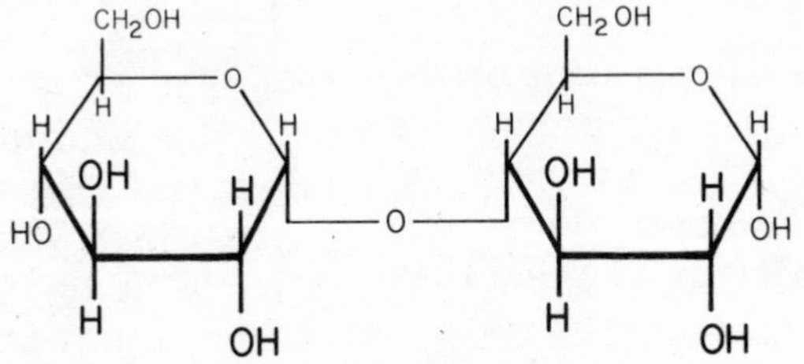

α-D-Maltose

(α-4-[α-D-Glucopyranosyl]-D-Glucopyranoside)

## Chemical Properties

*Action of Acid*

One of the most useful reactions in structural studies of carbohydrates is the acid-catalyzed hydrolysis of glycosidic bonds (acetals and ketals). For oligosaccharides or polysaccharides composed of neutral sugars all in the pyranose ring form, hydrolysis usually may be completed in 15 min to 1 hr with 1*N* HCl or $H_2SO_4$ at 100°C. Furanoside linkages are considerably more labile to acid, whereas the pyranoside linkages of acidic or basic sugar derivatives (uronic acids, amino sugars) are hydrolyzed more slowly than the corresponding pyranoside linkages formed from neutral sugars.

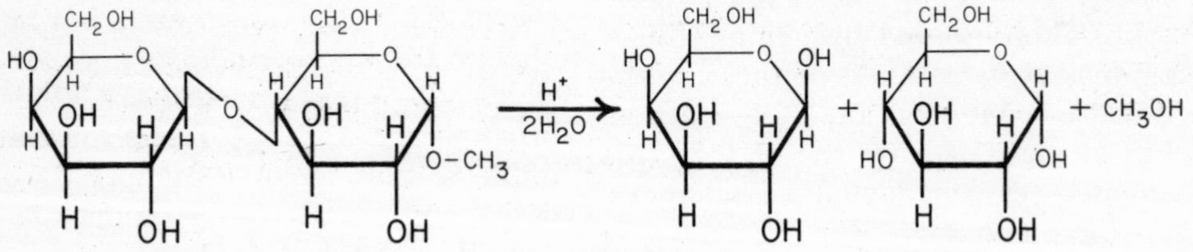

Concentrated acids (4–6*N* at 100°C; lower temperatures with more concentrated acids) cause the intramolecular dehydration of monosaccharides. Pentoses yield furfural, whereas hexoses yield first hydroxymethylfurfural, which is further degraded to levulinic acid on longer treatment:

$\xrightarrow[2H_2O]{H^+}$

α-Methyl-D-lactoside → D-Galactose (β-form written) + D-Glucose (α-form written) + Methanol ($CH_3OH$)

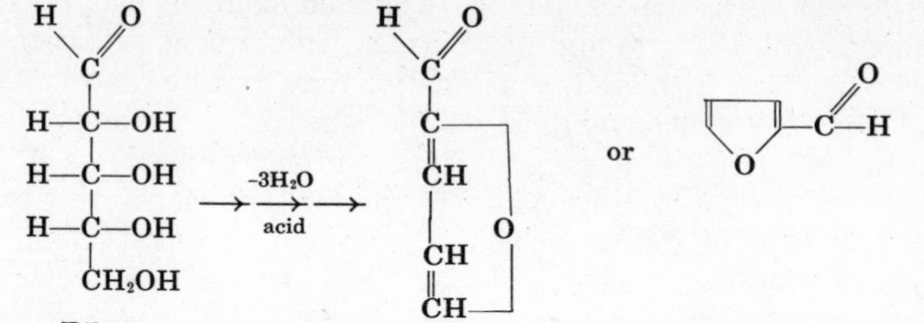

D-Ribose $\xrightarrow[\text{acid}]{-3H_2O}$ (furfural ring structure) or Furfural

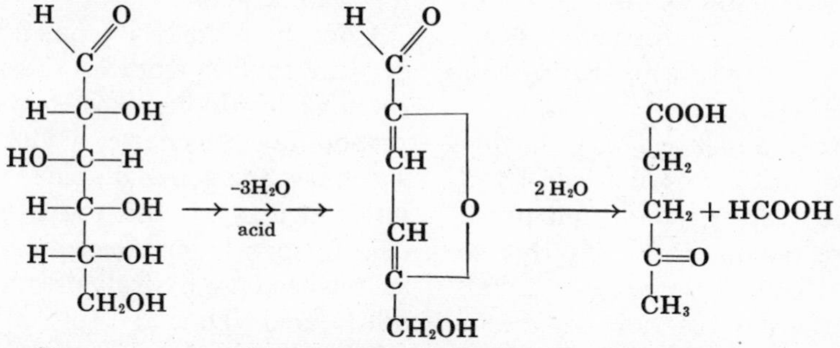

D-Glucose $\xrightarrow[\text{acid}]{-3H_2O}$ 5-Hydroxymethylfurfural $\xrightarrow{2\,H_2O}$ Levulinic acid ($COOH{-}CH_2{-}CH_2{-}C{=}O{-}CH_3$) + $HCOOH$

The relative rate of this reaction varies among sugars and also depends on the nature and strength of acid used. Accordingly this reaction can serve as the basis for a series of specific color reactions of great analytical value. The furfural or hydroxymethyl furfural produced by the dehydration of carbohydrates (note that polysaccharides are first hydrolyzed to monosaccharides) will condense with a large number of phenols and aromatic amines to yield colored products. The careful choice of acid strength, heating time, temperature, and phenol or aromatic amine can be used to limit color formation to a specific sugar or class of sugars (see Experiment 5, p. 20).

### *Action of Alkali*

In the presence of dilute alkali [for example, $0.035N$ NaOH or saturated aqueous $Ca(OH)_2$], monosaccharides are isomerized at room temperature, probably by way of an enolic intermediate. This type of interconversion is referred to as a Lobry de Bruyn-Van Eckenstein transformation and is illustrated below for glucose.

Alkaline Interconversion of Glucose, Mannose, and Fructose

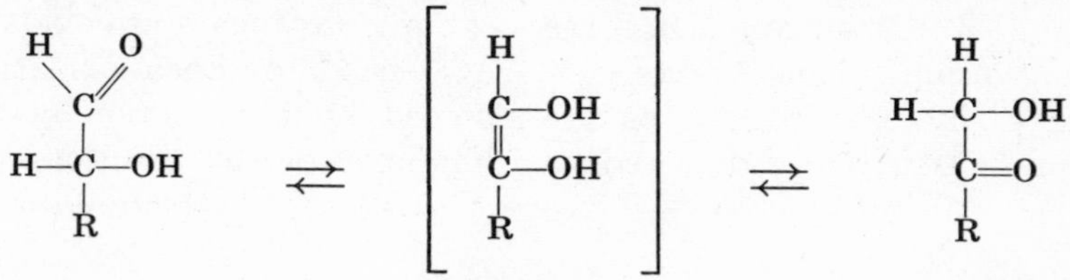

D-Glucose $\rightleftarrows$ [Common enol] $\rightleftarrows$ D-Fructose

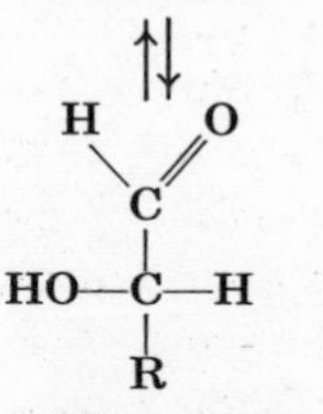

Common enol $\rightleftarrows$ D-Mannose

By increasing the concentration of alkali and/or the temperature, the rate and degree of isomerization are increased, and the double bond of the ene-diol moves down the carbon chain, giving rise to a variety of isomers and eventually to cleavage of carbon–carbon bonds. In the presence of $O_2$ this isomerization is accompanied by extensive oxidation, with the formation of such products as formaldehyde, glycolaldehyde, glyceraldehyde, lactic acid, methylglyoxal, and reductone (CHOH═COH—CHO).

In a similar manner oligosaccharides and polysaccharides are isomerized by alkali at the free reducing end; and unless precautions are taken to prevent the isomerization in these polymers, an alkali-catalyzed elimination reaction will occur, producing a stepwise cleavage of glycosidic bonds proceeding down the polysaccharide chain from the free reducing end [2].

### *Oxidation*

Sugars containing a free reducing group (that is, hemiacetal or hemiketal carbon or free aldehyde or ketone) are subject to oxidation by a variety of general oxidizing agents, whereas sugars lacking such groupings (for example, those containing only glycosidic linkages, as in sucrose) are resistant to these agents. The initial attack of the general oxidizing agents (for example, $Cu^{+2}$, $Bi^{+2}$, $Ag^{+}$, $Fe(CN)_6^{-3}$ and aromatic nitro compounds) occurs at the reducing group of the sugar (hence the name reducing group). Thus sugars containing these easily oxidized groups are called reducing sugars, since they reduce the added oxidizing agent. After the initial attack, subsequent oxidation and isomerization may continue down the carbon chain in a nonstoichiometric manner, yielding a large number of products of low molecular weight. In contrast to this type of general oxidation mechanism, alkaline hypohalite oxidizes aldehydes and hemiacetals (not ketoses and hemiketals) to the corresponding aldonic acids.

Concentrated $HNO_3$ has a unique property in that it oxidizes the two terminal carbons of aldohexoses, forming the corresponding saccharic acids. In general, saccharic acids are soluble in dilute $HNO_3$; but mucic acid, the particular saccharic acid formed from galactose (either free or as released by hydrolysis), is insoluble in dilute acid. Thus the formation of an insoluble saccharic acid when concentrated $HNO_3$ is heated with a carbohydrate and subsequently diluted signifies the earlier presence of galactose (see Experiment 5).

Oxidation of carbohydrates by the periodate ion ($IO_4^-$) follows a specific and stoichiometric course. Thus a study of the quantity of $IO_4^-$ consumed and the products formed from periodate oxidation of sugars often provides information of use in structural studies of carbohydrates. Periodate selectively cleaves a molecule between adjacent carbons which have any combination of hydroxyls, aldehydes, ketones, or primary amine groups. Thus the C—C bonds of *cis-* and *trans-*glycols, hydroxyaldehydes, and other combinations common to sugars are oxidatively cleaved by periodate. Each oxidation is accompanied by a reduction of $IO_4^-$ to $IO_3^-$. Examples of such reactions are

$$\begin{array}{l} \text{H—C—OH} \\ \text{H—C—OH} \end{array} + IO_4^- \longrightarrow \text{—C(H)=O} + \text{O=C(H)—} + IO_3^- + H_2O$$

$$\begin{array}{c}\mathrm{H}\diagdown\mathrm{C}{=}\mathrm{O}\\ |\\ \mathrm{H{-}C{-}OH}\\ |\end{array} + IO_4^- \longrightarrow \begin{array}{c}\mathrm{HCOOH}\\ +\\ \mathrm{H}\diagdown\mathrm{C}{=}\mathrm{O}\\ |\end{array} + IO_3^-$$

$$\begin{array}{c}\mathrm{H}\\ |\\ \mathrm{H{-}C{-}OH}\\ |\\ \mathrm{C{=}O}\\ |\\ \mathrm{H{-}C{-}OH}\\ |\end{array} + 2IO_4^- \longrightarrow \longrightarrow \begin{array}{c}\mathrm{H}\\ |\\ \mathrm{C}\\ \mathrm{H}\diagup\quad\diagdown\mathrm{O}\\ +\\ \mathrm{CO_2}\\ +\\ \mathrm{H}\diagdown\mathrm{C}{=}\mathrm{O}\\ |\end{array} + 2IO_3^-$$

$$\begin{array}{c}\mathrm{H}\diagdown\mathrm{C}{=}\mathrm{O}\\ |\\ \mathrm{H{-}C{-}NH_2}\\ |\\ \mathrm{H{-}C{-}OH}\\ |\end{array} + 2IO_4^- + H_2O \longrightarrow \longrightarrow \begin{array}{c}\mathrm{HCOOH}\\ +\\ \mathrm{HCOOH}\\ +\\ \mathrm{H}\diagdown\mathrm{C}{=}\mathrm{O}\\ |\end{array} + NH_3 + 2IO_3^-$$

$$\begin{array}{c}\mathrm{COOH}\\ |\\ \mathrm{HC{=}O}\end{array} + IO_4^- \longrightarrow \begin{array}{c}\mathrm{CO_2}\\ +\\ \mathrm{HCOOH}\end{array} + IO_3^-$$

In contrast, however, the following groupings are oxidized extremely slowly by periodate ion, and in practice may be considered resistant to periodate:

$$\begin{array}{c}\mathrm{COOH}\\ |\\ \mathrm{COOH}\end{array} \qquad \begin{array}{c}\mathrm{COOH}\\ |\\ \mathrm{HC{-}OH}\\ |\end{array}$$

Sugar hemiacetals and hemiketals which are in equilibrium with their respective free aldehydes and ketones react as if they were the free aldehydes or ketones. The oxidation level of each carbon increases by one level (alcohol $\longrightarrow$ aldehyde or ketone $\longrightarrow$ acid $\longrightarrow$ $CO_2$) for each adjacent carbon–carbon bond cleaved. Thus primary alcohols are released as formaldehyde; keto groups of sugars, as in fructose, are oxidized to $CO_2$; and so on.

If the periodate-reactive functional group of either of two adjacent carbon atoms is tied up in another bond (for example, an ether or an acetal), no oxidation or cleavage occurs at that particular carbon–carbon bond.

$$\begin{array}{c}|\\ \mathrm{H{-}C{-}OH}\\ |\\ \mathrm{H{-}C{-}O{-}CH_3}\\ |\end{array} + IO_4^- \longrightarrow \text{no reaction}$$

$$\begin{array}{c}\mathrm{OCH_3}\\ |\\ \mathrm{H{-}C{-}O{-}R}\\ |\\ \mathrm{H{-}C{-}OH}\\ |\end{array} + IO_4^- \longrightarrow \text{no reaction}$$

Thus acetals of aldohexopyranoses are not cleaved between carbons 1 and 2, 4 and 5, or 5 and 6, but are cleaved between carbons 2 and 3 and 3 and 4.

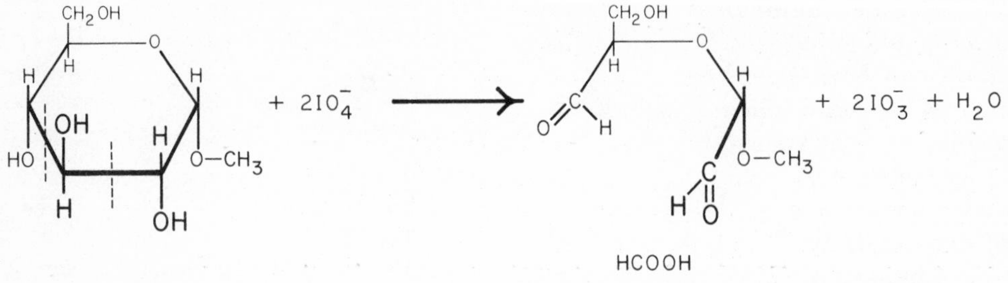

The rate of periodate oxidation of monosaccharides with $NaIO_4$ provides an interesting verification of the hemiacetal structure of sugars in solution. For example, during the oxidation of glucose with $NaIO_4$, there is an initial rapid consumption of 3 moles of $IO_4^-$ per mole of glucose. This is followed by a much slower consumption of two additional moles of $IO_4^-$ per mole of glucose until, eventually, the rate of consumption reaches the theoretical value of 5 moles of $IO_4^-$ per mole of glucose. Such results are best explained by considering the favored pyranose ring form of glucose. The initial rapid consumption of $IO_4^-$ probably corresponds to the oxidation of the three glycol groups in the molecule, leaving carbon 1 linked to carbon 5 as a formate ester. This residue is not susceptible to further oxidation by $IO_4^-$, but must await the gradual hydrolysis of the ester before the final two glycol bonds can be cleaved (see below).

In the neutral to slightly acidic solution in which this reaction is run, the hydrolysis of the formate ester proceeds very slowly; therefore the consumption of the final 2 moles of $IO_4^-$ is delayed. During the final stages of oxidation the rate-determining step is the hydrolysis of the ester rather than the oxidation of the product. This results in a slower rate of $IO_4^-$ consumption.

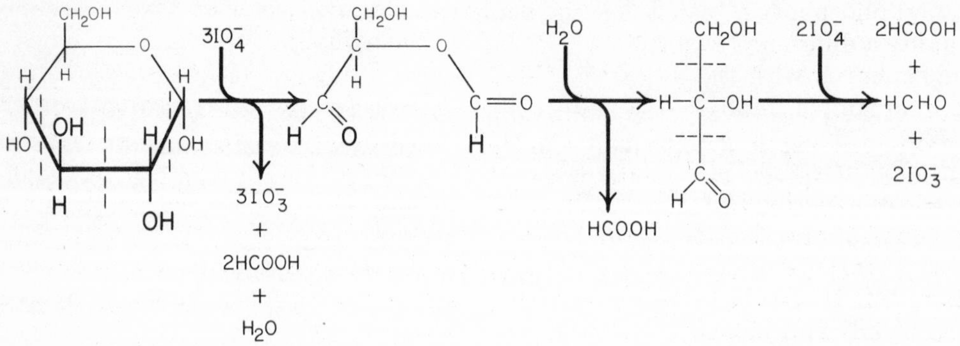

## Analysis of Carbohydrates

The carbohydrate experiments in this manual illustrate several of the physical and chemical properties useful in the isolation and characterization of carbohydrates. In addition to the principles illustrated in these experiments, other laboratory applications in carbohydrate chemistry are worthy of note here.

### *Quantitative Determination of Sugar Phosphates*

In the many quantitative colorimetric methods used to estimate inorganic phosphorus, the reagents employed do not react with organic phosphate [3]. Thus inorganic phosphate can be determined even in the presence of organic phosphate. Organic phosphates, such as sugar phosphates, can be acid hydrolyzed to form inorganic phosphate with varying ease, depending on the type of phosphate bond in question. These differences in ease of acid hydrolysis form the basis of a rough quantitative test for classes of organic phosphates [4]. Phosphate anhydrides, as in nucleoside triphosphate and pyrophosphate, and sugar phosphoacetals, as in glucose-1-phosphate, are extensively hydrolyzed by $1N$ HCl after seven min at 100°C (so-called 7-min phosphate; see Experiments 9 and 10). Phosphate esters require a longer period of heating; for example, the phosphate ester on the 1 position of fructose-1,6-diphosphate is virtually completely hydrolyzed after 30 min, and most phosphate esters [3-phosphoglyceric acid (PGA) is an exception] are hydrolyzed after 180 min in $1N$ HCl at 100°C. The quantity of PGA, or "resistant phosphorus," can be determined by making a difference comparison of 180-min phosphorus with the inorganic phosphate in a sample totally hydrolyzed with $10N$ $H_2SO_4$. Thus a comparison of the quantity of inorganic phosphate released after various periods of acid hydrolysis gives a first approximation of the concentration of various organic phosphates in an unknown. It should be noted that this method is only approximate, since the rate of hydrolytic cleavage in the various organic phosphates will differ under the conditions used (for example, some hydrolysis of fructose-1,6-diphosphate occurs during 7-min hydrolysis).

### *Differentiation of Polysaccharides by Iodine Color Complexes*

Certain polysaccharides (amylose, amylopectin, glycogen, and dextran) form characteristic color complexes when combined with molecular iodine [5]. The nature of these complexes is not well defined, but prevailing evidence points to an adsorptive complex between iodine and helically coiled polysaccharide chains. Chain lengths of at least eight linear sugar residues are required for the ready formation of helixes and intense iodine complexes. Thus linear or helically coiled polysaccharides (such as amylose and starches which are rich in amylose) form an intense blue-black color with iodine (see Experiment 9). Branched polysaccharides with interrupted helixes (for example, amylopectin) yield less intensely colored iodine complexes; and highly branched polysaccharides with short segments and hindered helix formation (for example, glycogen) yield only pale-brown color complexes with iodine.

### *Formation of Sugar Derivatives*

Once a tentative identification of a sugar (or any compound) has been made by determining various chemical and physical properties of the compound, it is desirable to confirm the identification by synthesizing a derivative and comparing it with the same derivative of an authentic sample.

Phenylhydrazones and osazones, both formed by reaction of phenylhydrazine or substituted phenylhydrazines with reducing sugars [6], have a limited usefulness in the characterization of sugars. In the formation of an osazone, one of the asymmetric centers of the parent sugar is destroyed. Thus D-glucose, D-mannose, and D-fructose all form the same osazone (see top of p. 10).

Furthermore, all osazones of sugars melt within the same temperature range. Thus the melting point of a particular osazone is not conclusive evidence in the characterization of a sugar. Conversion of osazones to osotriazoles, whose melting points have a wider range, is a better method [7].

Another derivative of an aldehyde sugar is the substituted benzimidazole [8]. This

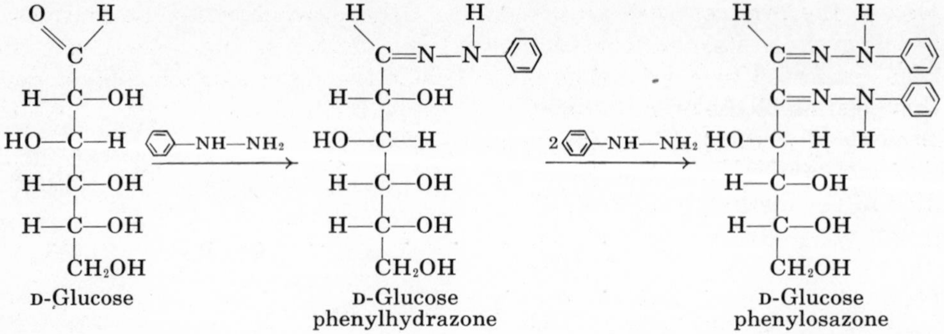

D-Glucose → ($C_6H_5$—NH—$NH_2$) → D-Glucose phenylhydrazone → (2 $C_6H_5$—NH—$NH_2$) → D-Glucose phenylosazone

compound is formed by a two-step process. The aldose is first oxidized with hypohalite to form an aldonic acid which is then condensed with *o*-phenylenediamine:

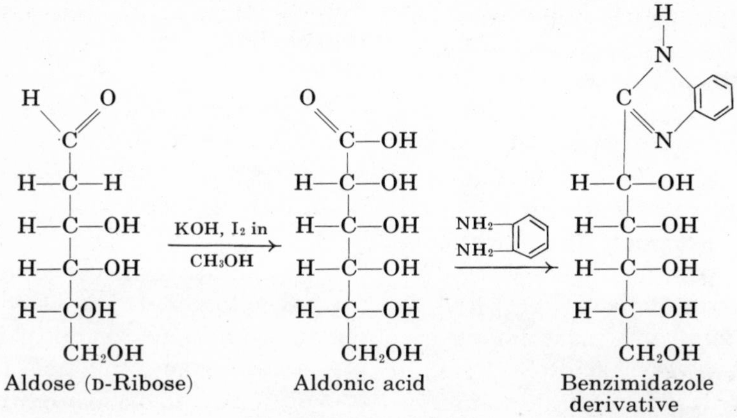

Aldose (D-Ribose) → (KOH, $I_2$ in $CH_3OH$) → Aldonic acid → (o-phenylenediamine) → Benzimidazole derivative

The melting points and specific rotations (see Experiment 4) of these compounds are sufficiently different for different sugars to aid in the identification of unknowns.

Another series of sugar derivatives is obtained by formation of methyl ethers of sugars and glycosides. Exhaustive methylation—that is, methylation of all free OH groups and hemiacetal OH groups with $(CH_3)_2SO_4$ and NaOH—has proved very useful in confirming identities and elucidating the structures of many carbohydrates, particularly the oligo- and polysaccharides [9]. For example, the position of the alcoholic hydroxyl involved in the glycosidic bond between the glucose residues of maltose may be ascertained by examining the products obtained upon successive oxidation, exhaustive methylation, and acid hydrolysis of maltose.

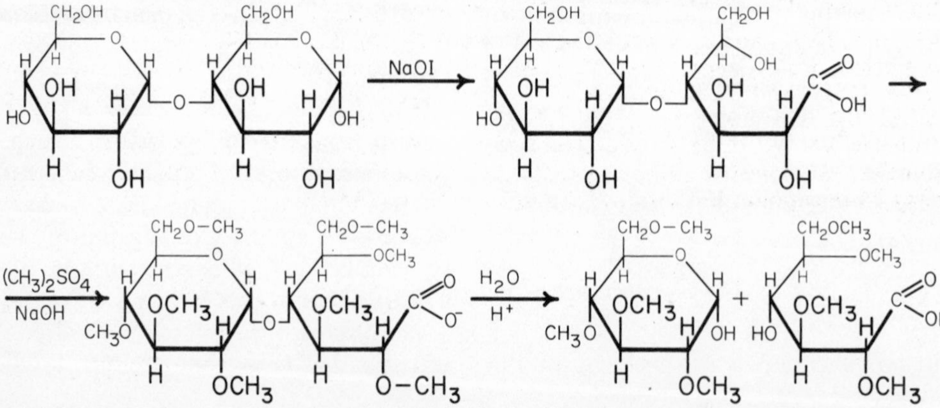

Maltose → (NaOI) → maltobionic acid → ($(CH_3)_2SO_4$, NaOH) → methylated derivative → ($H_2O$, $H^+$) → 2,3,4,6-tetra-O-methylglucose + 2,3,5,6-tetra-O-methylgluconic acid

Since one of the products has an unsubstituted OH group at its 4 position, it must have been connected by a glycosidic bond from this position to the other residue.

Treatment of reducing carbohydrates with dry methanolic HCl results in the formation of the methylglycosides.

Other carbohydrate derivatives sometimes useful in characterizing sugars include the sugar alcohols [10], or pentaacetates (for hexoses) [11] or involve the formation of a sugar with one more carbon atom by means of the Kiliani synthesis [12].

α-Methyl pyranoside

β-Methyl pyranoside

α-Methyl furanoside

β-Methyl furanoside


**General References**

Pigman, W. W. (Editor), *The Carbohydrates: Chemistry, Biochemistry, Physiology*, Academic, New York, 1957.

Colowick, S. P., and N. O. Kaplan (Editors), *Methods in Enzymology*, Academic, New York, Vols. 1 and 2, 1955; Vols. 3 and 4, 1957; Vol. 5, Vol. 6, 1962, in preparation.

**Specific References**

1. Reeves, R. E., *J. Am. Chem. Soc.*, **72,** 1499 (1950).
2. Whistler, R. L., and J. N. DeMiller, *Adv. in Carbohydrate Chem.*, **13,** 289 (1958).
3. Fiske, C. H. and Y. Subbarow, *J. Biol. Chem.*, **66,** 375 (1925).
4. Umbreit, W. W., R. H. Burris, and J. F. Stauffer, *Manometric Techniques*, Burgess, Minneapolis, 1957, pp. 272–279.
5. Bates, F. L., D. French, and R. E. Rundle, *J. Am. Chem. Soc.*, **65,** 142 (1943).
6. Shriner, R. L., R. C. Fuson, and D. Y. Curtin, *The Systematic Identification of Organic Compounds*, Wiley, New York, 1956, p. 131.
7. von Peckmann, H., *Ann.*, **262,** 265 (1891).
8. Dimler, R. J., and K. P. Link, *J. Biol. Chem.*, **150,** 345 (1943).
9. Haworth, W. N., *J. Chem. Soc.*, **107,** 13 (1915).
10. Blix, G. et al., *Acta Soc. Med. Uppsaliensis*, **61,** 1 (1956).
11. Montgomery, E., R. M. Haun, and C. S. Hudson, *J. Am. Chem. Soc.*, **59,** 1124 (1937).
12. Hudson, C. S., O. Hartley, and C. B. Purvis, *J. Am. Chem. Soc.*, **56,** 1248 (1934).

EXPERIMENT

# 1. Sugars: Reactions of Reducing Sugars

## Theory of the Experiment

The presence of carbohydrate in a preparation may be detected by either the Molisch test or the anthrone test. Both tests are based on the hydrolyzing and dehydrating action of concentrated $H_2SO_4$ on carbohydrates. In these tests the strong acid catalyzes the hydrolysis of any glycosidic bonds present in the sample and the dedehydration to furfural (pentoses) or hydroxymethyl furfural (hexoses) of the resulting monosaccharides. These furfurals then condense with $\alpha$-naphthol (Molisch test) or anthrone to give a colored product.

The Molisch test is a qualitative test for the presence of carbohydrate in a sample of unknown composition. To determine the amount and specific nature of the carbohydrate, further tests are required.

The most common method for detecting the presence of free reducing groups in a carbohydrate involves a determination of the capacity of the carbohydrate-containing sample to reduce $Cu^{+2}$ in alkaline solution. As stated previously, the carbohydrate is oxidatively degraded in a nonstoichiometric manner, with a corresponding reduction of the oxidizing agent. In spite of the nonspecific course of the reaction, it is found that when the conditions for the copper reduction are rigorously controlled, the amount of $Cu^{+2}$ reduced to $Cu^{+}$ is directly proportional to the amount of reducing sugar in the sample analyzed. But equimolar amounts of different reducing sugars differ in the rate at which they will reduce $Cu^{+2}$.

The $Cu^{+}$ formed in the reaction precipitates as the rust-colored $Cu_2O$. The amount of $Cu_2O$ formed may be determined gravimetrically, titrametrically, or (as in the present experiment), colorimetrically. In the Nelson test, the amount of $Cu_2O$ formed is determined by addition of arsenomolybdic acid, which is quantitatively reduced to arsenomolybdous acid by the $Cu^{+}$. The intense blue color of the arsenomolybdous acid is then measured colorimetrically.

## Experimental Procedure

### MATERIALS

Unknown sugar
5% Alcoholic $\alpha$-naphthol
Nelson's reagent A
Arsenomolybdate reagent
1% glucose
Conc. $H_2SO_4$
Nelson's reagent B
Glucose standard (100 $\mu$g/ml)

### *Molisch Test*

Prepare a 1% solution (10 mg/ml) of the unknown sugar, and transfer 0.5 ml to a 50-mm test tube. Prepare another tube with 1% glucose, and a third with water to serve as a control. Add two drops of 5% alcoholic $\alpha$-naphthol to each tube, and mix the contents. Then incline each tube and carefully allow 1 ml of concentrated $H_2SO_4$ to flow down the side of the tube so that it forms a layer beneath the aqueous solution. Note the color formed at the interface after a short time.

### *Nelson's Test*

Prepare Nelson's alkaline copper reagent by mixing 12.5 ml of Nelson's reagent A with 0.5 ml of Nelson's reagent B. Prepare a solution of unknown sugar containing 100 $\mu$g/ml (for example, 10 mg/100 ml), and then add the following quantities, in the order listed, to nine carefully cleaned and optically uniform tubes (see p. 187).

| Substance in milliliters | Tube Number | | | | | | | | |
|---|---|---|---|---|---|---|---|---|---|
| | 1 | 2 | 3 | 4 | 5 | 6 | 7 | 8 | 9 |
| $H_2O$ | 1.0 | 0.9 | 0.7 | 0.5 | 0.3 | 0 | 0.7 | 0.5 | 0 |
| Glucose standard (100 μg/ml) | 0 | 0.1 | 0.3 | 0.5 | 0.7 | 1.0 | 0 | 0 | 0 |
| Unknown sugar (100 μg/ml) | 0 | 0 | 0 | 0 | 0 | 0 | 0.3 | 0.5 | 1.0 |

Tube 1 contains no sugar and is therefore the blank. Add 1.0 ml of Nelson's alkaline copper reagent to each tube, and shake well. Place the tubes simultaneously in a boiling water bath (a 500-ml beaker or larger), and heat for exactly 20 min. Remove the tubes simultaneously, and place them in a beaker of cold water to cool. When the tubes are cool (25°C), add 1.0 ml of arsenomolybdate reagent to each, and shake well occasionally during 5-min period to dissolve the precipitated $Cu_2O$ and to reduce the arsenomolybdate. After the $Cu_2O$ has dissolved, add 7.0 ml of distilled water to each tube, and mix thoroughly. Next remove any stains and water droplets from the outside of the tubes with a clean cloth, and then read the optical densities in the colorimeter at 540 mμ after setting the instrument to zero optical density (100% transmission) with the blank (tube 1).

## Report of Results

Plot the optical density (ordinate) against the concentration of glucose in micrograms and micromoles (abscissa) to determine whether Beer's law (see Appendix I) is obeyed. Plot also the three determinations of your unknown sugar (see Appendix I). Calculate and compare extinction coefficients for glucose and the unknown sugar.

A positive Molisch test indicates that the material tested is a carbohydrate. A positive Nelson test indicates that the suspected carbohydrate material contains reducing groups which are not combined in glycosidic linkages. What conclusions can you draw from your data?

## Discussion

Frequently, the reducing test must be preceded by acid hydrolysis in order to release "potential" reducing centers by cleavage of glycosidic bonds. Preliminary hydrolysis was unnecessary in this experiment, since the sample is a monosaccharide.

Nelson's is only one of several widely used reducing tests. Others are discussed below.

*Benedict's test* (qualitative). Cupric ion, $Cu^{+2}$ is reduced to $Cu_2O$ in an alkaline citrate solution. The citrate buffer stabilizes the $Cu^{+2}$ during the reaction by keeping it from precipitating as black, insoluble CuO.

*Barfoed's test* (qualitative). Here again $Cu^{+2}$ is reduced to $Cu_2O$ in weak acid solution. Experimentally it has been shown that monosaccharides reduce more readily in a weak acid solution than do disaccharides. Hence this test will distinguish between these two types of sugar if carried out carefully.

*Nylander's test* (qualitative). The ion $Bi^{+3}$ is reduced to a precipitate of bismuth metal. This test is often used in sugar analysis of urine because it is not affected by other substances in the urine, such as uric acid or creatinine.

*Hagedron-Jensen test* (quantitative). Ferricyanide reduction is measured colorimetrically.

*Dinitrosalicylate method* (quantitative). The reduction of 3,5-dinitrosalicylate is measured colorimetrically.

These methods, as clinical tests, have been of great importance in medical diagnosis for many years. They have assisted in the diagnosis of diabetes, which leads to abnormally high levels of sugar in the blood and urine, and of other metabolic abnormalities.

**Exercises**

1. Why do mono-, oligo-, and polysaccharides all give a positive Molisch test?
2. Which of the following carbohydrates are reducing sugars: galactose, β-methylgalactoside, maltose, mannose, xylose, fructose, rhamnose, ribose, glucosamine?
3. If an unknown which gives a positive Molisch test fails to give a positive reducing test, what might you conclude? How would you go about testing your conclusion?
4. Why is green light (540 mμ) used to assay arsenomolybdous acid?
5. Why is it necessary to control rigorously the period of heating in Nelson's test?

**References**

Nelson, N., *J. Biol. Chem.*, **153,** 375 (1944).
Trevelyan, W. E., and J. S. Harrison, *Biochem. J.*, **50,** 298 (1958).

EXPERIMENT

# 2. Sugars: Specific Oxidation by Periodate

## Theory of the Experiment

Treatment of carbohydrates with periodate ($IO_4^-$) often yields considerable information about the structure and general polyhydroxyl character of the carbohydrate in question. The quantity of $IO_4^-$ consumed by complete oxidation of a carbohydrate is related to the original structure of the carbohydrate. Accordingly, knowledge of the number of moles of $IO_4^-$ consumed per gram of sample limits the number of possible structures. A full understanding of the reactions of periodate with sugars (see p. 6) will prove useful to you in evaluating your results.

Upon periodate oxidation of glycols, $IO_4^-$ is reduced to $IO_3^-$:

$$\begin{array}{c} | \\ \text{H—C—OH} \\ | \\ \text{H—C—OH} \\ | \end{array} + IO_4^- \longrightarrow \begin{array}{c} | \\ \text{H—C=O} \\ \\ \text{H—C=O} \\ | \end{array} + IO_3^- + H_2O$$

The extent of this reaction at any given time can be determined by conversion (in acid) of both the unreacted $IO_4^-$ and the $IO_3^-$ formed by the reaction to $I_2$:

$$8H^+ + IO_4^- + 7I^- \xrightarrow{H^+} 4I_2 + 4H_2O$$

$$6H^+ + IO_3^- + 5I^- \xrightarrow{H^+} 3I_2 + 3H_2O$$

As the above reactions indicate, $IO_4^-$ and $IO_3^-$ yield different amounts of $I_2$ when quenched (reduced) with acidic iodide. Therefore one can calculate the amount of $IO_4^-$ which had reacted with the sample by first quenching and then titrating the resulting $I_2$ with a water blank for comparison.

The $I_2$ formed is determined by titration with thiosulfate; the starch-$I_2$ complex is used as the indicator:

$$I_2 + 2Na_2S_2O_3 \longrightarrow 2NaI + Na_2S_4O_6$$

Further useful information can be obtained by studying the organic products formed by periodate oxidation of carbohydrates. Formic acid (HCOOH) is a prominent product of the oxidation of simple sugars. Complete periodate oxidation of fructose yields 3 moles of formic acid per 5 moles of $IO_4^-$ consumed, and complete oxidation of glucose with $IO_4^-$ yields 5 moles of HCOOH for each 5 moles of $IO_4^-$ consumed. The contrasts in values for the ratio

$$\frac{\text{moles HCOOH formed}}{\text{moles } IO_4^- \text{ consumed}}$$

are even more pronounced with the aldo- and ketopentoses. Formic acid can be readily determined by titration with dilute base after previous reduction of excess periodate with ethylene glycol.

## Experimental Procedure

MATERIALS

Unknown monosaccharide
20% KI
Ethylene glycol
0.5*M* $NaIO_4$
0.1*N* $Na_2S_2O_3$
0.05*N* NaOH
1*N* $H_2SO_4$
1% Soluble starch
1% Phenolphthalein

### *Periodate Reaction*

Weigh accurately (to ±1.0 mg) a 250- to 300-mg sample of the unknown, and transfer it to a 50-ml volumetric flask. Dissolve the sample in 10 ml of water, and add 25 ml of 0.5*M* $NaIO_4$ solution; finally, add water to bring the volume up to the 50-ml mark. Mix the solution thoroughly, and record the time at which you add the periodate. Prepare a blank having the same periodate concentration as the sample. Incubate both reaction mixtures at 37°C.

### *Quantitative Measurement of Periodate Consumption*

At timed intervals (for example, after 15 min, 1 hr, 24 hr, 48 hr), remove a 1-ml aliquot of each reaction mixture (experimental and blank), and add both to a 125-ml Erlenmeyer flask containing 5 ml of 1*N* $H_2SO_4$ and 5 ml of 20% KI. A large quantity of iodine will form at once. Titrate the iodine *immediately* with 0.1*N* $Na_2S_2O_3$ solution, adding $S_2O_3^{-2}$ slowly so that a yellow color persists during the titration, until only a pale yellow color remains. Add 1 ml of starch indicator, and continue adding $S_2O_3^{-2}$ until the blue color just disappears. (A total of 12–20 ml of 0.1*N* $S_2O_3^{-2}$ will probably be consumed.) Record the amount of $S_2O_3^{-2}$ required to titrate the iodine. The last two titration values will probably be identical, indicating that the reaction has reached completion.

### *Determination of Formic Acid*

After 48 hours, or when the consumption of periodate has stopped, add a 5-ml aliquot of the sample to 1 ml of ethylene glycol. Allow the mixture to stand at room temperature for 5 min, occasionally swirling it, while the unreacted periodate is consumed. Then add one drop of 1% phenolphthalein, and titrate the solution with 0.05*N* NaOH until a definite pink color appears. Perform a similar analysis on the blank.

## Report of Results

### *Moles of Periodate Consumed*

The moles of periodate consumed during oxidation of the sugar is inversely related to the amount of $I_2$ formed on quenching. The amount of $I_2$ present can therefore be determined by the titration with $S_2O_3^{-2}$, for the difference between the titration of the sample and that of the blank is a direct measurement of the amount of periodate consumed in the reaction. Accordingly, the moles of oxidant ($IO_4^-$) consumed can be calculated by the equation

$$\text{Moles } IO_4^- \text{ consumed} = \frac{\text{vol of sample in ml}}{\text{vol of aliquot in ml}} \times \frac{\left(\text{ml to titrate blank} - \text{ml to titrate aliquot}\right)\left(N \text{ of thiosulfate}\right)}{2 \times 1000}$$

Tabulate your data to show the moles of periodate consumed versus time of oxidation. Determine the moles of periodate consumed per mole of unknown, assuming first that the unknown is a hexose and then that it is a pentose.

### *Moles of Formic Acid Formed*

The moles of formic acid formed is directly related to the number of equivalents or moles of base consumed during titration. Therefore the relationship

$$\frac{(\text{ml of base})(N \text{ of base})(\text{total vol of sample})}{(\text{vol of aliquot})(1000)}$$

gives the moles of HCOOH formed.

Obtain the ratio

$$\frac{\text{moles HCOOH formed}}{\text{moles } IO_4^- \text{ consumed}}$$

for your unknown sugar.

In practice, the quantities, millimoles of $IO_4^-$ consumed and millimoles of HCOOH formed, are frequently used to avoid small numbers. Omission of the factor 1000 from the denominator of either of the above equations yields results in millimoles.

## Discussion

The results obtained from this experiment should verify the polyhydroxyl character of the unknown. If the analysis is precise, it is possible to differentiate ketoses from aldoses, for the ratio

$$\frac{\text{HCOOH formed}}{IO_4^- \text{ consumed}}$$

indicates whether the sugar is an aldose or a ketose. Several other methods which aid in defining the structure of an unknown monosaccharide are discussed in Experiments 3 and 4.

### Exercises

1. What would be the products of complete periodate oxidation of each of the following compounds: (a) galactose, (b) fructose, (c) lactose, (d) xylose, (e) sucrose, (f) trehalose, (g) raffinose, (h) $\alpha$-methyl galactose, (i) glucosamine, (j) amylose, (k) inulin, and (l) adenosine?
2. How many moles of periodate would be consumed in the periodate oxidation of compounds (a)–(i) of question 1?
3. Why does the number 2 appear in the denominator of the equation relating the $S_2O_3^{-2}$ titration with the moles of $IO_4^-$ consumed?
4. Why is ethylene glycol used in the procedure for measuring the HCOOH formed upon $IO_4^-$ oxidation? Would glycolaldehyde work as well? Explain your answer.

### Reference

(*Most standard texts in organic chemistry discuss the methods of this experiment.*)

Dyer, J. R., *Methods of Biochem. Anal.*, **3**, 111 (1956).

EXPERIMENT

# 3. Sugars: Chromatography

## Theory of the Experiment

Paper chromatography is one of the most useful methods for the separation and characterization of small quantities of compounds (see Appendix II). Frequently, by comparing the migration of the unknown relative to that of the knowns in a particular solvent system, a tentative identification of the unknown can be made.

Specific color reactions can also be used to detect and identify compounds on paper chromatograms. For example, reducing sugars give colored products when reacted with aniline-acid-oxalate (aldohexoses form brown spots, and aldopentoses form reddish spots). Periodate-reactive compounds are identifiable with the periodate-benzidine spray. In the latter method, papers are initially sprayed with periodate reagent and then, after a few minutes, with the benzidine. The benzidine is oxidized by the periodate to a blue color in areas where the periodate has not been previously consumed by reaction with sugar. Thus sugars appear as white spots in a blue field.

## Experimental Procedure

### MATERIALS

Isopropanol:acetic acid:$H_2O$ (3:1:1)
0.5% $NaIO_4$
Whatman No. 1 paper
1% Glucose, galactose, mannose, fructose, ribose, xylose, and arabinose
Aniline-acid-oxalate reagent
0.5% Benzidine reagent
4-Liter amber glass chromatography jars
Metric scale

Place in a chromatography jar sufficient isopropanol:acetic acid:$H_2O$ (3:1:1) to form a layer 1 cm deep. Seal the top, and allow the solvent to equilibrate for 10 min. While waiting, mark an origin, or starting line, 3 cm from the edge of a sheet of Whatman No. 1 paper (22 × 22 cm) with nine evenly distributed pencil marks (2 cm apart), placed so as to leave at least a 2-cm margin at the edges. Using a capillary tube, spot the paper with samples 5–7 mm in diameter (that is, approx. 0.005 ml) of each of the 1% sugar solutions listed above plus two spots of the unknown—one spot to a mark.

Staple the paper in the form of a cylinder with the starting line at one end. Lower the cylinder (starting line down!) into the previously equilibrated chromatography jar containing isopropanol:acetic acid:$H_2O$ (3:1:1). Remove the paper when the solvent front nears the top of the cylinder. Mark the solvent front and allow the paper to dry. Spray the dried chromatogram with either aniline-acid-oxalate or periodate-benzidine as follows.

### *Aniline-acid-oxalate Spray for Reducing Sugars*

Spray the chromatogram with the reagent until fully dampened, avoiding excessive application of spray, as this may cause movement of the compounds on the paper. Then heat the paper at 100–105°C for 5–15 min to develop the colored spots resulting from reaction of the reducing sugar with the reagent.

### *Periodate-benzidine Spray for Polyols*

Spray the chromatogram with 0.5% $NaIO_4$ until just damp. Allow the $IO_4^-$ to react with the compounds on the paper for about 5 min at room temperature, and then spray the paper with 0.5% benzidine reagent. The periodate sprayed on the areas lacking sugar will react with the benzidine, forming a blue color; the areas containing sugar will have previously consumed the periodate and will appear as white spots on the blue field.

## Report of Results

After detecting the sugars, calculate their $R_f$ values, and tentatively identify each unknown by comparing its $R_f$ value and color reaction with those of the knowns.

## Discussion

The benzidine portion of the periodate-benzidine spray system forms the basis of an excellent method for detecting reducing sugars (see Horrocks in references). When a spray of 0.5% benzidine is used, reducing sugars appear as brown spots. Thus insufficient exposure to the periodate (that is, not enough spray, or too high a concentration of sugar on a spot) followed by the benzidine spray will produce brown spots instead of white spots in a blue field.

In practice, the identification of the unknown should be confirmed by comparing it with known compounds in a variety of solvent systems. This is not done here because of time limitations. Final characterization of the unknown requires more substantial data than that obtained from chromatography.

### Exercises

1. What is the chemical basis for the aniline-acid-oxalate and periodate-benzidine sprays that lead to color tests?
2. Would these reagents be equally satisfactory in locating the following compounds on paper: (a) mannose, (b) ribose, (c) riboflavin, (d) mannitol, (e) fructose, and (f) ribitol?

### References

(*Also see References at end of Appendix II.*)

Horrocks, R. H., *Nature,* **164,** 444 (1949).

Cifonelli, J. A., and J. Smith, *Anal. Chem.,* **26,** 1132 (1954).

Partridge, S. M., *Biochem. J.,* **42,** 238 (1948).

EXPERIMENT

# 4. Sugars: Polarimetry

## Theory of the Experiment

The magnitude and direction of rotation of the plane of polarized light by an asymmetric compound is a specific physical property of the compound that may be used in its characterization. The measurement of this property of a compound is called polarimetry. When plane polarized light is passed through a solution of a pure compound, the degree to which the plane of light is rotated is found to be directly proportional to the number of asymmetric molecules of the compound through which the light passes. Therefore the rotation observed will depend on the nature of the asymmetric compound, the concentration of the compound in solution, and the length of the light path through the solution. In addition, the observed rotation will vary with the wavelength of plane polarized light used and with the temperature. When these variables are controlled, one may relate, by the formula

$$\left[\alpha\right]_D^T = \frac{[\alpha]_{\text{obs}}}{l \times c},$$

the degree of rotation actually observed to the characteristic capacity of the compound to rotate polarized light. Here $[\alpha]_{\text{obs}}$ is the observed rotation, $c$ is the concentration in g/ml, and $l$ is the length of the light path in decimeters. The quanity $[\alpha]_D^T$

is called the specific rotation at the temperature $T$ when the $D$ line of the sodium spectrum is used as the light source. Frequently this formula is written

$$\left[\alpha\right]_D^T = \frac{\alpha \times 100}{l \times c}$$

where the units of $c$ are g/100 ml instead of g/ml. Every optically active compound has a characteristic specific rotation which may be used as a physical constant to distinguish it from similar compounds.

The usefulness of this type of measurement for identifying reducing sugars is enhanced by the fact that it is often possible to obtain a characteristic initial specific rotation when the sugar is first dissolved and a characteristic equilibrium specific rotation after the sugar has mutarotated. These two values permit a determination of the anomeric form originally present as well as the identification of the sugar in question.

## Experimental Procedure

MATERIALS

Unknown carbohydrate
Polarimeter
$0.01M$ HCl

Adjust the polarimeter to zero degrees using water (the blank). If your polarimeter does not have a readily accessible zero adjustment, use as a starting point for your determinations the average of several readings on a water blank. Then carefully weigh about 1 g ($\pm 0.01$ g) of unknown, and dissolve it in about 20 ml of $H_2O$ in a 25-ml volumetric flask. Fill the flask to the mark, mix the contents, and immediately transfer the solution to the polarimeter tube, taking care to avoid introducing air bubbles. Determine the observed rotation in the polarimeter. Then combine the contents of the polarimeter tube with the original solution and add a drop of dilute HCl to the flask. Using the acidified sugar solution repeat the determination of the rotation at 5-minute intervals until you obtain an equilibrium value.

## Report of Results

Making corrections for the water blank if necessary, calculate the specific rotation of the original solution (that is, before mutarotation) and of the final equilibrium mixture. Compare your values with the specific rotations of the common sugars given in Table I. From your results, make a tentative identification of your compound.

**TABLE I.** Specific Optical Rotations $[\alpha]_D^{20}$ of Sugars (in degrees)

| Sugar | α Form | Equilibrium mixture | β Form |
|---|---|---|---|
| D-Ribose | −23.1 | −23.7 | —— |
| L-Arabinose | +54.0 | +104.5 | +175.0 |
| D-Xylose | +92.0 | +19.0 | −20.0* |
| D-Glucose | +113.4 | +52.2 | +19.0 |
| D-Galactose | +144.0 | +80.5 | +52.0 |
| D-Fructose | −21.0* | −92.0 | −133.5 |
| D-Mannose | +34.0 | +14.6 | −17.0 |
| L-Rhamnose | −7.7 | +8.9 | +54.0* |
| Lactose | +90.0 | +55.3 | +35.0 |
| Maltose | +168.0* | +136.0 | +118.0 |
| Sucrose | —— | +66.5 | —— |
| Raffinose | —— | +105.2 | —— |
| Trehalose | —— | +178.3 | —— |

*Calculated value.

## Discussion

In practice it is frequently of value to confirm a tentative polarimetric identification by running known compounds. This check will permit the correction of errors in your first observation and will confirm the identification.

### Exercises

1. The equilibrium mixture of D-glucose has a specific rotation of $+52.2°$, whereas the specific rotation of the $\alpha$ and $\beta$ forms are $+113.4°$ and $+19.0°$, respectively. What percentage of a glucose solution exists in the $\beta$-form at equilibrium?
2. Will the compound $\beta$-methylglucose mutarotate? Explain your answer.

### References


Gilman, H., *Organic Chemistry* (2nd ed.), Wiley, New York, Vols. 1 and 2, 1943; Vols. 3 and 4, 1953.

Bates, F. J. et al., *Polarimetry, Saccharimetry and the Sugars*, National Bureau of Standards, Circular C 440, Washington, D.C., 1942.


EXPERIMENT

# 5. Sugars: Qualitative Tests and Derivatives

## Theory of the Experiment

The various procedures applied in Experiments 1–4 usually indicate the structure of the unknown carbohydrate. This experiment serves to confirm the postulated identity. This confirmation is achieved in two ways. The first consists in subjecting the unknown and a series of knowns to one or more qualitative tests (Seliwanoff's test, Bial's test, mucic acid test), each of which is specific to a particular type of sugar or to a single sugar.

Seliwanoff's test is a timed color reaction that is specific for ketoses. In concentrated HCl solution ketoses undergo dehydration to yield furfural derivatives more rapidly than do aldoses. Further, most furfural derivatives will form complexes with resorcinol to yield color. Consequently, the relative rates of color development in a solution containing sugar, HCl, and resorcinol provide evidence for the aldose or ketose nature of the sugar in question.

Bial's test is a color reaction that is specific for pentoses. Under carefully controlled conditions of temperature, time, and HCl concentration pentoses are rapidly converted to furfural, while some hydroxymethyl furfural is formed from hexoses present. In the presence of ferric ion and orcinol (5-methylresorcinol), furfural condenses rapidly to yield a colored product.

The mucic acid test is based on the formation of a crystalline saccharic acid that is insoluble in dilute $HNO_3$—a reaction that is unique to galactose and galactose-containing compounds (see p. 00).

The second step in this confirmation consists in synthesizing derivatives (methyl glycosides) of both the unknown and the known. When the cyclic hemiacetal of an aldose is treated with anhydrous alcohol in the presence of an acid catalyst, the hydrogen of the hemiacetal is replaced by the alkyl group of the alcohol, and the alkyl glycoside is formed. Keto sugars such as fructose undergo this reaction less readily. Since there are normally two cyclic forms of aldo sugars (pyranose and furanose) and since two anomeric glycosides ($\alpha$ and $\beta$) can be formed, four possible isomers of the glycoside can be obtained. Under controlled conditions, it is possible to favor the synthesis of a specific isomer. Thus under the conditions usually employed, one or two isomeric forms predominate. The methylglycosides obtained by reaction of the unknown with methanol are compared by chromatography with the products obtained from a similar reaction with a known which is suspected to be identical to the unknown. Similar $R_f$ values of methylated products of the two samples provide confirmatory evidence of identity.

## Experimental Procedure

MATERIALS

Unknown sugar
$n$-Butyl alcohol
Dry Dowex 50 in $H^+$ form
Ethylacetate:$n$-propanol:$H_2O$ (5:3:2)
Whatman No. 1 paper
Crystalline sugars (glucose, galactose, xylose, arabinose, fructose, mannose, and ribose)
Seliwanoff's reagent
Conc. $HNO_3$
0.5% $NaIO_4$
Microscope
Bial's reagent
Dry methanol
0.5% Benzidine reagent

### *Formation of Methylglycoside*

It is necessary to prepare the methylglycoside of both the unknown and the known. Since this reaction requires a period of refluxing, start the glycoside formation before performing the other tests. Reverse this order only if previous tests indicate a ketose. In that event, use Seliwanoff's test as confirmation only, and do not attempt methyl-ketal formation.

Dissolve or suspend 1 g of suspected aldehydo sugar in 100 ml of dry methanol in an Erlenmeyer or round-bottom flask. To this add first 2 g of dry Dowex 50 cation-exchange resin in $H^+$ form as a catalyst and then 2 or 3 boiling chips to reduce bumping. Reflux the reaction mixture for 3 hrs, cool the flask, and filter off the resin. Concentrate the filtrate to a syrup on a steam bath. In a like manner prepare the glycosides of appropriate known sugars.

Apply 3 spots (1, 2, and 5 $\mu$l on spots 2, 4, and 7 mm in diameter) for each of the glycosides from the known and unknown sugars and 5 $\mu$l of a 1% solution of the corresponding free sugars to a sheet of Whatman No. 1 paper, and chromatograph in one dimension, using ethyl acetate: $n$-propanol:$H_2O$ (5:3:2) as the solvent. Locate the compounds on the paper with periodate-benzidine spray (see p. 17).

### *Seliwanoff's Test*

Prepare 0.5% solutions of glucose, fructose, and the unknown. Add 1 ml of each solution and 1 ml of $H_2O$ to four tubes. Add 9 ml of Seliwanoff's reagent to each tube, and heat them in a boiling water bath. Compare the times required for color development in the knowns and the unknown. Ketoses react readily in this reaction, whereas aldoses react more slowly. This can be used as the basis of a quantitative test if an appropriate heating period is selected.

### *Bial's Test*

Place 3 ml of Bial's reagent in each of four tubes. Add 0.1 ml of a 0.5% solution of: aldopentose to the first tube, aldohexose to the second, and an unknown to the third. To the fourth tube add water. Heat the solutions gently over a Bunsen burner until the first bubbles rise to the surface. Dilute the solutions to 10 ml with $H_2O$, add 5 ml of $n$-butanol, and shake the tubes. Observe and record your results.

### *Mucic Acid Test*

Place about 50 mg of galactose, glucose,

and unknown in separate test tubes, and add to each 1 ml of $H_2O$ and 1 ml of concentrated $HNO_3$. Heat the tubes in a boiling water bath for 1 hr, then allow the bath and tubes to cool slowly. Scratch the tubes with a clean stirring rod to hasten crystallization. Observe the crystals under a microscope, and check their solubility by adding 2 ml of $H_2O$ to each tube. Mucic acid will not dissolve under these conditions.

## Report of Results

Tabulate your results from Experiments 1–5 in the form shown below, and compare your unknown with the knowns. From your data, make your decision as to the identity of your sugar, and report it to the instructor.

Any major discrepancies between the unknown and the suspected known should be resolved by repeating the appropriate tests.

| Assay method | Unknown sugar | α-D-Galactose |
|---|---|---|
| Molisch test | + | + |
| Nelson's test | + | + |
| Moles $IO_4^-$/moles sugar | 4.8 | 5.0 |
| HCOOH formed/$IO_4^-$ consumed | 1.1 | 1.0 |
| $R_f$ of sugar | 0.53 | 0.55 |
| $\alpha_D$ initially | +137 | +144 |
| $\alpha_D$ at equilibrium | +80.0 | +80.5 |
| $R_f$ of methylglycosides observed | 0.51, 0.34, 0.10 | 0.48, 0.33, 0.11 |
| Seliwanoff's test | − | − |
| Bial's test | − | − |
| Mucic acid test | + | + |

## Discussion

The presence of 2–4 glycosides identified on the chromatograms after glycoside formation from a single sugar serves as a good demonstration of the various anomeric forms of any given monosaccharide.

The series of tests performed in Experiments 1–5 are typical of the steps performed to identify unknown biological materials. The unknown is first classified by simple qualitative tests; then tentatively identified by more specific methods, such as chromatography and polarimetry; and finally identified by the preparation of a characteristic derivative.

### Exercises

1. Why are there four possible forms of methylglycosides of aldohexoses?
2. Draw the structures of the saccharic acids expected by the action of $HNO_3$ on (a) glucose, (b) ribose, (c) galactose, and (d) mannose?
3. Why is it necessary to run water blanks and known sugars in Seliwanoff's test and in Bial's test?
4. Two α-methylglycosides (*A* and *B*), subjected to neutral periodate oxidation, are found to yield the same dialdehyde (*C*). Compound (*A*) consumed 2 moles of periodate per mole of sugar, concomitantly forming 1 mole of formic acid. Compound (*B*) consumed 1 mole of periodate per mole of sugar, but did not produce any formic acid. Oxidation of (*C*) with $Br_2$ water yielded (*D*):

$$R{-}\overset{O}{\overset{\|}{C}}{-}H \xrightarrow[Br_2]{\text{aqueous}} R{-}\overset{O}{\overset{\|}{C}}{-}OH.$$

which, on subsequent acid hydrolysis, yielded glyoxylic acid,

$$\left( \overset{O}{\underset{H-C}{\|}} - \overset{O}{\underset{C-OH}{\|}} \right)$$

L-glyceric acid, and methanol. Draw possible Haworth formulas for (*A*), (*B*), and (*C*). Give names for the (*A*) and (*B*) you have drawn.

**References** (*consult standard texts in organic chemistry*)

Bial, M., *Deut. med. Wochschr.*, **29,** 477 (1903).

Seliwanoff, T., *Ber.*, **20,** 181 (1887).

Kent, W. H., and B. Tollens, *Ann.* **227,** 221 (1885).

EXPERIMENT

# 6. Glycogen: Isolation

## Theory of the Experiment

Living organisms frequently contain stored carbohydrates which apparently act as reserve materials. These substances are stored in the form of polysaccharides (polyglycosides), such as the starch and inulin of higher plants and the glycogen of higher animals. Such polysaccharides have chemical and physical properties which differ from those of the simpler sugars studied in Experiments 1–5. In Experiments 6–8 we shall examine some of these properties by characterizing the polysaccharide glycogen, which we will obtain from rat liver.

The physical and chemical properties of many neutral polysaccharides are sufficiently different from those of other naturally occurring substances to permit their ready isolation. Thus when rat liver is homogenized in trichloroacetic acid (TCA), many high-molecular-weight compounds, such as proteins and nucleic acids, are readily precipitated, while the polysaccharide, glycogen, remains in solution. Since polysaccharides are considerably less soluble than sugars in aqueous alcohol, glycogen can be separated from sugars and other water-soluble compounds by precipitation with alcohol. Purified glycogen is obtained from aqueous solution by subsequent reprecipitation with alcohol.

## Experimental Procedure

MATERIALS

Rat
10% TCA
Absolute ethanol
Washed sand
Clinical centrifuge
NaCl
Dissecting tools
95% Ethanol
Diethyl ether
Ice
5% TCA

## Isolation

Decapitate a rat, and remove the liver. Place the liver on ice immediately. When cool, quickly weigh the liver to the nearest 0.1 g, cut it into small pieces, and drop it into a precooled mortar containing one volume (1 ml per gram of tissue) of 10% TCA. Add 0.5 g of washed sand, and, keeping the mortar on ice, grind the liver with the pestle. (**Caution:** Take care that TCA does not splash on hands.) Centrifuge the homogenate in balanced, heavy-walled centrifuge tubes for 5 min at 2000 × *g* (top speed in the International Clinical Centrifuge).

Decant the opalescent supernatant into a 100-ml graduated cylinder. To achieve a more complete extraction of liver glycogen, rinse the mortar and pestle with one more volume of 5% TCA. Pour these washings into the centrifuge tubes, which contain the residue obtained from the first extraction. Stir the mixtures with a glass rod, allow the tubes to stand for 5 minutes, and then centrifuge at 2000 × *g*. Discard the precipitates. Add the supernatants to the first TCA extract in the graduated cylinder, and record the total volume.

While stirring the combined TCA extracts, slowly add 2 volumes of 95% ethanol per volume of TCA extract. Stir the mixture, and allow it to stand until the precipitate flocculates. If no flocculation occurs add a pinch of sodium chloride, and warm the tube gently by placing it in a beaker of warm water until a precipitate forms. Transfer the suspension to centrifuge tubes, balance them, and after the precipitate has begun to settle, centrifuge the material for three minutes at 2000 × *g* Discard the supernatant. Dissolve the white precipitate in about 5 ml of $H_2O$, and reprecipitate it by adding 2 volumes (10 ml) of 95% ethanol.

Collect the precipitate by centrifugation, and wash it in the centrifuge tube with 3 ml of 95% ethanol by dispersing it with a glass rod. After centrifugation, wash the precipitate first in 3 ml of absolute EtOH and then in 3 ml of diethyl ether. Air-dry the preparation by spreading the white precipitate in a thin layer over a watch glass. Using an analytical balance, weigh the preparation in a tared, stoppered bottle, and record the yield of glycogen. Label the sample, and store it for use in Experiment 7.

## Report of Results

In addition to indicating the total yield of glycogen, calculate the content of glycogen in the liver as grams of glycogen per 100 grams of fresh liver. Compare your results with those of other students to see if liver size affects glycogen content.

## Discussion

The method of isolation used in this experiment is based on those properties of glycogen which are distinct from those of other materials present in rat liver. This procedure gives a quantitative extraction of the total liver glycogen.

### Exercises

1. What factors might influence the percentage of glycogen in a rat liver?
2. Why is it necessary to wash this glycogen preparation with alcohol and ether?
3. Why are time and temperature important in the initial stages of the isolation but not in the later steps?

### Reference

Stetten, M. R. et al., *J. Biol. Chem.*, **222** 587 (1956).

EXPERIMENT

# 7. Glycogen: Acid and Enzymatic Hydrolysis

## Theory of the Experiment

Glycogen has a branched structure with linear chains of consecutive glucose residues joined by α-1,4 linkages and with α-1,6 linkages at the branch points. The result is a fanlike structure with one terminal reducing end and many nonreducing ends (Fig. 1).

The hydrolysis of these glycosidic bonds is catalyzed by either acids or enzymes. In the acid-catalyzed hydrolysis there is a random cleavage of bonds, with the intermediate formation of all the various possible oligosaccharides and with the final conversion of these oligosaccharides to glucose. Several enzyme catalyzed hydrolyses are more specific with respect to bonds cleaved —for example, α-amylase of human saliva. This enzyme is also found in a variety of plant tissues and animal pancreatic juices. The α-amylases catalyze the rapid, random hydrolysis of internal α-1,4 bonds. They do not, however, hydrolyze α-1,6 linkages, regardless of molecular size, nor do they hydrolyze maltose. Thus glycogen is initially split by α-amylase action into branched dextrins of medium molecular weight; only small amounts of maltose are formed. Further action of α-amylase decreases the molecular weight of these dextrins, yielding hepta-, hexa-, penta-, tetra-, and trisaccharides. The final degradation products of the action of α-amylase on glycogen are glucose, maltose, and isomaltose (6-α-D-glucopyranosyl-α-D-glucopyranoside). The glucose is formed by the relatively slow end cleavages of the oligosaccharides which yield glucose and an oligosaccharide one residue shorter.

The polysaccharide nature of glycogen is verified in this experiment by demonstrating the increase in the number of reducing groups during the acid-catalyzed

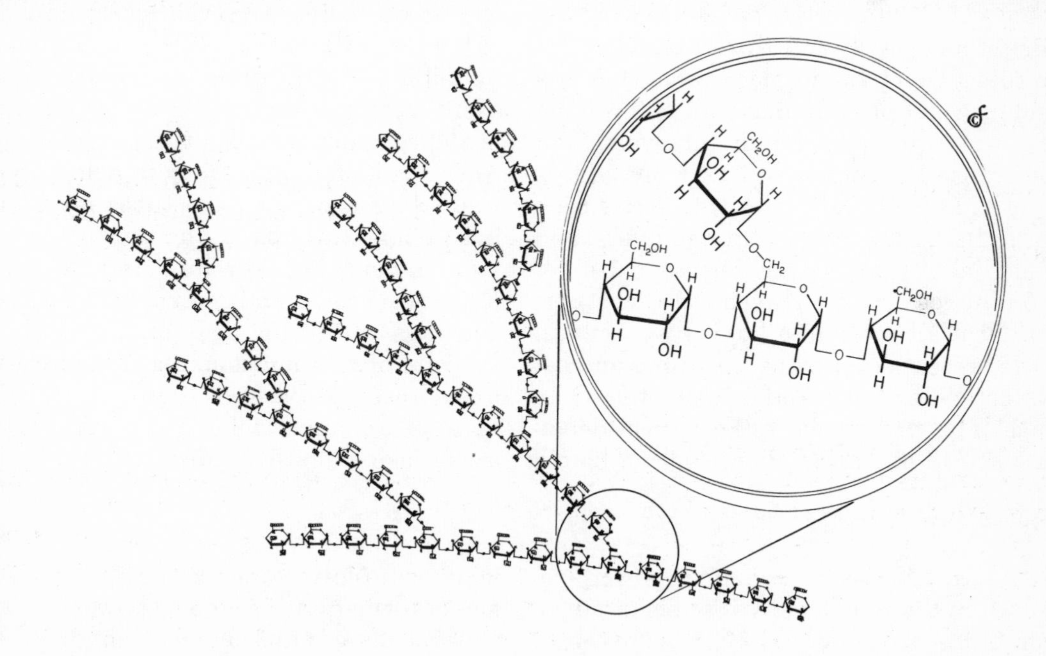

FIG. 1. *Glycogen structure.*

and α-amylase-catalyzed hydrolyses of glycogen. The progress of hydrolysis can be observed by measuring the increase in reducing sugar as assayed by 3,5-dinitrosalicylate reduction. The extent or percentage of hydrolysis is determined by comparing the amount of reducing sugar formed at any time with the amount of sugar present after total acid hydrolysis.

The α-amylase concentration varies among samples of saliva according to the physiological state and genetic constitution of the donor. Therefore greater or lesser dilution of saliva may be necessary so that the concentration of enzyme (catalyst) will catalyze the hydrolysis at a measurable rate.

## Experimental Procedure

### MATERIALS

Rat-liver glycogen
1.2*N* NaOH
0.02*M* Sodium phosphate buffer (*p*H 6.9)
2*N* HCl
3,5-Dinitrosalicylate reagent
0.005*M* NaCl
0.02 Sodium phosphate buffer (*p*H 6.9), 0.005*M* NaCl

### *Acid Hydrolysis*

Dissolve 40 mg of rat-liver glycogen in 5 ml of $H_2O$ (8 mg/ml), and add an 0.4-ml aliquot of the solution to each of seven numbered tubes (tubes 2–8). Fill tube 1 with 0.4 ml of water, and use it as a blank. Add 0.6 ml of 2*N* HCl to each tube, and record the time. Immediately add 1 ml of 1.2*N* NaOH to tubes 1 and 2, and place tubes 3–8 in a boiling water bath. Remove tubes 3–7 from the water bath at 3-min intervals, and neutralize the contents of each by adding 1 ml of 1.2*N* NaOH. After tube 8 has been in the bath a total of 25 min, remove it, and add 1 ml of 1.2*N* NaOH. Then add 1 ml of 3,5-dinitrosalicylate reagent to each tube, and heat all tubes in a boiling water bath for 5 minutes. Immerse the tubes in cool water. When cool, add 17 ml of $H_2O$ to each tube, and measure the optical densities at 540 mμ, using tube 1 as the blank.

### *Enzymatic Hydrolysis*

Collect about 2 ml of saliva in a small beaker. When ready to do the enzymatic hydrolysis (below), prepare a 1:80 dilution of this saliva by adding 0.25 ml of saliva to 19.75 ml of $H_2O$. Use the dilute saliva *immediately* for the hydrolysis.

Dissolve 32 mg of rat-liver glycogen in 4 ml of 0.005*M* NaCl, 0.02*M* sodium phosphate buffer (*p*H 6.9), and add a 0.4-ml aliquot to each of eight tubes. (The chloride ion "activates" α-amylase.) Prepare a water blank also. Add 0.5 ml of dilute saliva to each tube, and allow the tubes to incubate at room temperature. Immediately after adding the saliva, stop the reaction in one of the tubes containing saliva and glycogen by adding 1 ml of dinitrosalicylic acid reagent (zero time sample). Stop the reaction in the remaining tubes at 2-min intervals. Add 1 ml of dinitrosalicylic acid to the water blank, heat all the tubes for 5 min in a boiling water bath, and then cool them by immersion in cold water. Add 18.1 ml of $H_2O$ to each tube, and determine the optical densities at 540 mμ against the blank. If the rate of hydrolysis is too fast to give a linear rate for the first 4 min, or too slow to give an appreciable rate, repeat the assay, using another saliva level.

## Report of Results

The free glucose, or reducing power, released in the complete hydrolysis of glycogen (tube 8) represents 100% conversion of glycogen to glucose. With this as a standard, plot the percentage of hydrolysis (or formation of reducing power) versus time for the acid and enzymatic hydrolyses.

## Discussion

A second enzyme, β-amylase, also catalyzes the hydrolysis of glycogen. It catalyzes the successive hydrolyses of the second α-1,4 glycosidic bond from the free nonreducing ends of glucose chains, releasing maltose units. But β-amylase does not hydrolyze α-1,6 bonds, nor does it hydrolyze α-1,4 bonds of glucose chains beyond an α-1,6 branch residue. Thus, after all the nonreducing-end glucose chains have been trimmed back to the branch residues, the final products of the action of β-amylase on glycogen are maltose and the remaining "limit dextrin."

There are other methods for measuring the hydrolysis of glycogen and other polysaccharides, such as measurement of viscosity and observation of changes in the color of the polysaccharide iodine complexes. Each of these methods measures a different property of the polysaccharide. Reducing power is a measure of the number of free reducing groups in the reaction mixture; viscosity depends on the size and shape of the substrate molecules, and is decreased as hydrolysis proceeds; the color of the iodine complex becomes less intense as the chain length is decreased during hydrolysis.

### Exercises

1. A sample of glycogen is divided into two equal parts: one is treated with α-amylase; the other, with β-amylase. (a) After the reaction is complete, which preparation has the greater viscosity? The greater reducing power? The more intensely colored iodine complex? (b) Assume that the concentration of each enzyme is adjusted such that both enzymes yield reducing power at the same rate during the initial stages of hydrolysis. Are the rates of loss of viscosity also the same?
2. Would you expect treatment of glycogen with 1*N* NaOH at 100°C to yield results similar to those obtained from the acid hydrolysis of this experiment?

### References

Sumner, J. B., *J. Biol. Chem.*, **65,** 393 (1925).

Nord, F. F. (editor), *Advances in Enzymology*, Vol. XII, Interscience, New York, 1951, pp. 379–395.

Meyer, K. H. et al., *Helv. Chim. Acta.*, **31,** 2158 (1948).

EXPERIMENT

# 8. Glycogen: Periodate Oxidation

## Theory of the Experiment

The structure of glycogen and similar polysaccharides can be described as a branched, fanlike arrangement of glucopyranose residues with α-1,4 linkages along the chains and α-1,6 linkages at the branch points. As a result of this structure, only the terminal reducing end and all of the nonreducing ends yield formic acid on periodate oxidation. Furthermore, in such molecules the amount of formic acid formed from the single reducing end is insignificant in comparison with that formed from the many terminal nonreducing sugar residues. Accordingly, the number of nonreducing ends of glycogen is approximately equal to the number of molecules of formic acid formed upon periodate oxidation. The *average* segment length and the percentage of branching of the polysaccharide can in turn be calculated from the number of end groups, if the total number of glucose residues has been determined.

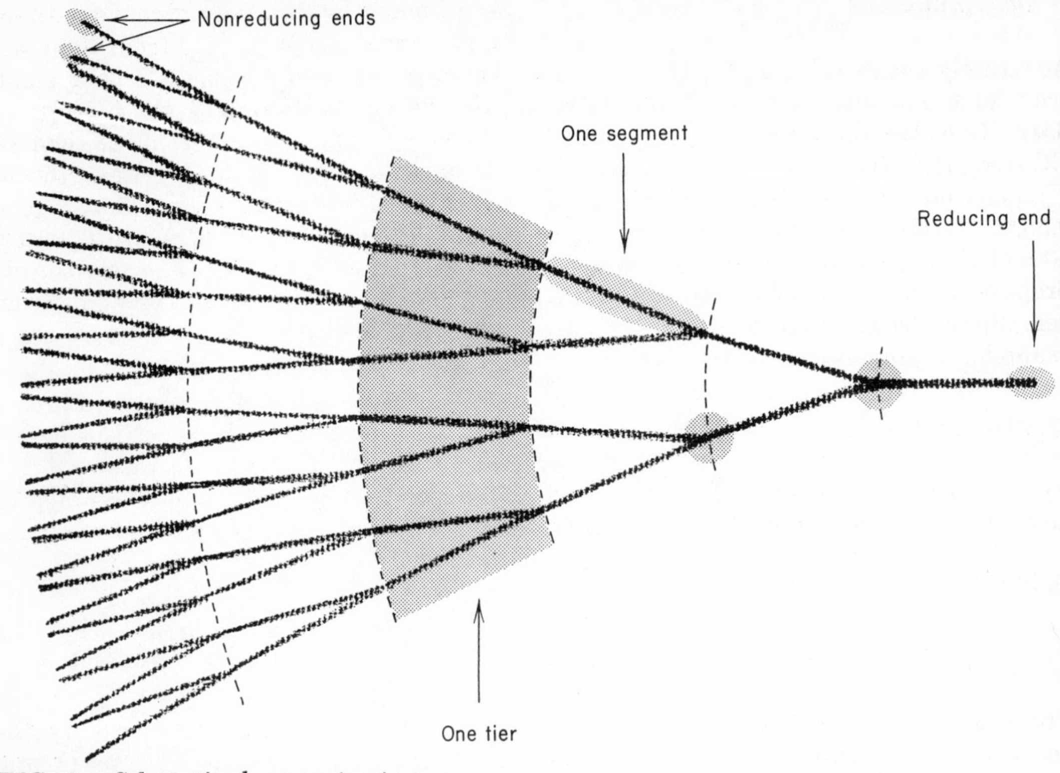


FIG. 2. *Schematic glycogen structure.*

The schematic diagram of the glycogen molecule (Fig. 2) illustrates that the number of branching $\alpha$-1,6 bonds is always one less than the number of nonreducing ends, $n$, of glycogen. Since $n$ is a large number, $n \cong n - 1$; that is, the number of $\alpha$-1,6 bonds is equal to the number of nonreducing ends. Therefore the percentage of branching, or the percentage of all the glucose residues from which branches originate, is determined by the equation

$$\%\ \text{branching} = \frac{\begin{pmatrix}\alpha\text{-1,6 bonds in}\\ \text{glycogen sample}\end{pmatrix}}{\begin{pmatrix}\text{glucose residues}\\ \text{in glycogen sample}\end{pmatrix}} \times 100$$

$$= \frac{\begin{pmatrix}\text{moles HCOOH formed by}\\ IO_4^- \text{ oxidation of sample}\end{pmatrix}}{\begin{pmatrix}\text{moles glucose in}\\ \text{glycogen sample}\end{pmatrix}} \times 100$$

The schematic diagram of glycogen also indicates that the number of 1,4-linked segments equals $2n - 1$. Since $2n$ is a large number, $2n \cong 2n - 1$. Thus it is possible to calculate the average number of glucose residues in a segment by use of the equation

$$\text{Glucose units in a segment} = \frac{\begin{pmatrix}\text{moles glucose}\\ \text{in glycogen sample}\end{pmatrix}}{\begin{pmatrix}\text{segments in}\\ \text{glycogen molecule}\end{pmatrix}}$$

$$= \frac{\begin{pmatrix}\text{moles glucose in}\\ \text{glycogen sample}\end{pmatrix}}{2 \times \begin{pmatrix}\text{moles HCOOH formed by}\\ IO_4^- \text{ oxidation of sample}\end{pmatrix}}$$

In the equation for glycogen, glucose exists in the glycoside form ($C_6H_{10}O_5$) with a molecular weight of 162.

## Experimental Procedure

### MATERIALS

Rat-liver glycogen
$0.35M$ $NaIO_4$
Standard $0.01N$ NaOH
Boiled, $CO_2$-free $H_2O$
Ethylene glycol
1% Phenolphthalein

### Determinations

Accurately weigh 250 mg of glycogen, and transfer the sample to a 50-ml volumetric flask. Dissolve the glycogen in 20 ml of $CO_2$-free $H_2O$ (you may need to heat the glycogen on a steam bath), add 10 ml of 0.35*M* $NaIO_4$, and adjust the volume to 50 ml by adding boiled $H_2O$. For the blank prepare a similar sample, omitting glycogen. Place the solutions in the dark (for example, in your desk). At approximately 1-, 4-, and 8-hr intervals, remove a 10-ml aliquot from the blank and from the reaction mixture, and place each in a small Erlenmeyer flask. Add 2 or 3 drops of ethylene glycol to each, and place them in the dark for 10–15 min. Titrate both aliquots (reaction mixture and blank) to a phenolphthalein end point with 0.01*N* NaOH, using an ascarite tube over the air inlet of the burette to exclude $CO_2$ from the NaOH solution.

## Report of Results

By plotting the amount of formic acid formed versus time, determine when the periodate reaction reaches completion. Calculate the average number of glucose residues per segment and the percentage of branching from the total amount of formic acid formed.

## Discussion

The studies in Experiments 7 and 8 are not sufficient to serve as a complete characterization of glycogen, but they do point out some of the important features of the glycogen structure. The observation that unhydrolyzed glycogen is essentially free of reducing power is consistent with the hypothesis that glycogen, as a polysaccharide, would have few reducing groups per molecule. The results of the acid hydrolysis show that glycogen is a polysaccharide containing many glycosidic bonds. The partial hydrolysis catalyzed by the specific enzyme $\alpha$-amylase indicates the nature of some of the internal linkages of the glycogen molecule. The apparent lack of reducing power before hydrolysis and the large amounts of formic acid produced upon periodate oxidation are indications that glycogen is a branched structure. The calculations of per cent branching and average chain length are in agreement with this hypothesis.

### Exercises

1. Describe one method of demonstrating the existence of 1,6-glycosidic bonds in glycogen. What further operations could be performed to characterize the glycogen structure?
2. Assuming a molecular weight of $6 \times 10^6$ for the glycogen assayed, use your data to calculate the number of tiers in your glycogen. (Note that the total number of segments is $2T - 1$, where $T$ is the number of tiers.)
3. What would be the approximate average segment lengths of amylose and amylopectin obtained by $IO_4^-$ oxidation and HCOOH titration?
4. Assuming a fixed segment length of 5 residues in a glycogen molecule, show that the amount of HCOOH produced from the reducing end of a symmetrical 10-tier molecule of glycogen by $IO_4^-$ oxidation is very small compared with that produced from the nonreducing ends.

### References

Bobbitt, J. M., *Advances in Carbohydrate Chem.*, **11**, 1 (1956).
Dyer, J. R., *Methods of Biochem. Anal.*, **3**, 111 (1956).

EXPERIMENT

# 9. Glucose-1-Phosphate: Enzymatic Formation from Starch

## Theory of the Experiment

Starch, a carbohydrate widely distributed in plants, is a mixture of two polysaccharides: amylose, a straight-chain polymer of glucose units joined by $\alpha$-1,4 linkages; and amylopectin, a branched-chain glucose polymer which differs from glycogen primarily in its larger number of $\alpha$-1,4-linked glucose units between the $\alpha$-1,6 branch points. Both glycogen and starch are acted upon by the enzyme phosphorylase, which catalyzes the phosphorolysis of glycogen or starch to glucose-1-phosphate.

In this experiment, soluble starch is incubated with phosphate and a phosphorylase preparation from potatoes. At the end of 24–48 hr, the reaction is stopped by heating the mixture to destroy the enzyme. After removal of the denatured enzyme, the unreacted inorganic phosphate is removed from the filtrate by precipitation as magnesium ammonium phosphate. The glucose-1-phosphate is then isolated by ion exchange chromatography and is subsequently crystallized.

Glucose-1-phosphate is determined by measurement of the phosphate released by acid hydrolysis. Phosphoacetals of all sugars can be hydrolyzed by treatment with 1$N$ acid for 7 min at 100°C, but the various sugar phosphate esters (for example, glucose-6-phosphate) ordinarily require longer heating times or more concentrated acid for complete hydrolysis.

The preparation and isolation of glucose-1-phosphate can be conveniently completed within three laboratory periods. The enzyme incubation can be started in the first period. The removal of cations with Dowex 50 should be completed in the second period and the product stored at 0–5°C. The ion exchange can then be completed and the crystallization begun in the third period. Thus Experiment 10 may be started in the next period, after the collecting and drying of the crystals of dipotassium glucose-1-phosphate dihydrate. Deionized or distilled water should be used throughout and no solutions should be discarded until a product is obtained. Store all solutions at 0–5°C between laboratory periods to avoid bacterial or chemical degradation.

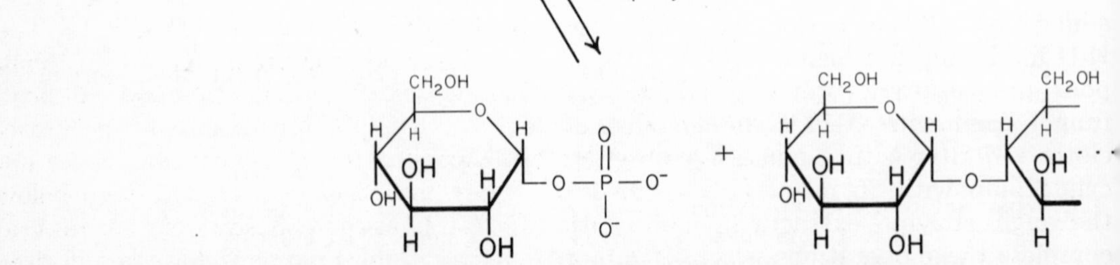

Nonreducing end

Phosphorylase

## Experimental Procedure

MATERIALS

Phenylmercuric nitrate slurry
0.8*M* Potassium phosphate buffer *p*H 6.7
Soluble starch
Filter aid
14% $NH_4OH$
2*N* NaOH
$Mg(Ac)_2 \cdot 4H_2O$
IR-45 ($OH^-$ form)
Absolute methanol
Cheesecloth
Cylindrical glass tubes
Acid molybdate reagent
Phosphate standard
2*N* HCl
Potatoes
Dowex 50 ($H^+$ form)
5% KOH
Blendor
Glass Wool
Charcoal
Reducing reagent
0.01*N* KI, 0.01*N* $I_2$

### *Preparation of Starch*

Using a minimum of $H_2O$, make a smooth slurry of 10 g of soluble starch. Add this to 70 ml of boiling $H_2O$, and heat and stir until the solution is nearly clear. Then stop the heating, and add 180 ml of cold $H_2O$ to help cool the solution to room temperature. Do not add the enzyme until the solution has cooled to room temperature (heat inactives the enzyme).

### *Preparation of Enzyme*

Peel a medium-sized potato (precooled for 24 hours at 1–5°C), and cut it into $\frac{1}{2}$-in. cubes. Blend 150 grams of these cubes, added over a 30-sec period, with 150 ml of $H_2O$ for 2 min in a blendor. Then quickly pour the resultant brei onto a Büchner funnel lined with 8–10 layers of cheesecloth, and filter with vacuum, washing the crude pulp with 25 ml of $H_2O$ to insure thorough enzyme extraction. Failure to complete these operations within 2 min of blending may result in loss of enzyme activity. After filtration, add 100 mg of phenylmercuric nitrate, as a powder or slurry, to inhibit the action of other enzymes and bacterial growth. Allow the preparation to stand for 5 min; then decant the solution and any floating phenylmercuric nitrate from the accumulated precipitate. Adjust the extract to a volume of 250 ml with $H_2O$.

### *Incubation of the Enzyme with Starch*

Add the 250 ml of enzyme solution to 250 ml of starch solution. Then add 250 ml of 0.8*M* phosphate buffer solution, record the total volume, and store the solution in a stoppered Erlenmeyer flask in your desk.

During the incubation period (24–48 hr) the reaction mixture will form a red and then a dark blue or purple color, owing to the action of other enzymes present in the crude extract. The colored materials will be removed during later procedures.

### *Removal of Inorganic Phosphate*

After 24–48 hr of incubation, stop the enzymatic reaction by rapidly heating the solution to 95°C and then slowly cooling it. Remove the coagulated protein by suction filtration in a large Büchner funnel containing filter paper precoated with filter aid. Then remove the excess phosphate by dissolving 0.2 mole of magnesium acetate [44 g of $Mg(Ac_2 \cdot 4H_2O)$] in the filtered solution and then adjusting the *p*H to 8.5 with 14% ammonia (use *p*H paper first, then a meter). Cool the solution in a salted ice bath for 10 min, and remove the precipitated magnesium ammonium phosphate by suction filtration. Remove duplicate 0.1-, 0.2-, and 0.5-ml aliquots of the filtered solution for the inorganic phosphate and 7-min phosphate assays described below. If the phosphate assay reveals that an excess of inorganic phosphate is still present in the incubation filtrate (that is, if an intense blue color forms in the unhydrolyzed 0.1-ml aliquot), add 1 g of $Mg(Ac)_2 \cdot 4H_2O$, adjust to $p$H 8.5 with 14% ammonia, cool the solution, and filter it as above before repeating the phosphate assays. Calculate the number of micromoles of inorganic phosphate and 7-min phosphate in the entire volume of filtered solution. The incubation filtrate should be essentially free of inorganic phosphate before proceeding.

*Note:* In the above procedure, filter clogging due to gelatinized starch may be encountered in the filtration steps. A short treatment of the solution with $\alpha$-amylase (1–2 ml of saliva) will aid in filtration.

### *Assay of Inorganic Phosphate and Glucose-1-Phosphate. Range-finding for Phosphate Assay*

In the Fiske-Subbarow colorimetric method for the determination of phosphate, the color yield is directly proportional to the amount of inorganic phosphate only when the aliquot taken for analysis contains between 0.1 and 1.0 micromole of phosphate. Since the efficiency of the enzymatic formation of glucose-1-phosphate may vary somewhat, depending upon the sources of the enzyme and the starch, as well as the length of the incubation period, it is possible that the aliquot sizes suggested may not lie within the range in which the assay is valid. In this event it will be necessary to analyze several different aliquot sizes until one is found which falls within the accurate range of the assay. In this experiment, as in all isolation procedures, it is necessary to obtain an accurate measurement of the amount of the desired compound in each fraction.

Therefore, you will need to determine, and use, appropriate aliquot ranges for analysis before proceeding to the next steps in the experiment. Further, you must keep accurate records of aliquot sizes and protocols used in order to evaluate your data correctly.

### *Seven-min Hydrolysis*

Set aside one of the duplicates of each pai of samples to be assayed for inorgani phosphate. To the other sample add ar equal volume of $2N$ HCl; 1 ml of $1N$ HC is preferable with aliquots of less thar 0.2 ml. Now place the acidified samples ir a boiling water bath for 7 min. Remove the samples, cool and neutralize them to $p$H 6.5–7.5 by adding (pipet) a stoichiometric amount of $2N$ NaOH. Dilute all the tubes (including the unhydrolyzed aliquots) to 3 ml with $H_2O$.

### *Inorganic Phosphate Determination (Modified Fiske-Subbarow Method)*

For each phosphate analysis prepare tubes containing (1) a water blank, (2) the unhydrolyzed aliquot(s), (3) the hydrolyzed aliquot(s), and (4) two phosphate standards (0.4 and 0.8 micromoles of inorgani phosphate). Adjust the volume of all tubes to 3.0 ml with water, and add, in order

1 ml of acid molybdate reagent,
1 ml of reducing reagent (3% $NaHSO_3$, 1% Elon),
5 ml of $H_2O$.

Mix the solutions by inverting the tubes and allow the color to develop for 20 mi before reading the optical density at 66 m$\mu$. Calculate the quantity of inorgani phosphate and glucose-1-phosphate in th aliquot and in the entire reaction mixture

### *Use of Cation-exchange Resin*

Decolorize the resultant solution, freed o inorganic phosphate, by stirring with 2 of charcoal and then removing the char coal by vacuum filtration using filter aid This procedure yields a clear to yellow solution containing glucose-1-phosphate unreacted starch, and many salts whic were present either in the original potat extract or in the reagents added durin the course of the experiment.

Remove the cations by treatment with Dowex 50 in the following manner: Add 300 ml of moist Dowex 50 in $H^+$ form to the decolorized solution and stir gently for 5 min before separating by vacuum filtration. If the *p*H of the filtrate is not acidic (*p*H 1–3), add 100 ml of moist Dowex 50 in $H^+$ form, then stir and filter the solution as before. Repeat this procedure until the *p*H of the solution is 1–3. Record the volume of the resultant solution, and remove aliquots identical to those previously used in the assay of the $MgNH_4PO_4$ supernatant. Assay these for 7-min phosphate. If necessary the remaining solution may be stored at 0–5°C for several days without loss of glucose-1-phosphate.

### *Use of Anion-exchange Resin*

The next step in the purification procedure is the column chromatography of the Dowex 50 filtrate on Amberlite IR-45 in $OH^-$ form. This involves removing the anions in the acid solution from all other contaminating materials. Thus when the acidic filtrate (*p*H 1–3) contacts the IR-45 in $OH^-$ form, the glucose-1-phosphate is ionically absorbed on the resin while unionized materials in the acid solution, such as acetic acid and unreacted starch, pass through the column. When the resin is eluted with strong alkali (5% KOH), the adhering anions are displaced by the $OH^-$ ions and are obtained in the eluate from the column.

Prepare a column about 4 cm in diameter and 20–30 cm long using a rubber stopper, screw clamp, and glass wool plug (Fig. 3). Mix 250 ml of $H_2O$ with 250 ml of moist IR-45 ($OH^-$), and pour the resultant slurry into the column in such a manner as to exclude air pockets from the settled resin. Then wash the resin with $H_2O$ until the effluent is about *p*H 9 (*p*H paper). Drain or pipet off the excess fluid until the fluid surface just covers the top of the resin bed. Cover the surface of the resin with a layer of glass wool.

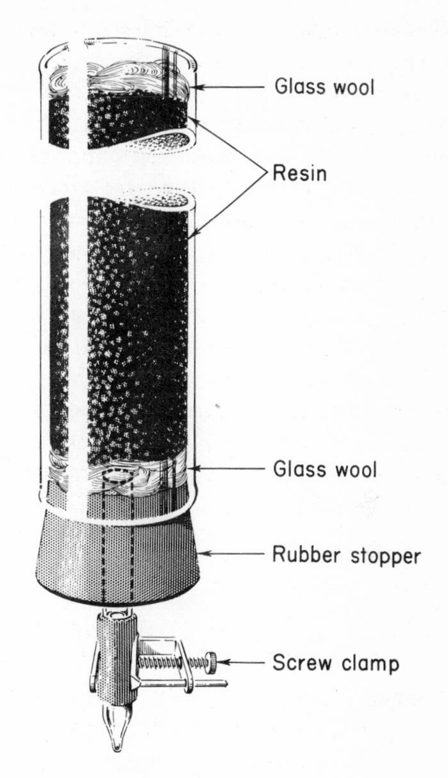


FIG. 3. *Anion exchange column.*

To effect adsorption of the glucose-1-phosphate by the IR-45, gently pour the acidic solution treated with Dowex 50 onto the column, avoiding the introduction of air bubbles into the resin bed. Open the screw clamp, and adjust the flow rate to about 15 ml/min. Pass the entire solution through the IR-45 resin without permitting air to enter the column. Collect the entire effluent, and assay aliquots for 7-min phosphate. If appreciable 7-min phosphate appears (> 30% of G-1-P of Dowex 50 filtrate), and if time permits, add an additional 50 g of IR-45 to the column, and pass the effluent through the column again. After the original solution has passed into the column and the effluent has been satisfactorily freed of the 7-min phosphate, wash the column with deionized $H_2O$ until the effluent no longer gives a positive test for starch (that is, a blue color upon mixing a drop of effluent with a drop of $0.01N$ $I_2$ in $0.01N$ KI).

To elute the glucose-1-phosphate from the resin, pass 5% KOH through the column, adjusting the flow rate to 20 ml/min. Collect separate, successive 50-ml fractions in a graduate. Continue collecting fractions for about 200 ml after the *p*H of the effluent becomes markedly alkaline (*p*H 11–13) as observed with *p*H paper. (The usual total volume of all fractions equals 500 ml.) Then test 0.1-ml aliquots from each fraction for 7-min phosphate. Some fractions will require smaller aliquots to fall within the linear range of the standard curve, but 0.1-ml aliquots will serve to find the "peak" for glucose-1-phosphate. Combine the fractions containing 80–90% of the recovered glucose-1-phosphate, and adjust the solution to *p*H 8 or higher by adding a few drops of 5% KOH. Add three volumes of absolute methanol to the combined fractions, and leave them at 0–5°C for at least 12 hr for crystallization of dipotassium glucose-1-phosphate dihydrate. In this temperature range this compound will remain stable indefinitely. Finally, collect the crystals by centrifugation or filtration of the cold solution. Wash the crystals with 5 ml of absolute methanol, then dry, weigh, and store the crystals for use in Experiment 10.

## Report of Results

1. Prepare a flow sheet of the steps in the isolation procedure, indicating the purpose of each step.
2. Prepare a table (see sample below) showing the percentage of the glucose-1-phosphate recovered at each step in the procedure.
3. Prepare a graph of the elution pattern of the glucose-1-phosphate, plotting micromoles of glucose-1-phosphate/50 ml as the ordinate and the fraction numbers, or milliliters of effluent, as the abcissca.
4. Calculate the maximum yield of glucose-1-phosphate that would be expected if the reaction had reached equilibrium. (The quantity of starch is altered only slightly during the incubation, therefore the starch concentrations in the numerator and de

**TABLE II.** Percentage of Glucose-1-Phosphate Recovered During Purification.

| Step | Vol of solution | μMoles of 7-min phosphate/ml | Total μMoles of 7-min phosphate | Percentage recovered |
|---|---|---|---|---|
| $Mg(NH_4)PO_4$ supernatant | | | | 100 |
| Dowex 50 supernatant | | | | |
| Original IR-45 ($OH^-$) washes | | | | |
| Selected KOH effluent fractions | | | | |
| Glucose-1-phosphate crystals* | —— | —— | | |

*Assume that the crystals are dipotassium glucose-1-phosphate dihydrate (molecular weight 372).

nominator are roughly equivalent and cancel out. Use as your constant for calculation

$$\frac{\text{glucose-1-phosphate}}{\text{inorganic phosphate}} = 0.088$$

Compare your yield with the theoretical yield expected. Suggest reasons for any discrepancy between your results and the theoretical value calculated from the phosphorylase constant.

## Discussion

Since the reaction catalyzed by phosphorylase is a phosphorylysis of starch rather than a hydrolysis, no free reducing groups are formed in the reaction. But even though one may have no knowledge concerning the nature of the product of such an enzymatic reaction, one can tell whether the starch is being depolymerized by observing changes in viscosity, iodine color, and so on in the polysaccharide solution as the reaction proceeds.

This experiment demonstrates the use of an assay procedure to follow the progress of fractionation in the isolation of the product of an enzyme-catalyzed reaction. Here again, as discussed in the glycogen hydrolysis experiment, it is apparent that a study of the reaction products at various stages of the starch breakdown gives information concerning the chemical nature of the enzyme-catalyzed reaction. Further steps, necessary to characterize the product as glucose-1-phosphate, will be performed in Experiment 10.

**Exercises**

1. Consider the following set of data for inorganic phosphate determinations on unhydrolyzed and 7-min, hydrolyzed aliquots taken from a 700-ml volume of the $MgNH_4PO_4$ supernatant obtained in this experiment.

| Vol of aliquot | Optical density at 660 mμ | |
|---|---|---|
| | Unhydrolyzed | 7-Min hydrolyzed |
| 0.1 ml | 0.004 | 0.225 |
| 0.2 ml | 0.010 | 0.435 |
| 0.5 ml | 0.022 | 0.860 |
| Standards | | |
| 0.4 μmole inorganic phosphate | 0.150 | |
| 0.8 μmole inorganic phosphate | 0.300 | |

(a) Which optical density values are usable for further calculations?
(b) Assume 100% recovery of the glucose-1-phosphate and inorganic phosphate at the beginning of the anion exchange step. How many milliequivalents of anion exchange resin must be used to handle a $p$H 3 solution of these anions ($p\text{Ka}_1 = 1$, and $p\text{Ka}_2 = 6$ for glucose-1-phosphate)?
(c) Assume 80% recovery of the glucose-1-phosphate as crystals. How many grams of dipotassium glucose-1-phosphate dihydrate would be isolated from the above solution?

2. Propose a series of steps, using ion exchange resins, for the isolation of glucose-6-phosphate from a solution containing methylamine, sodium acetate, glucose, and glucose-6-phosphate.
3. What steps would you perform to convert the rather insoluble barium salt of glucose-6-phosphate to the dipotassium salt?

**References**

Fiske, C. H., and Y. Subbarow, *J. Biol. Chem.*, **66,** 375 (1925).

McCready, R. M., and W. Z. Hassid, *J. Am. Chem. Soc.*, **66,** 560 (1944).

———, *in* Colowick, S. P., and N. O. Kaplan (Editors), *Methods in Enzymology*, Academic, New York, Vol. 3, 1957, p. 137.

EXPERIMENT

# 10. Glucose-1-Phosphate: Chemical Characterization

## Theory of the Experiment

It is necessary to establish the identity of any newly isolated compound. In the present experiment, the first step consists in showing that the product presumed to be glucose-1-phosphate is in fact composed of equal quantities of glucose and phosphate. It is also necessary to distinguish between the various isomers of glucose-phosphate. The second step is particularly pertinent to the present experiment, since it is known that glucose-1-phosphate may be converted to glucose-6-phosphate by the enzyme phosphoglucomutase, which is present in the crude potato extracts of Experiment 9.

The characterization of glucose-1-phosphate is based on the relative stability of various sugar phosphates in acid solution. Sugar phosphoacetals are nonreducing sugars, and release equal amounts of reducing sugar (characterizable by chromatography) and phosphate upon 7-min hydrolysis in 1*N* acid at 100°C. In contrast, glucose-6-phosphate and other phosphate esters require more concentrated acid or more prolonged heating for complete hydrolysis. It is possible, therefore, to char acterize glucose-1-phosphate by qualitativ (chromatographic) and quantitative analy sis of the material(s) present before an after "7-min hydrolysis." It is also poss ble to estimate the purity of the product b careful evaluation of the results.

Chromatographic characterization c sugar phosphates is aided by our abilit to detect the phosphate present, althoug other means such as the aniline-acid-oxalat and periodate-benzidine also detect glu cose-6-phosphate. Phosphate detection i usually accomplished by means of a com bined acid-molybdate spray reagent. Th acid of the spray reagent hydrolyzes th phosphate esters or acetals, resulting i the release of inorganic phosphate, whic then combines with the molybdic acid an is reduced to yield a blue spot (phospho molybdous acid). Though frequently use for the detection of sugar phosphates, thi type of spray is effective for all organi phosphates.

## Experimental Procedure

MATERIALS

Nelson's reagents A and B
Glucose standard
1*N* NaOH
Reducing reagent
10*N* $H_2SO_4$
0.5% $NaIO_4$
1% Glucose-1-phosphate
1% Glucose
U.V. lamp
Arsenomolybdate reagent
1*N* HCl
Phosphate ($P_i$) standard (1 $\mu$mole/ml)
2.5% Ammonium molybdate
Oil or sand bath
Whatman No. 1 paper
0.5% Benzidine reagent
1% Glucose-6-phosphate
$CH_3OH:NH_4OH:H_2O$ (6:1:3)
30% $H_2O_2$

### *Characterization of Compounds Present Before and After Hydrolysis*

A sample of the suspected glucose-1-phosphate can be checked for purity by evaluation of the following quantities:

(a) reducing power before 7-min hy drolysis (tube 1 below);
(b) reducing power after 7-min hydroly sis (tube 2 below);

(c) inorganic phosphate present before 7-min hydrolysis (tube 3 below);
(d) inorganic phosphate present after 7-min hydrolysis (that is, 7-min phosphate tubes 4 and 4′ below);
(e) phosphate present after total hydrolysis (tube 5 below).

To determine these values, prepare a solution of the isolated, suspected glucose-1-phosphate (1.60–2.20 mg/ml to two decimal places), and place 0.1-ml aliquots in each of 6 tubes. Number the tubes 1, 2, 3, 4, 4′, and 5.

#### 7-MIN HYDROLYSIS

Add 1.0 ml of 1*N* HCl to tubes 2, 4, and 4′. Heat them for 7–8 min in a boiling water bath; then cool them in a beaker of cold water. Finally, neutralize the contents by adding 1 ml of 1*N* NaOH to each of the tubes.

#### TOTAL HYDROLYSIS

Add 0.5 ml of 10*N* $H_2SO_4$ to tube 5. Heat the tube for 30–60 min at 130–160°C in an oil or sand bath. When the solution is clear, cool the tube, and slowly (**Caution:** Concentrated $H_2SO_4$) add 2 ml of $H_2O$. Finally, heat this diluted solution for 10 min in a boiling water bath to hydrolyze any pyrophosphates formed during the digestion.

#### PREPARATION FOR ASSAY OF UNHYDROLYZED PRODUCT

Add 3 ml of $H_2O$ to tubes 1 and 3, and save this for the assay of unhydrolyzed product.

#### NELSON'S TEST

Analyze tube 1 (untreated) and tube 2 (7-min hydrolyzed) for reducing sugar by Nelson's test (see Experiment 1) as follows: To the sample add 1 ml of Nelson's combined reagents A and B (25:1). Heat the solution for 20 min in a boiling water bath, cool, add 1 ml of arsenomolybdate reagent, shake the tube for 5 min, and dilute the contents to 10 ml with $H_2O$. Read the optical densities at 540 mμ. Treat blank and standards (50 and 100 μg of glucose) similarly.

#### DETERMINATION OF INORGANIC PHOSPHATE

Using the modified Fiske-Subbarow method, assay the following for inorganic phosphate: tube 3 (no hydrolysis), tubes 4 and 4′ (7-min hydrolysis), tube 5 (total hydrolysis), three standards (0.3, 0.5, and 0.7 micromoles of phosphate), and a blank. Tube 5 already has sufficient acid for the assay; therefore, add 0.5 ml of 10*N* $H_2SO_4$ to tubes 3, 4, and 4′ (not 5), to the standards, and to the blank. Then add 1 ml of 2.5% ammonium molybdate and 1 ml of reducing reagent to all tubes, and make up to 10 ml with $H_2O$. Allow them to stand for 20 min, then read the optical density of each at 660 mμ against the blank.

### *Chromatographic Characterization*

Prepare a small quantity of a 1% solution of the suspected glucose-1-phosphate, and hydrolyze a 0.1-ml aliquot of the solution with an equal volume of 2*N* HCl at 100°C for 7 min. This procedure yields the organic compound released by 7-min hydrolysis. Then, to an ascending chromatogram, apply the following spots, 5–7 mm in diameter: 1 spot of the hydrolyzed material; 1 spot of 1% authentic glucose-1-phosphate; 2 spots of 1% authentic glucose-6-phosphate; 1 spot of 1% isolated, suspected glucose-1-phosphate; and 1 spot of 1% glucose. Develop the chromatogram in methanol:ammonium hydroxide: $H_2O$ (6:1:3). After drying the developed chromatogram, cut the paper into vertical strips. Spray the strips containing glucose-1-phosphate (authentic and suspected) and one of the strips containing glucose-6-phosphate with the phosphate spray (3%

$HClO_4$, 1% $(NH_4)_2MoO_2$, 0.1% Versene in 0.1*N* HCl), and allow them to air dry before developing the phosphate-containing spots by irradiation with an ultraviolet lamp or bright sunlight for a few minutes.

Spray the remaining strips with the periodate-benzidine spray. After the spots have developed, determine the $R_f$ values of the compounds. (*Note:* Acid hydrolysis of glucose-1-phosphate may yield some hydroxymethylfurfural, which will appear as a second spot. Further, glucose-1-phosphate does not give a strong response to the periodate-benzidine spray.)

## Report of Results

Prepare a table similar to the one shown below to aid in the proper evaluation of your results (your concentration may differ from that shown). Average the values you obtain from duplicates 4 and 4′.

Determine the number of micromoles of glucose-1-phosphate and glucose-6-phosphate in a 0.1-ml aliquot of your sample. Further, calculate the percentage of purity of the glucose-1-phosphate by comparing your data with the expected results from an equivalent amount of glucose-1-phosphate dipotassium dihydrate. Confirm your results by examining your chromatogram.

Determine the extent of hydration of your isolated compound by performing the following operations:

1. Calculate the percentage of 7-min phosphorous (not phosphate) in your sample (for example, μg P/100 μg sample).
2. Calculate the expected percentage of 7-min phosphorous from dipotassium glucose-1-phosphate.
3. Calculate the expected percentage of 7-min phosphorus from dipotassium glucose-1-phosphate dihydrate.
4. Compare your value for the percentage of phosphorus (1) with the two theoretical values (2 and 3), and decide which formula best fits your data.

Weigh and label the remaining glucose-1-phosphate, and turn it in to the instructor.

| 0.1 ml of material (180 μg/0.1 ml) | μMoles reducing power | μMoles inorganic phosphate |
|---|---|---|
| Initial sample | 0.05 | 0.00 |
| 7-Min hydrolyzed sample | 0.47 | 0.42 |
| Total-hydrolysis sample | — | 0.47 |

## Discussion

Glucose-1-phosphate was first characterized by Cori in 1937, and hence is often referred to as the Cori ester. During the same decade, a number of other phosphate esters of sugars were isolated from tissues and identified (for example, Robison's ester, or glucose-6-phosphate, 1931; Harden-Young ester, or fructose-1,6-diphosphate, 1932). The study of the enzymatic interconversions of these sugar phosphates, which are important intermediates in carbohydrate metabolism, will be considered in Experiment 30.

### Exercises

1. Describe chemical tests for determining whether a given pure sample of unknown is (a) glucose-6-phosphate, (b) glucose-1-phosphate, (c) β-methylglucoside, (d) glucose, (e) fructose-1,6-diphosphate, or (f) sorbitol.
2. Which of the following compounds yield inorganic phosphate upon 7-min hydrolysis: acetyl phosphate, 3-phosphoglyceraldehyde, ribose-5-phosphate, adenosine-5′-phosphate, ribose-1-phosphate, and pyrophosphate?

**References**


Cori, C. F. et al., *J. Biol. Chem.*, **121,** 465 (1937).

Nelson, N., *J. Biol. Chem.*, **153,** 375 (1944).

Umbreit, W. W., R. H. Burris, and J. F. Stauffer, *Manometric Techniques* (Rev. ed.), Burgess, Minneapolis, 1957.

Hanes, C. S., and F. A. Isherwood, *Nature*, **164,** 1107 (1949).


EXPERIMENT

# 11. Other Carbohydrates: Isolation and Characterization of Trehalose

## Theory of the Experiment

The object of Experiments 11 and 12 is to isolate other carbohydrates from biological sources and then to confirm their structures.

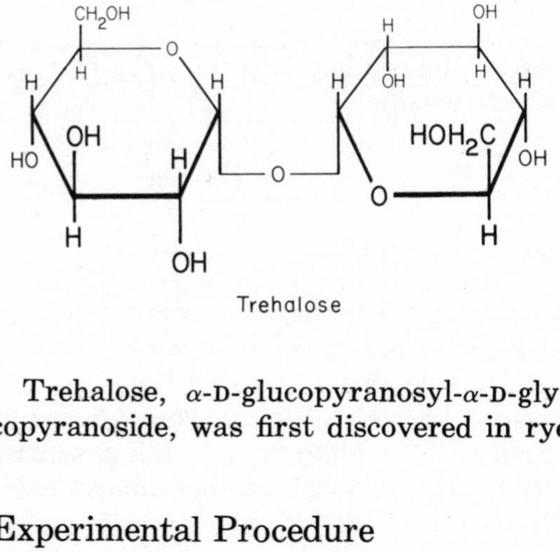

Trehalose

Trehalose, α-D-glucopyranosyl-α-D-glycopyranoside, was first discovered in rye ergot in the early nineteenth century. Since that time it has been isolated from young mushrooms, higher plants, seaweeds, the hemolymph of insects, and many other sources. Trehalose was first isolated from yeast in 1925 by Koch and Koch. During the course of investigations to discover the nature of "bios," a yeast growth factor, these workers allowed an alcoholic extract of yeast to stand for several months. At the end of this time, they found crystals clinging to the side of the flask. These crystals were shown to be trehalose, and had no bios activity. The method used in this experiment involves an alcohol extraction of a yeast, deproteinization of the extract, and eventual crystallization of trehalose from the alcoholic solution.

## Experimental Procedure

### MATERIALS

Active, dried baker's yeast
Saturated $Ba(OH)_2$
Filter aid
95% Ethanol
Charcoal
0.1*M* HCl
20% $ZnSO_4$
1% Phenolphthalein
70% Ethanol

### *Trehalose Isolation*

Make a paste of 32 g of dried baker's yeast with 68 ml of $H_2O$. Add 250 ml of 95% ethanol, and allow the mixture to stand for 30 min with intermittent stirring. Filter the mixture on a Büchner funnel, and wash the precipitate with three 30-ml portions of 70% ethanol. Combine the washes and the original extract. Now add 20 ml of 20% $ZnSO_4$, then 1 ml of 1% phenolphthalein and sufficient (about 50 ml) saturated $Ba(OH)_2$ to make the mixture pink to phenolphthalein. Add 2 g of activated charcoal. Heat the entire mixture to 70°C on a steam bath, and filter, while it is hot, through a Büchner funnel precoated with filter aid. Adjust the deproteinized solution to about *p*H 7 with 0.1*M* HCl, and concentrate it by gentle heating under vacuum in a flask until about 10 ml of syrup remains. Slowly mix 80 ml of 95% EtOH into the syrup; stopper the flask, and store it in the desk. Sometimes crystals will form overnight, but occasionally a week or more is required. Such slow crystallization often produces single crystals over ½ inch long. Crystallization can be hastened by cooling the sample to 0°C or by adding more 95% ethanol.

*Characterization*

Perform one or more of the following tests on an aqueous solution of the sugar to characterize it as trehalose:

1. Chromatography with known sugars, using a solvent system that will yield a distinctive $R_f$ value for trehalose.
2. Polarimetry (the $[\alpha]_D^{20}$ of trehalose is +178.3°).
3. Reducing-power assay, before and after acid hydrolysis, using Nelson's test or 3,5-dinitrosalicylate reduction.
4. Chromatography of the hydrolysis products obtained by maintaining a 0.5% solution of trehalose in 1*N* HCl at 100°C for 20 min.
5. Periodate assay, using a procedure similar to that of Experiment 2, but measuring $IO_4^-$ consumption over only a 3- to 4-hour interval.

## Report of Results

Express the results of the isolation in terms of the percentage yield from the original dried yeast. Discuss your data and conclusions concerning the structure of the product.

## Discussion

In the isolation of trehalose, most of the materials originally extracted from the yeast are systematically removed before the trehalose is finally precipitated. Thus insoluble material and various aromatic compounds are removed by the initial filtration and charcoal treatment. Proteins are removed by heat coagulation and formation of their insoluble zinc salts. Phosphorylated sugars are removed as their insoluble barium salts. Thus during the final precipitation, trehalose is essentially the only ethanol-insoluble component in the extract.

### Exercises

1. Why is only one specific rotation value given for trehalose rather than three values, as given for most sugars in Table I (p. 19)?
2. In the experimental procedure, the solution is adjusted to an alkaline *p*H before being heated to denature the proteins. Why isn't the heating done at acid *p*H instead?

### References

Steward, L. C. et al., *J. Am. Chem. Soc.*, **72,** 2059 (1950).

Koch, E. M., and F. C. Koch, *Science*, **61,** 570 (1925).

Payen, R., *Can. J. Research.*, **27B,** 749 (1949).

EXPERIMENT

# 12. Other Carbohydrates: Isolation and Characterization of Glucosamine

## Theory of the Experiment

The exoskeletons of marine invertebrates are composed largely of calcium salts and the structural polysaccharide chitin. Chitin consists of long chains of $\beta$-1,4-linked N-acetyl-D-glucosamine units.

The monomeric units of chitin can be isolated from hydrolysates of invertebrate exoskeletons—for example, crab shells. Such an isolation requires a preliminary removal of salts (for example, $CaCO_3$) and other contaminating materials by extraction with acid alone or with acid and then alkali. The N-acetyl-D-glucosamine can then be obtained by a controlled acid or enzymatic hydrolysis. Complete acid hydrolysis, as in this experiment, ruptures not only the glycosidic bonds but also the amide bonds of chitin, yielding the monosaccharide D-glucosamine as the hydrochloride.

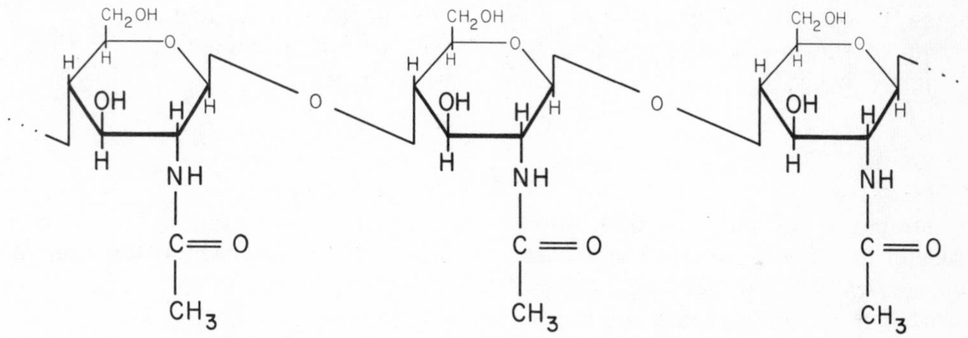

Chitin

## Experimental Procedure

MATERIALS

Crab or lobster shells
Activated charcoal
10% HCl
95% Ethanol
Conc. HCl (37%)
Glucosamine hydrochloride

### *Isolation*

Scrape and wash about 40 g of crab or lobster shells free of *all* adhering tissues (including eyes), break the shells into small pieces, and soak them in 10% HCl (5 ml/g of shells) for 24 hours. The next day, decant the HCl, wash the pieces of shell with water several times, and squeeze them to remove most of the water. After checking to be sure that no tissue is adhering to the shells, reflux the extracted shells in concentrated HCl (10 ml of HCl/1.5 g of extracted shells) for 2–3 hours or until the whole mass becomes a thick, black syrup. Because this reaction tends to bump, mount the reflux condenser on an angle or on a side-arm adaptor. Have your instructor check the apparatus before starting the refluxing. Add an additional 40 ml of concentrated HCl four times during the hydrolysis to maintain the acid strength. Dilute the resultant syrup with 3 volumes of $H_2O$, warm the syrup with about 5 g of activated charcoal, and filter with vacuum filtration. If the filtrate is not light-colored and clear, repeat the charcoal step. Concentrate the filtrate under vacuum until crystals appear in the flask. Dissolve these crystals by adding a minimum amount of water, and then add 1 volume of 95% ethanol. Crystallization occurs overnight at room temperature, but a higher yield is obtained if the flask is held at lower temperatures. Recrystallize the crude glucosamine-HCl by dissolving it in a minimum of warm water and adding 1 volume of 95% ethanol. Dry and weigh the purified product, and characterize as described below.

### *Characterization*

Knowing the structure of authentic glucosamine-HCl, you will be able to predict the chemical properties of the molecule. Devise a series of chemical tests, physical tests, chromatographic separations, and so on to determine the identity of the unknown material. Obtain known samples of glucosamine-HCl, glucose, or any other carbohydrate desired, from your instructor. Give the product remaining after characterization to the instructor.

## Report of Results

Express the quantity of material obtained in terms of the percentage of yield from the original weight of dry shells. Record your data and your arguments for chemical structure in a form similar to that used in Experiment 5.

## Discussion

Glucosamine is of considerable interest because of its wide occurrence in a variety of biochemically important complex molecules. It is found in muramic acid, a compound of the cell walls of bacteria. Glucosamine is also found as a component of blood-group substances, certain oligosaccharides present in milk, and many plant and animal polysaccharides.

### Exercises

1. Why is glucosamine isolated in the form of its hydrochloride rather than in the form of the free glucosamine?
2. Suggest some other sources of chitin for the glucosamine isolation.

### References

Wolfrom, M. L., and M. J. Cron, *J. Am. Chem. Soc.*, **74,** 1715 (1952).
Hudson, C. S., and J. K. Dale, *J. Am. Chem. Soc.*, **38,** 1434 (1916).

# PART TWO **LIPIDS**

## Introduction

Many substances exist in nature which are soluble in fat solvents (ether, chloroform, benzene) or yield such materials on hydrolysis. To these substances we apply the general term *lipid*. Definition of this term is difficult due to the heterogeneous nature of the substances it embraces. Bloor has characterized lipids as (1) insoluble in water, but soluble in fat solvents; (2) actual or potential esters of fatty acids; and (3) utilizable by living organisms. These three criteria are generally acceptable, although there are some exceptions to each.

Contrary to the first part of the definition, many universally accepted lipids are insoluble in some of the common fat solvents. Sphingomyelins and cerebrosides are insoluble in ethyl ether, and the phosphatides are insoluble in acetone. Nor are all lipids actual or potential esters of fatty acids. For example, the fatty-acid moiety of the cerebrosides is present as an amide. Furthermore, certain lipids may contain ether linkages rather than ester linkages. The last point in the definition also has limited usefulness. For example, mineral oils (which usually are not considered as lipids) are utilized by strains of bacteria found around oil wells!

### Structure and Classification

Lipids are conveniently classified as simple lipids (which yield fatty acids and alcohols on hydrolysis) and complex lipids (which may contain phosphate, various carbohydrates, and nitrogenous components such as sphingosine, ethanolamine, serine, hexosamines). The following scheme includes the types which occur generally in plant and animal tissues.

*Simple lipids*
- Triglycerides (fats and oils)
- Waxes
- Sterol esters (sterols)

*Complex lipids*
- Glycerophosphatides (phospholipids) (derived from $\alpha$-glycerophosphoric acid). Lecithin, phosphatidyl ethanolamine (cephalin), phosphatidyl serine, phosphatidyl inositol, plasmalogens.
- Sphingolipids (derived from sphingosine and phytosphingosine). Ceramides, sphingomyelins, cerebrosides, gangliosides, and other complex oligosaccharide derivatives of ceramides (for example, phytoglycolipid).

#### *Triglycerides and Related Simple Lipids*

Triglycerides—that is, glycerylesters of fatty acids (e.g., triolein and $\alpha$-palmityl-$\alpha'$-$\beta$-diolein below)—are by far the most abundant lipids in higher plants and animals, accumulating in fat depots and seeds as a storage form of food energy.

$$\begin{array}{l} \alpha\ CH_2\text{—O—oleyl} \\ \quad\ \ | \\ \beta\ CH\text{—O—oleyl} \\ \quad\ \ | \\ \alpha'\ CH_2\text{—O—oleyl} \end{array}$$

Triolein

$$\begin{array}{l} \alpha\ CH_2\text{—O—palmityl} \\ \quad\ \ | \\ \beta\ CH\text{—O—oleyl} \\ \quad\ \ | \\ \alpha'\ CH_2\text{—O—oleyl} \end{array}$$

$\alpha$-Palmityl-$\alpha'$-$\beta$-diolein

$$(\text{Oleyl} = CH_3(CH_2)_7\text{—}CH\text{=}CH\text{—}(CH_2)_7\text{—}\overset{O}{\overset{\|}{C}}\text{—};\ \text{palmityl} = CH_3(CH_2)_{14}\overset{O}{\overset{\|}{C}}\text{—})$$

Natural fats generally contain a variety of fatty acids, and are best represented as "mixed" fats (for example, α-palmityl-α′-β-diolein above).

More recently it has been recognized that diglycerides and monoglycerides exist in various tissues, usually in small amounts, and that diglycerides play an important role in the biosynthesis of some of the more complex lipids [1].

Other compounds related to the triglycerides are the alkoxydiglycerides, which contain ether linkages. These compounds are important constituents of certain marine organisms (for example, starfish, sharks) [2].

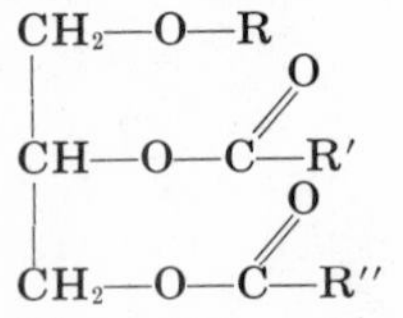

$CH_2$—O—R

$CH$—O—C(=O)—R′

$CH_2$—O—C(=O)—R″

Alkoxydiglyceride

### *Waxes*

The term *wax* is usually applied to esters of fatty acids with higher aliphatic alcohols. Paraffin wax, a hydrocarbon, is therefore not a true wax. The alcohols and acids usually vary from $C_{26}$ to $C_{30}$, but some are in the $C_{34}$ to $C_{36}$ range. The acids are frequently hydroxy acids. The waxes, which are inert materials, are often secreted as a protective coating, but occasionally waxes may serve as the main storage lipid of a plant or an animal. For example, the seed lipid of the shrub *Simmondsia californica* consists of 50% "oil," which is composed of wax esters and devoid of triglycerides. Waxes replace triglycerides to a considerable extent in aquatic animals, such as the crustaceans and whales.

### *Sterols and Sterol Esters*

The sterols are hydroxylated derivatives of the cyclopentanoperhydrophenanthrene (steroid) nucleus.

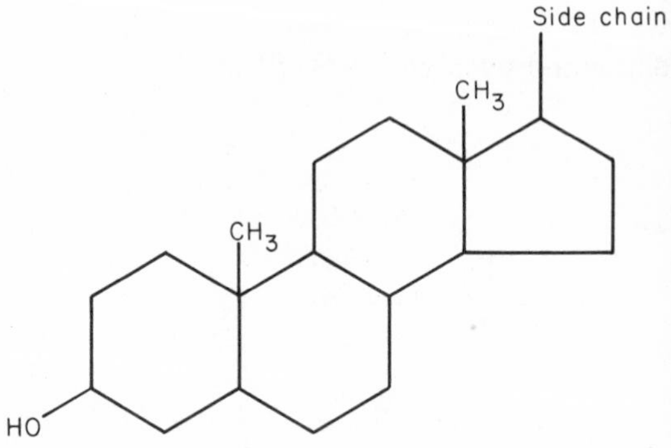

Side chain

$CH_3$

$CH_3$

HO

Typical sterol

All sterols are capable of forming esters with fatty acids, but in general, the free sterols are more abundant than their esters. Examples of common sterols are: cholesterol, typical of higher animals; β-sitosterol, common in higher plants; and ergosterol, typical of fungi (see below).

When cholesterol occurs in the esterified forms, it usually is combined with common fatty acids, such as palmitic, oleic, or linoleic acids. The plant sterols are occasionally esterified with acids, such as acetic and cinnamic acids, or are incorporated into glycosides (saponins).

It should be noted that the simple lipids are neutral esters and contain no ionic groups or highly polar substituents. Thus they are less polar in character than the complex lipids and tend to be more soluble in hydrocarbon solvents (which is of importance in their preparation) and in acetone.

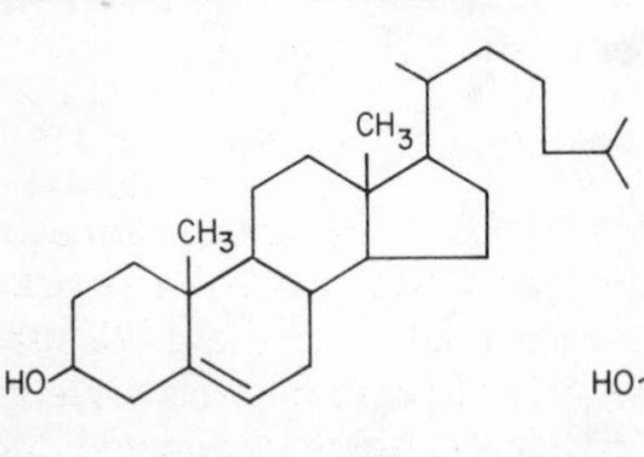

$CH_3$ $CH_3$ HO

Cholesterol

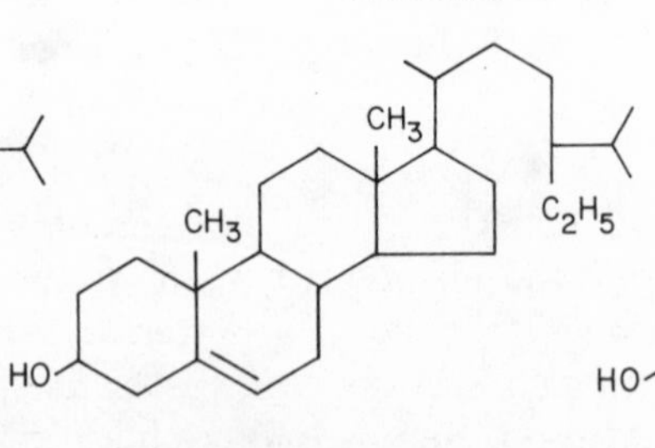

$CH_3$ $CH_3$ $C_2H_5$ HO

β-Sitosterol

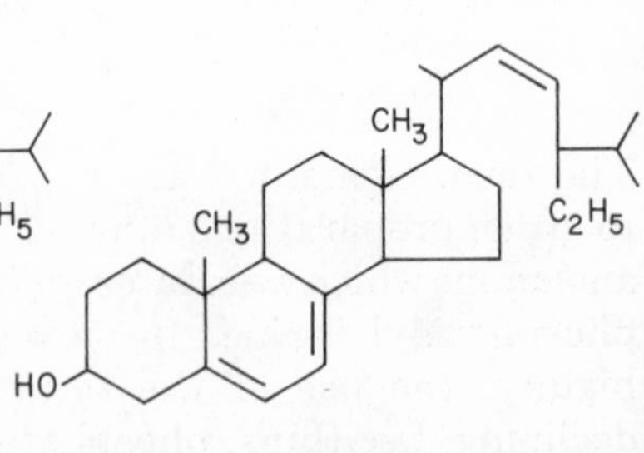

$CH_3$ $CH_3$ $C_2H_5$ HO

Ergosterol

### *The Glycerophosphatides (often called phospholipids)*

These lipids are acyl derivatives of $\alpha$-glycerophosphoric acid. The simplest glycerophosphatides are the phosphatidic acids, which contain $\alpha$-glycerophosphoric acid esterified with two fatty acids.

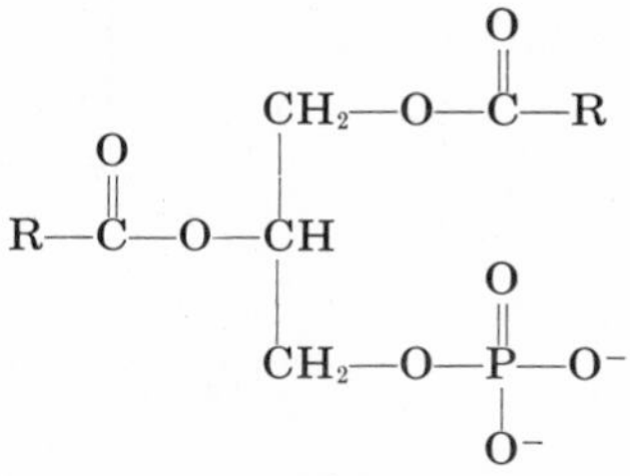

```
                      O
                      ‖
             CH2—O—C—R
    O        |
    ‖        |
R—C—O—CH
             |        O
             |        ‖
             CH2—O—P—O⁻
                      |
                      O⁻
```

A phosphatidic acid

Small quantities of phosphatidic acids have been isolated from a wide variety of plant and animal tissues. There is some doubt whether these compounds exist in appreciable amounts in tissues, since more complex glycerophosphatides are readily hydrolyzed by enzymes that are widely distributed in such tissues, yielding phosphatidic acids [3].

The major glycerophosphatides are esters of phosphatidic acid having the general formula shown below, in which R represents a long-chain alkyl group and R′ an amino alcohol or inositol.

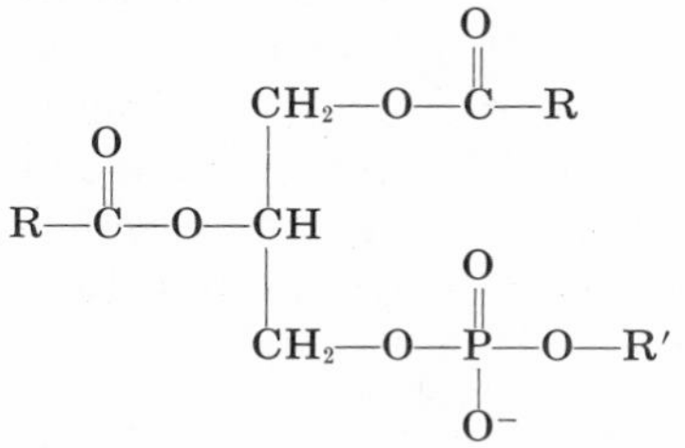

```
                      O
                      ‖
             CH2—O—C—R
    O        |
    ‖        |
R—C—O—CH
             |        O
             |        ‖
             CH2—O—P—O—R′
                      |
                      O⁻
```

The term cephalin was originally applied to crude preparations of phosphatidyl ethanolamine which were later found to contain phosphatidyl serine. In view of this ambiguity, the use of the term cephalin is declining. Lecithins, phosphatidyl ethanolamine, and phosphatidyl serine are very important "complex" lipids, comprising major fractions of the lipid materials isolated from nerve tissue of higher animals (see Experiment 16). The so-called "inositol phosphatides," or "phosphoinositides," appear to be esters of phosphatidic acid and inositol or inositol mono-, di-, and triphosphate. The structure of phosphatidyl inositol is relatively certain [4], and recent work has defined the structures for di- and triphosphoinositides [5, 6].

In lecithin (phosphatidyl choline)

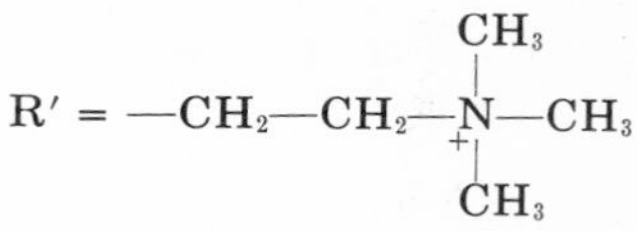

```
                       CH3
                       |
R′ = —CH2—CH2—N—CH3
                      +|
                       CH3
```

In phosphatidyl ethanolamine

$R' = —CH_2—CH_2—NH_3^+$

In phosphatidyl serine

```
                          O
                        //
R′ = —CH2—CH3—C
              |         \
              NH3        O⁻
```

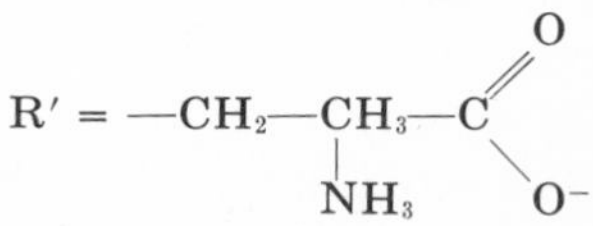

In phosphatidyl inositol (phosphoinositide)

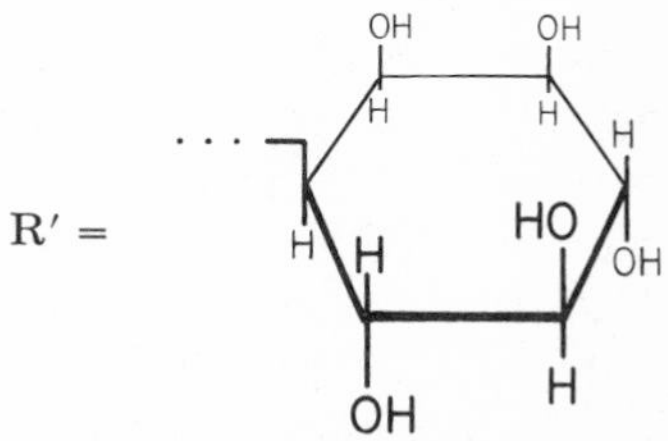

R′ = (inositol ring structure)



The plasmalogens represent an unusual type of glycerophosphatide which has been isolated from animal and plant tissues [7, 8]. Hydrolysis of these materials yields long-chain aldehydes plus monoacylated

$\alpha$-glycerophosphate. The long-chain aldehydes originate from vinyl ether structures rather than from the previously assumed acetal structure. The vinyl ether group appears usually (if not always) in the $\alpha$-position. The R′ phosphomonoester group may be either choline or ethanolamine, although ethanolamine usually predominates.

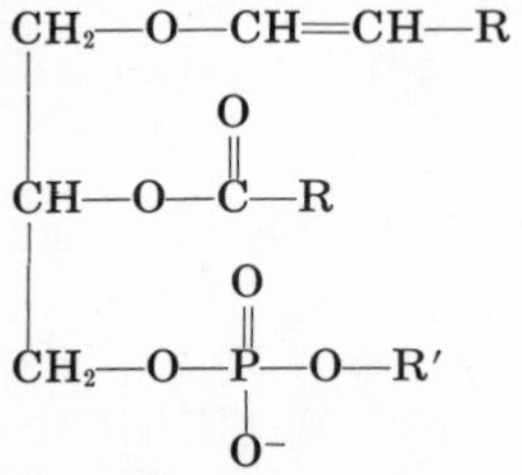

$$\begin{array}{l} CH_2{-}O{-}CH{=}CH{-}R \\ | \\ CH{-}O{-}\overset{\displaystyle O}{\overset{\|}{C}}{-}R \\ | \\ CH_2{-}O{-}\overset{\displaystyle O}{\overset{\|}{P}}{-}O{-}R' \\ \qquad\quad\;\; | \\ \qquad\quad\;\; O^- \end{array}$$

Structure of plasmalogens
(R = aliphatic chain;
R′ = choline or ethanolamine)

### *Sphingolipids*

The sphingolopids contain $C_{18}$ hydroxyamino alcohols. Four such long-chain hydroxyamino alcohols have been isolated from natural lipids. These are:

| | |
|---|---|
| Sphingosine | $CH_3(CH_2)_{12}CH{=}CH{-}\underset{OH}{CH}{-}\underset{NH_2}{CH}{-}\underset{OH}{CH_2}$ |
| Dihydrosphingosine | $CH_3(CH_2)_{14}{-}\underset{OH}{CH}{-}\underset{NH_2}{CH}{-}\underset{OH}{CH}$ |
| Phytosphingosine | $CH_3(CH_2)_{13}{-}\underset{OH}{CH}{-}\underset{OH}{CH}{-}\underset{NH_2}{CH}{-}\underset{OH}{CH_2}$ |
| Dehydrophytosphingosine | $CH_3(CH_2)_xCH{=}CH(CH_2)_{11-x}{-}\underset{OH}{CH}{-}\underset{OH}{CH}{-}\underset{NH_2}{CH}{-}\underset{OH}{CH_2}$ |

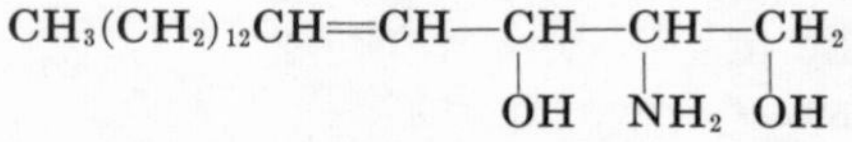

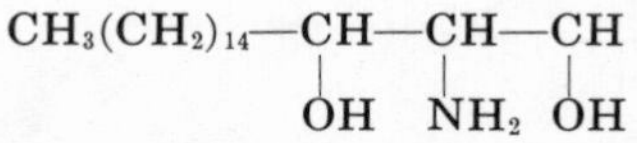

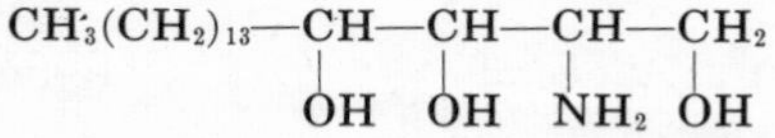

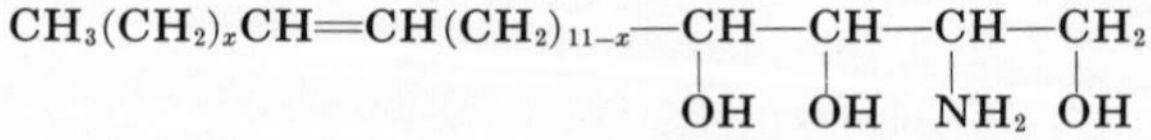

Sphingosine and dihydrosphingosine are typical of animal tissues (although the latter has been found recently in wheat flour [9]. Phytosphingosine and dehydrophytosphingosine have not been discovered in animal tissues [10].

The free sphingosine forms occur in nature in very small amounts if at all. Acyl amides (ceramides) of sphingosine are found in the spleen and liver.

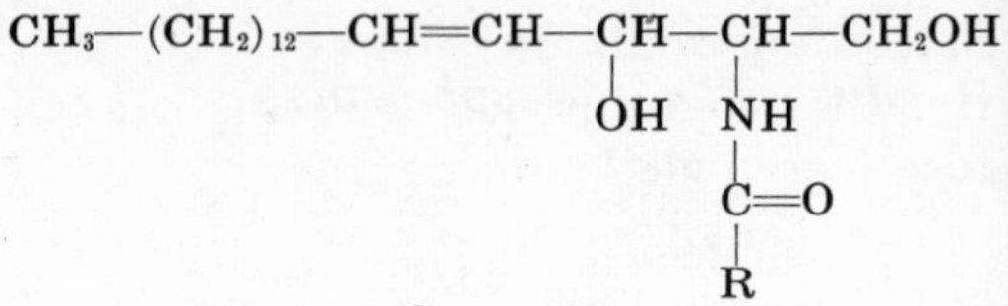

$$CH_3{-}(CH_2)_{12}{-}CH{=}CH{-}\underset{OH}{CH}{-}\underset{\underset{\underset{R}{|}}{C{=}O}}{\underset{|}{NH}}{\!\!\!\!\!CH}{-}CH_2OH$$

Ceramide

Phytosphingosine ceramide occurs in yeasts.

The major sphingolipids, however, are more complicated structures consisting of ceramides linked to a phosphate ester or, glycosidically, to carbohydrate residues (mono- and higher) [11, 12]:

*ceramide*—O—*hexose*

Cerebroside

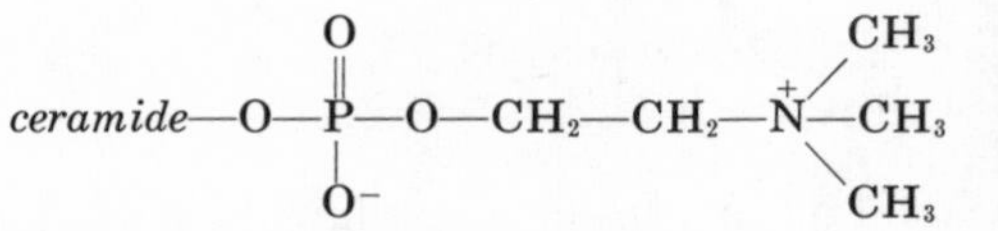

$$\textit{ceramide}{-}O{-}\overset{\displaystyle O}{\overset{\|}{\underset{\displaystyle O^-}{\underset{|}{P}}}}{-}O{-}CH_2{-}CH_2{-}\overset{+}{N}(CH_3)_3$$

Sphingomyelin

*ceramide*—O—*glucose*
|
*galactose*
|
*hexosamine*
|
*acetyl neuraminic acid*

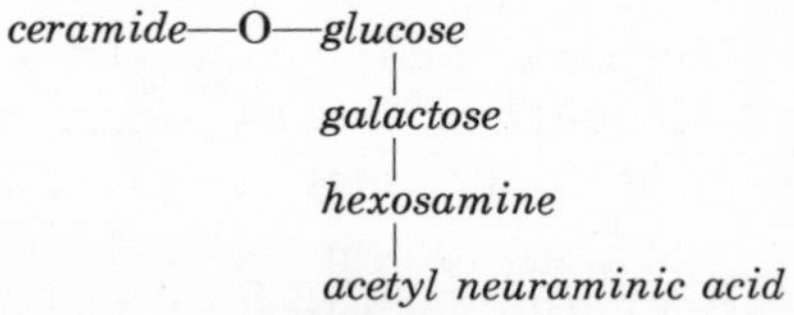

Ganglioside

The fatty acid components of the sphingolipids are usually saturated (some monounsaturated components are also present) with an unusually high proportion of long-chain $C_{24}$ acids such as lignoceric (*n*-tetracosanoic) and cerebronic ($\alpha$-hydroxylignoceric).

The carbohydrate component of brain cerebrosides is galactose. Glucose is present in small amounts in spleen cerebrosides.

Spleen also contains a disaccharide derivative that contains both glucose and galactose (ceramide lactoside).

The gangliosides, a more complicated group of ceramide oligosaccharides, contain—in addition to galactose and glucose—both N-acetylgalactosamine and N-acetylneuraminic acid.

Apparently N-acetylneuraminic acid is formed by the condensation of mannosamine and pyruvic acid or a closely related compound [13] (see opposite).

$$
\begin{array}{c}
\mathrm{O} \\
\parallel \\
\mathrm{C{-}OH} \\
| \\
\mathrm{C{-}H} \quad (\text{ring O to C-6}) \\
| \\
\mathrm{CH_2} \\
| \\
\mathrm{HO{-}C{-}H} \\
| \\
\mathrm{CH_3{-}\overset{\displaystyle O}{\overset{\parallel}{C}}{-}\overset{\displaystyle H}{\overset{|}{N}}{-}C{-}H} \\
| \\
\mathrm{C{-}H} \\
| \\
\mathrm{H{-}C{-}OH} \\
| \\
\mathrm{H{-}C{-}OH} \\
| \\
\mathrm{CH_2OH}
\end{array}
$$

N-Acetylneuraminic acid (sialic acid)

A phytoglycolipid has been isolated from corn and soybeans and other seeds and partially characterized by Carter and coworkers [14]. (The oligosaccharide contains glucosamine, glucuronic acid, galactose, arabinose, and mannose.) The proposed structure is

$$
\mathrm{CH_3(CH_2)_{13}\underset{\displaystyle OH}{\underset{|}{CH}}{-}\underset{\displaystyle OH}{\underset{|}{CH}}{-}\underset{\underset{\displaystyle RCO}{|}}{\underset{\displaystyle NH}{\underset{|}{CH}}}{-}CH_2{-}O{-}\overset{\displaystyle O}{\overset{\parallel}{\underset{\displaystyle O^-}{\underset{|}{P}}}}{-}O{-}}\textit{inositol}{-}\textit{oligosaccharide}
$$

Phytosphingosine with fatty acid amide

# Isolation of Lipid Fractions

The first step in the preparation of a lipid is extraction from the tissue. This procedure is complicated for the following reasons:

1. Lipids are often bound to proteins and polysaccharides of tissues in complexes of widely varying degrees of stability. Thus, in order to break these complexes, it is often necessary to employ extraction conditions that denature or disrupt the associated proteins or polysaccharides. The extent of these complexes varies among tissues; for example, it is claimed that all of the lipid of egg yolk is present as lipoprotein in the native state. Structures involving complex lipids are usually more stable than those derived from simple lipids because of the ionic and polar attractions involved.

2. Lipid mixtures have remarkable ability to carry nonlipid materials into organic solvents, partly by virtue of ionic interactions. Thus hexane extracts of blood serum contain urea, sodium chloride, and amino acids; and hexane extracts of soybean contain raffinose and stachyose.

3. The glycerophosphatides usually contain substantial amounts of highly unsaturated fatty acids which undergo rapid autoxidation catalyzed by Cu or Fe salts. These salts tend to be carried along by lipids. Thus certain lipid extractions may have to be done in a nitrogen atmosphere and

may require the use of peroxide-free solvents.

4. Extraction of wet tissues often produces intractable emulsions; hence prior dehydration is usually essential. But dehydration of certain tissues appears to cause irreversible binding of some of the lipid components, decreasing the yield.

5. Solvents vary widely in their ability to extract different types of lipids, necessitating the use of a series of solvents for complete extraction.

6. Enzymes which hydrolyze lipids exist in various tissues. Certain of these are activated by organic solvents, the rate of activation increasing as the solvent is warmed. At least one enzyme is active in an ether dispersion. Thus consideration must be given to the possibility of enzymatic degradation, particularly when dealing with plant sources.

No single satisfactory procedure has been discovered which avoids all of these complications. The extraction conditions must be adapted to the tissue employed and to the quantity and type of lipid desired.

Acetone is a very useful solvent. It provides a mild but rapid method of dehydrating tissue, and as the water content decreases, acetone extracts fats, sterol esters, sterols, and other simple lipids. The complex lipids are relatively insoluble in acetone and are converted into friable powders by this solvent, thus facilitating extraction with other solvents. Following extraction with acetone, extraction with ether or hexane will remove glycerophosphatides (more or less completely); hot ethanol will extract the sphingolipids. (Ethanol denatures proteins and disrupts lipoprotein complexes.) Thus partial fractionation of the lipids can be achieved by a selective series of solvent extractions.

For small-scale work, chloroform-methanol (3:1) extraction of the wet tissue is very effective. This method removes almost all of the lipid, some in the form of lipid-protein complexes (proteolipids). Equilibration of the extract with a water layer removes nonlipid material. Addition of salts (magnesium chloride) to the aqueous layer minimizes the loss of lipid to the aqueous layer. For analytical and small-scale isolation work this procedure has several advantages.

Unfortunately, separations based on solubility differences are almost always only partially satisfactory, because the solubility properties of the constituents of lipid mixtures are quite different from those of the pure lipids. For example, phosphatides are partly soluble in acetone in the presence of triglycerides, and sphingolipids have some solubility in ether solutions of lipids. Phosphatidyl ethanolamine is not extracted by alcohol from crude lipid preparations, although the pure material is quite soluble in alcohol. Thus solvent fractionations must be repeated several times, and even so, do not usually give clean separations.

Although there has been some success with the use of countercurrent distribution, the most promising method for lipid purification after initial extraction is chromatography with silicic acid columns. An effective procedure has been described in detail by Hirsch and Ahrens [15], and adequate, similar macro methods are available for practical preparation of relatively homogeneous lipids.

It should be emphasized that terms such as lecithin, phosphatidyl ethanolamine and sphingomyelin do not denote pure compounds, but represent mixtures of compounds with varying amounts of different fatty acids. These mixtures usually cannot be separated from one another by ordinary methods of purification. Hence the term "pure complex lipid" means a lipid whose non-fatty acid moiety is homogeneous.

## Analysis of Lipids

Lipids are usually waxes or oils whose properties make characterization by physical constants difficult if not impossible. For that reason detection and determination of the lipid constituents in a tissue extract must depend on a variety of analytical approaches, both on the extracted intact lipids and on various partial or total hydrolysis products. Nitrogen and phosphorus analyses and N/P ratios once afforded a main approach, resulting in the use of such terms as monoamino-monophosphatide and diaminomono-phosphatide. These analyses are still useful in determining minor phosphatide contamination in fats and oils. Unsaturation can be measured quantitatively in terms of the iodine number, whereas molecular size of lipids and fatty acids can be measured in terms of saponification numbers and neutralization equivalents respectively. However, modern developments have provided more precise tools for characterization of lipids and lipid mixtures.

### *Characterization of Intact Lipids*

#### PAPER CHROMATOGRAPHY

Intact lipid mixtures can often be separated by paper chromatography either directly or on "reverse phase" papers. In the latter method, the filter paper is converted to a nonpolar stationary phase by impregnation with silicone or rubber latex, and the aqueous phase becomes the mobile phase. Glass filter paper can also be used to advantage. Resolution of complex mixtures is often poor, with considerable trailing, but partially purified fractions often give good results, permitting positive identification of lipid components.

Lipids can be visualized on papers with the aid of a variety of spray reagents. Sudan black [16] serves as a general lipid stain, whereas various spray reagents are used to stain specific lipids; for example, ninhydrin locates phosphatidyl ethanolamine and phosphatidyl serine (see p. 96), acidic phosphomolybdate locates phosphate esters (see p. 37), and periodate-benzidine (see p. 17) locates phosphoinositides.

#### SILICIC ACID COLUMNS

Small lipid samples can be fractionated in a reproducible way on silicic acid columns. Triglycerides and fatty acids are eluted by hexane or hexane-ether mixtures. Increasing the polarity of the eluant with hexane-ether or chloroform, or varying concentrations of methanol in chloroform, brings increasingly polar lipids through the column. Infrared examination of the eluted fractions can then provide both qualitative and rough quantitative data concerning the constituent lipids.

### *Characterization of Lipid Components*

Analysis of lipid components by chromatographic or other means usually requires a preliminary hydrolysis step in which the conditions employed must be selected with regard to the materials used and the object in view.

#### ALKALINE HYDROLYSIS

Alkaline hydrolysis (for example, the Schmidt-Thannhauser procedure employs 1*N* KOH at 37°C for 24 hr) cleaves most ester bonds but leaves the amide groups in sphingolipids intact. Alcoholic alkali may be necessary with lipid mixtures which do not disperse readily into aqueous solutions. Potassium hydroxide is preferable, since it is more soluble in alcohol than NaOH; similarly, potassium salts of fatty acids are more soluble than their equivalent sodium salts.

The behavior of glycerophosphatides during alkaline hydrolysis requires some further comment. Hydrolysis of the acyl groups from lecithin leaves glycerylphosphorylcholine (GPC). This, however, is not present in the usual alkaline hydrolysates; instead $\alpha$- and $\beta$-glycerophosphate and choline are produced. Extensive study of this and similar reactions has shown that phosphate diesters with an $\alpha$-hydroxyl group undergo ready cleavage involving participation of the hydroxyl group through a cyclic intermediate [17] (see also p. 127).

Mild alkali will cleave GPC considerably more rapidly than it will cleave glycerophosphorylethanolamine (GPE), but both are readily degraded as shown below. Originally these results were believed to indicate the presence of $\beta$-glycerophosphatides, but this view has now been abandoned on the basis of a variety of evidence; for example, enzymatic or very mild alkaline hydrolysis of lecithin (Dawson procedure of short exposure to $0.1N$ NaOH in methanol [18]) yields only $\alpha$-GPC.

Since glycerophosphatides never give glycerol on mild alkaline hydrolysis, the production of glycerol under these conditions indicates the presence of mono-, di-, or triglycerides.

The amide bonds of the sphingolipids can be cleaved under drastic alkaline conditions, the divalent bases such as $Ba^{+2}$, $Ca^{+2}$ being somewhat more effective in this hydrolysis than the monovalent ones. Refluxing a cerebroside for 6–12 hr with saturated barium hydroxide cleaves the amide bond but not the glycosidic linkage, giving galactosylsphingosine (psychosine) as the product.

Alkaline hydrolysis is usually preferable when the fatty acids are desired, since strong acids may cause some alterations of unsaturated acids. Alkaline reagents are also useful when preservation of glycosidic or vinyl ether bonds is desired. Alkaline reagents degrade serine slowly, and under drastic conditions cause some hydrolysis and destruction of glycerophosphate.

### ACIDIC HYDROLYSIS

Acid hydrolyses are usually carried out using $6N$ aqueous hydrochloric acid (constant boiling) or 5–10% solutions of HCl in methanol (to promote solubility). Most glycerophosphatides are acid hydrolyzed to fatty acids, glycerophosphate and base, just as with alkali, but inositol phosphatides yield inositol phosphate and diglycerides upon acid hydrolysis. Hydrochloric acid has the advantage of being easily removed by vacuum, facilitating paper-chromatographic examination of the hydrolysis products. Ethanolamine and serine are relatively stable in acidic reagents.

Cerebrosides in methanolic HCl (or $H_2SO_4$) are rapidly cleaved to sphingosine,

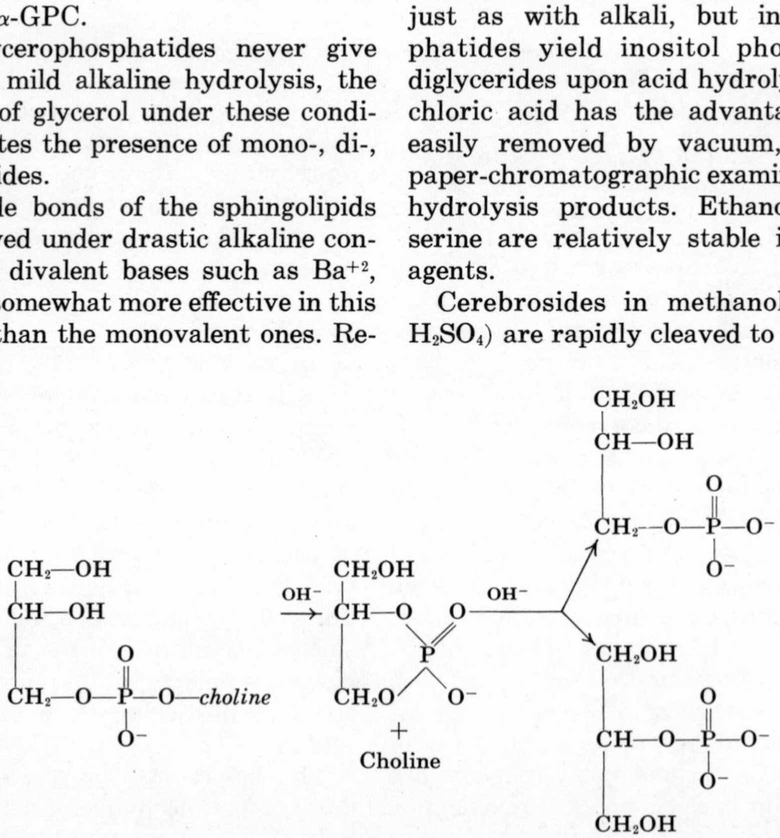

$$CH_2OH-CH(OH)-CH_2-O-P(=O)(O^-)-O\text{-}choline \xrightarrow{OH^-} \text{cyclic } CH_2OH-CH(-O-)-CH_2O-P(=O)(O^-) + \text{Choline} \xrightarrow{OH^-} CH_2OH-CH(OH)-CH_2-O-PO_3^{2-} \;\text{and}\; CH_2OH-CH(O-PO_3^{2-})-CH_2OH$$

methyl glycoside, and fatty acid plus some fatty acid methyl ester. Further refluxing for 4 hr in 5% methanolic hydrogen chloride is adequate to complete the hydrolysis of the fatty acid ester.

The vinyl ether group of plasmalogens is very readily cleaved by weak acids. Mercuric chloride catalyzes this reaction. Cleavage is complete in about an hour in 90% acetic acid containing $HgCl_2$.

### IDENTIFICATION OF HYDROLYSIS PRODUCTS

A number of paper-chromatographic systems have been developed for detection of the constituents of lipids. Fatty acids can be separated on paper chromatograms and located by means of *p*H indicators or dyes that fluoresce after subsequent washing. Recent advances in vapor-phase chromatography of fatty acid methyl esters now offer a more powerful tool for characterization of fatty acids than paper chromatography [19]. These methods allow the separation and characterization of as little as 10–100 μg of fatty acid methyl esters (formed by refluxing fatty acids in methanol with an acid catalyst).

Many of the other products of lipid hydrolysis are characterized on paper chromatograms. Choline is detected on paper chromatograms by conversion to choline phosphomolybdate (see p. 64); serine, ethanolamine, and sphingosine are located by means of the ninhydrin reaction (see p. 96); and carbohydrates are detected by the periodate-benzidine or aniline-acid-oxalate spray systems (see p. 17).

## General References


Deuel, H. J., Jr., *The Lipids: Their Chemistry and Biochemistry*, Interscience, New York, 1951.

Fieser, L. F., and M. Fieser, *Steroids*, Reinhold, New York, 1959.

Hanahan, D. J., *Lipid Chemistry*, Wiley, New York, 1960.

Lovern, J. A., *The Chemistry of Lipids of Biochemical Significance*, Wiley, New York, 1955.


## Specific References


1. Grien, D. E., *Scientific American*, **202,** 46 (1960).
2. Karnovsky, M. L., and A. F. Brumm, *J. Biol. Chem.*, **216,** 689 (1955).
3. Kates, M., *Can. J. Biochem. and Physiol.*, **33,** 575 (1955).
4. Pizer, F. L., and C. E. Ballou, *J. Am. Chem. Soc.*, **81,** 915 (1959).
5. Dittmer, J. C., and R. M. C. Dawson, *Biochim. et Biophys. Acta*, **40,** 379 (1960).
6. Grado, C., and C. E. Ballou, *J. Biol. Chem.*, **236,** 54 (1961).
7. Rapport, M. M., and R. E. Frauzl, *J. Neurochem.*, **1,** 303 (1957).
8. Wagenkencht, A. C., *Science*, **126,** 1288 (1957).
9. Carter, H. E. et al., *J. Biol. Chem.*, **236,** 1912 (1961).
10. Carter, H. E. et al., *J. Biol. Chem.*, **206,** 613 (1954).
11. Rouser, G. et. al., *J. Am. Chem. Soc.*, **75,** 310, 313 (1953).
12. Svennerholm, L., *Nature*, **177,** 524 (1956).
13. Comb, D. G., and S. Roseman, *J. Am. Chem. Soc.*, **80,** 497 (1958).
14. Carter, H. E. et al., *J. Biol. Chem.*, **233,** 1309 (1958).
15. Hirsch, J., and E. J. Ahrens, *J. Biol. Chem.*, **233,** 311 (1958).
16. Marinetti, G. V., and E. Stotz, *J. Am. Chem. Soc.*, **77,** 6668 (1955).
17. Bear, E., and M. Kates, *J. Biol. Chem.*, **185,** 615 (1950).
18. Dawson, R. M. C., *Biochem. J.*, **62,** 689 (1956).
19. Wilkens Instrument and Research, Inc., "Fatty Acid Methyl Ester Analysis," in *Gas Chromatography Notebook*, Walnut Creek, Calif., 1958.

EXPERIMENT

# 13. Fats and Oils: Chemical Properties

## Theory of the Experiment

One of the means of analyzing lipids involves hydrolysis and subsequent isolation and characterization of the products. Hydrolysis of simple lipids is most often accomplished by hot alkali (saponification). Since the lipid materials are frequently insoluble in water, hydrolysis in aqueous alkali is slow. The rate of hydrolysis is accelerated by the use of appropriate solvents (for example, ethanol or ethylene glycol).

From the sample equation

$$\mathrm{R{-}\overset{\displaystyle O}{\overset{\|}{C}}{-}O{-}R'} + \mathrm{KOH} \longrightarrow \mathrm{R{-}\overset{\displaystyle O}{\overset{\|}{C}}{-}O^-}\ \mathrm{K^+} + \mathrm{R'OH}$$

we can see that 1 mole of alkali is consumed for every mole of ester saponified. Therefore the quantity of alkali consumed by complete hydrolysis of a lipid sample is related to the number of ester bonds originally present in the sample.

The consumption of alkali by a particular lipid is commonly expressed in terms of the saponification number, which is defined as the number of milligrams of KOH consumed in the complete saponification of 1 g of fat or oil. The products of the saponification of a neutral fat or oil are glycerol an a mixture of the salts of the componen fatty acids. The free fatty acids of hig molecular weight, in contrast to their salts are insoluble in $H_2O$, but are soluble in non polar solvents such as ether. Hence thes fatty acids can be separated from glycero by acidification of the saponified solution followed by filtration or ether extraction

An estimate of the average molecula weight of the fatty acid fraction can b made from the neutralization equivalen of the sample. The neutralization equiva lent is defined as the number of grams o acid required to neutralize 1 equivalent o alkali. In practice, the neutralizatio equivalent is determined by titrating an hydrous organic acids with standard alka and then using the equation

$$\text{N.E.} = \frac{\text{weight of sample in grams} \times 1000}{\text{volume of alkali in milliliters} \times \text{normalit}}$$

Glycerol, released by alkaline hydroly sis, can be detected by several tests, whic involve either internal dehydration (acro lein test) or oxidation of glycerol and con densation of the product with phenoli compounds to give chromogenic products Consult an organic chemistry textboo for the details of these reactions.

## Experimental Procedure

MATERIALS

Unknown oil
About $0.5N$ standardized HCl
$KHSO_4$
Glycerol
Conc. $H_2SO_4$
95% Ethanol
$0.1N$ NaOH
5.6% Alcoholic KOH
1% Alcoholic phenolphthalein
2% NaOCl (Clorox)
5% $\alpha$-Naphthol
Ether
Conc. HCl

### *Determination of Saponification Number*

Using a good triple beam balance, weig out about 10 g of the unknown oil (to th nearest 0.1 g), and place it in a 500-m Erlenmeyer flask. Add 100 ml of alcoholi KOH, and reflux the mixture on a stea bath until saponification is complete– that is, when the mixture is clear (on phase). This requires about 30 minutes Prepare a blank by refluxing 50 ml of th alcoholic potassium hydroxide in exactly the same way. Cool both flasks, and titrate with standard (5.0*N*) hydrochloric acid, using 1 ml of 1% phenolphthalein as the indicator.

### *Determination of Neutralization Equivalent*

After titrating the saponified lipid, remove the alcohol from the solution by evaporation over a steam bath in the hood. Then add about 20 ml of water, and adjust the *p*H to 1–2 (*p*H paper) by adding concentrated HCl drop by drop and swirling the mixture. Next, extract the acidified aqueous mixture three times with 10 ml portions of ether (using a separatory funnel or by decantation). Place the ether extracts into a tared 125-ml Erlenmeyer flask. Make an additional extraction with ether if any solid or oily free fatty acids remain in the aqueous phase. Save the aqueous phase (including washes) remaining after the various ether extractions for the detection of glycerol.

Remove the ether of the extract under reduced pressure, and determine the weight of the dry or oily residue. Next, wash the surface of the residue with 10 ml of $H_2O$ to remove any adhering HCl. Then add 10 ml of ether to the residue, and dry the solution again to remove the last traces of HCl. When the sample is free of ether, add 50 ml of warm 95% EtOH, and titrate it to a phenolphthalein end point with 0.1*N* NaOH. Gentle heating will hasten solution of the solid. Similarly titrate a solvent blank containing an equal volume of alcohol.

### *Detection of Glycerol*

For either or both of these tests, use the aqueous extract of the saponified lipid remaining after ether extraction.

#### *Acrolein Test*

Use the acrolein reaction to test the following substances: 1.5 ml of ether-extracted, aqueous, saponified lipid that has been concentrated under vacuum to 0.5 ml; several drops of nonsaponified lipid; and several drops of glycerol.

Add about 0.5 g of $KHSO_4$ to the sample in a dry test tube. Heat the tube *carefully* and *slowly* (to avoid charring) over a Bunsen burner. The characteristic odor of acrolein indicates the presence of glycerol.

#### *Colorimetric Test*

Test the ether-extracted, aqueous, saponified lipid (1 ml) by the following procedure, using 1 ml of water as a blank and aqueous glycerol (1 ml of 10% solution) as a standard. Add 1 ml of sodium hypochlorite solution (Clorox). After 2–3 min add 3 or 4 drops of concentrated HCl to the tubes, and boil for 1 min to remove the chlorine (preferably in the hood). Finally add 0.2 ml of 5% $\alpha$-naphthol and then 4 ml of concentrated $H_2SO_4$. Shake the tubes *carefully*. An emerald-green color indicates the presence of glycerol.

## Report of Results

Calculate both the saponification number of the saponified lipid and the mean molecular weight of the lipid, assuming that it is a triglyceride. Then calculate the mean neutralization equivalent of the fatty acids of the lipid and the molecular weight of the original lipid, using the neutralization equivalent found by direct titration of the isolated fatty acids obtained from the triglyceride.

## Discussion

This experiment provides a crude characterization of a triglyceride. From the determination of the saponification number of the triglyceride or the neutralization equivalent of the isolated fatty acids, you can obtain the average molecular weight of the intact triglyceride or the component fatty acids. Some fat constants for common fats are given in Table III. Compare your results with the values in the table.

Information on the identity of the constituent fatty acids may be obtained by such methods as (1) the determination of the volatility of the fatty acids, (2) the determination of the iodine number, (3) the isolation of the fatty acids by chromatography, or (4) the quantitative determination of the esterified acids by gas chromatography.

**TABLE III.** Physical and Chemical Properties of Natural Oils

| Oil | Class* | Saponification number | Iodine number | Solidification point (°C) |
|---|---|---|---|---|
| Almond | nondrying | 183–207 | 93–103 | −15 to −20 |
| Corn | semidrying | 187–193 | 111–128 | −10 to −20 |
| Cottonseed | semidrying | 194–196 | 103–111 | +12 to −13 |
| Linseed | drying | 188–195 | 175–202 | −19 to −27 |
| Olive | nondrying | 185–196 | 79–88 | turbid +2, ppt. −6 |

*Nondrying oils are still fluid after 18–20 days of exposure to the air. Semidrying oils form a sticky film after 1 week of exposure to the air. Drying oils, when spread thinly, form an acetone-insoluble film after 2–6 days.

### Exercises

1. What are the possible sources of error in the determination of the saponification number and the neutralization equivalent?
2. What additional tests would enable you to identify your lipid sample (assuming that it is one of the oils listed in Table III)?
3. An unknown, simple, saturated triglyceride is observed to have a saponification number of 189. What is the molecular weight of the compound? What is the name of the compound?
4. Write the reactions of glycerol with $KHSO_4$ and NaOCl.
5. Assume that a neutral fat is saponified and then extracted with ether while it is still alkaline. Would you expect most of the fatty acids to be in the ether extract? Would the result be different if the saponified solution had been acidified before extraction?

### Reference

Hilditch, T. P., *Ann. Rev. Biochem.*, **22,** 125 (1953).

EXPERIMENT

# 14. Fats and Oils: Enzymatic Hydrolysis

## Theory of the Experiment

Higher animals break down ingested triglycerides in the small intestine by the action of hydrolytic enzymes (lipases) present in the secretion of the pancreas. Lipases belong to the general class of enzymes known as esterases, which catalyze the general reaction

$$R-\overset{O}{\overset{\|}{C}}-O-R' + H_2O \rightleftharpoons R-\overset{O}{\overset{\|}{C}}-OH + R'OH$$

The simple esterases catalyze only the hydrolysis of esters of simple alcohols (e.g., methyl caproate, ethyl butyrate). Lipases catalyze the hydrolysis of esters of glycerol, but do not hydrolyze simple esters. The compounds acted upon by lipase (that is, the substrates of lipase) are diglycerides and triglycerides; the hydrolysis of monoglycerides by these enzymes is very slow. Thus, starting with a triglyceride, the reactions are as below.

According to the work of Mattson and Beck, and that of Savary and Desnuelle (see References), the triglyceride is preferentially hydrolyzed at either the $\alpha$ or $\alpha'$ (1 or 3) positions to yield an optically active 1,2-diglyceride. In a second reaction this is converted to the final product, an optically inactive 2-monoglyceride. The rates of hydrolysis are not appreciably affected by variations in the number of carbon atoms (12–18) and in the number of double bonds (0–2) in the fatty acid R group.

Hog, sheep, or beef pancreas is a rich and readily available source of lipase. Crude preparations of lipase are frequently made by extraction of an acetone powder (a pulverized, dehydrated preparation) of pancreas. Such a preparation is used in this experiment.

A number of requirements are involved in the demonstration of optimum lipase activity. First, it is necessary to increase the effective concentration of the substrate. High-molecular-weight di- and triglycerides are practically insoluble in aqueous solutions and are therefore unavailable to the enzyme. The effective concentration of these substrates is increased by emulsification with such agents as soaps, pro-

$$\begin{array}{l} \alpha'\text{-1}\ CH_2-O-\overset{O}{\overset{\|}{C}}-R_1 \\ \quad\ \ | \\ 2\ \ CH-O-\overset{O}{\overset{\|}{C}}-R_2 \\ \quad\ \ | \\ \alpha'\text{-3}\ CH_2-O-\overset{O}{\overset{\|}{C}}-R_3 \end{array} + H_2O \rightleftharpoons \begin{array}{l} 1\ CH_2OH \\ \quad | \\ 2\ CH-O-\overset{O}{\overset{\|}{C}}-R_2 \\ \quad | \\ 3\ CH_2-O-\overset{O}{\overset{\|}{C}}-R_8 \end{array} + R_1COOH,$$

$$H_2O + \begin{array}{l} CH_2OH \\ | \\ CH-O-\overset{O}{\overset{\|}{C}}-R_2 \\ | \\ CH_2-O-\overset{O}{\overset{\|}{C}}-R_3 \end{array} \rightleftharpoons \begin{array}{l} CH_2OH \\ | \\ CH-O-\overset{O}{\overset{\|}{C}}-R_2 \\ | \\ CH_2OH \end{array} + R_3COOH.$$

teins, or bile salts. Second, it is necessary to add calcium ions to precipitate the free fatty acids as they are formed during hydrolysis. Third, as is true of most enzymes, the activity of lipase preparations is markedly affected by the hydrogen ion concentration. Optimum activity is obtained at *p*H 8–9. To maintain a favorable *p*H, an ammonium hydroxide-ammonium chloride buffer (*p*H 8) is used. The reactions involved are shown below.

In the course of hydrolysis, hydrogen ions are produced (reactions 1–2). These hydrogen ions are consumed by the buffer, with the concomitant formation of $NH_4^+$ ion and the weakening of the buffer capacity (reaction 4). The buffer capacity is restored during hydrolysis by the occasional addition of ammonium hydroxide, keeping the system pink to phenolphthalein. Thus the extent of the hydrolysis is reflected by the increase in $NH_4^+$. The ammonium ion concentration is determined by titration of aliquots of the reaction mixture in alcohol-ether with standard potassium hydroxide (reaction 5). The alcohol-ether serves two purposes: (1) to stop lipase activity and (2) to increase the apparent *p*K of the phenolphthalein by 2–3 *p*H units (phenolphthalein becomes a weaker acid). The *p*K of the ammonium buffer system is apparently unaffected, or may decrease (that is, ammonium ions may become more acidic). Thus the addition of alcohol-ether to an aliquot of the lipase reaction mixture decolorizes the phenolphthalein (the unionized form of this indicator is colorless), and it is possible to titrate $NH_4^+$ with the stronger base to the usual pink color of phenolphthalein.

$$(1)\quad \begin{array}{l} CH_2{-}O{-}C(=O){-}R_1 \\ | \\ CH{-}O{-}C(=O){-}R_2 \\ | \\ CH_2{-}O{-}C(=O){-}R_3 \end{array} + H_2O \longrightarrow R_3{-}C(=O){-}OH + \begin{array}{l} CH_2{-}O{-}C(=O){-}R_1 \\ | \\ CH{-}O{-}C(=O){-}R_2 \\ | \\ CH_2OH \end{array} \quad \text{Further hydrolysis}$$

$$(2)\quad R{-}C(=O){-}OH \longrightarrow R{-}C(=O){-}O^- + H^+$$

$$(3)\quad R{-}C(=O){-}O^- + \tfrac{1}{2}\,CaCl_2 \longrightarrow \tfrac{1}{2}\left(R{-}C(=O){-}O\right)_2 Ca + Cl^-$$

$$(4)\quad HCl + NH_4OH \longrightarrow NH_4^+ + Cl^- + H_2O$$

$$(5)\quad NH_4^+ + Cl^- + KOH \longrightarrow NH_4OH + 2\ KCl$$

$$(6)\quad HCl + KOH \longrightarrow KCl + H_2O$$

## Experimental Procedure

### MATERIALS

Artificial bile solution
2% Detergent (Dreft solution)
$0.05M$ $NH_4Cl$-$NH_4OH$ buffer (*p*H 8)
$0.1M$ $CaCl_2$
Glass beads
Oils or fats of Expt. 14
1% Phenolphthalein
95% Ethanol:ether (9:1)
$0.02N$ KOH
$5N$ $NH_4OH$
Pancreatin

### *Preparation of Enzyme*

A commercial preparation of dried pancreas, Pancreatin, is suitable for this experiment. Suspend 1 g of Pancreatin in 50 ml of $0.05M$ $NH_4Cl$-$NH_3$ buffer (*p*H 8), a 2% solution. After stirring, separate the insoluble material by centrifugation, and discard it. The supernatant extract should be kept at 0°C until used.

*Enzymatic Assay*

Weigh accurately 0.5 g of the same oil used in Experiment 14 into a 125-ml Erlenmeyer flask. Add 5 ml of artificial bile salt solution, then 1 ml of 2% Dreft solution, and several glass beads. Warm the solution on a steam bath, stopper the flask tightly, and shake vigorously until a homogeneous emulsion is obtained. Cool the mixture to room temperature while shaking the flask gently.

Add 10 ml of 0.05$M$ $NH_4Cl$-$NH_4OH$ buffer ($p$H 8) *dropwise with shaking.* Then add 10 ml of 0.1$M$ $CaCl_2$ in the same manner. Add 1 ml of 1% phenolphthalein and 4 ml of water, while shaking the flask, and bring the mixture to 37°C. Finally, add 4 ml of 2% Pancreatin. Mix the solution thoroughly by shaking, and withdraw *immediately* an initial 5-ml sample. Titrate this sample *immediately* as described below.

Allow the reaction mixture to incubate at 37°C with occasional shaking, and maintain a favorable $p$H (faint pink color of phenolphthalein in flask) by adding small amounts of 5$N$ $NH_4OH$. During the incubation, the calcium salts of the released fatty acids will precipitate.

At 15, 30, 45, and 90 min after starting the incubation, remove 5-ml aliquots from the reaction, and titrate them as described below, taking care not to vary the procedure.

Titrate each sample as follows: Add 75 ml of 95% ethanol:ether (9:1), and titrate the solution immediately with 0.02$N$ standardized KOH.

## Report of Results

Subtract the quantity of $NH_4^+$ ion present in the total reaction mixture at zero time (obtained from the titration of the zero-time aliquot) from the $NH_4^+$ ion present in the total volume at the time of each titration, making appropriate corrections for aliquot removal. Express these results graphically by plotting either the quantity of $NH_4^+$ ion present or the quantity of acid released versus time.

Use the saponification number (100% hydrolysis) for the oil in Experiment 13 to calculate the percentage hydrolysis of the oil after 90 min of enzymatic hydrolysis.

## Discussion

This experiment is an in vitro demonstration of the hydrolysis of neutral fat by lipase. Since the equilibrium state of this reaction is not completely in the direction of hydrolysis, the forward reaction is favored by the $Ca^{+2}$ ion "trap," which removes the released fatty acids as their insoluble soaps.

In vivo (that is, in the intestines of higher animals) there is no similar high concentration of $Ca^{+2}$ ion to trap the released fatty acids, but the hydrolysis is enhanced by the absorption of the products. However, two other lipase-catalyzed reactions, transesterification and net synthesis of ester bonds, also occur.

$$R'—\overset{O}{\overset{\|}{C}}—O—R + R''—\overset{O}{\overset{\|}{C}}—OH \underset{\text{Transesterification}}{\rightleftarrows} R''—\overset{O}{\overset{\|}{C}}—O—R + R'—\overset{O}{\overset{\|}{C}}—OH,$$

$$ROH + R'—\overset{O}{\overset{\|}{C}}—OH \underset{\text{Net synthesis}}{\rightleftarrows} R'—\overset{O}{\overset{\|}{C}}—O—R + H_2O.$$

The existence of the transesterification reaction implicates an alkyl-enzyme (Enz-R) as an intermediate in lipase-catalyzed hydrolysis of a fatty acid ester:

$$\text{Enz-H} + R'\text{—}\overset{\displaystyle O}{\overset{\|}{C}}\text{—O—R} \rightleftarrows \text{Enz-R} + R\text{—}\overset{\displaystyle O}{\overset{\|}{C}}\text{—OH}$$

$$\text{Enz-R} + \text{HOH} \rightleftarrows \text{Enz-H} + \text{ROH}$$

$$R'\text{—C—}\overset{\displaystyle O}{\overset{\|}{\text{OR}}} + H_2O \rightleftarrows R'\text{—}\overset{\displaystyle O}{\overset{\|}{C}}\text{—OH} + \text{ROH}$$

or in transesterification:

$$\text{Enz-H} + R'\text{—}\overset{\displaystyle O}{\overset{\|}{C}}\text{—O—R} \rightleftarrows \text{Enz-R} + R'\text{—}\overset{\displaystyle O}{\overset{\|}{C}}\text{—OH}$$

$$\text{Enz-R} + R''\text{—}\overset{\displaystyle O}{\overset{\|}{C}}\text{—OH} \rightleftarrows \text{Enz-H} + R''\text{—}\overset{\displaystyle O}{\overset{\|}{C}}\text{—OR}$$

The existence of the net synthesis reaction points out that enzymes can catalyze reactions in either direction. Like all catalysts, enzymes do not change the equilibrium states of reactions, but affect only the rate at which equilibrium is attained.

### Exercises

1. What factors would you expect to have an influence on the rate of hydrolysis observed in the experiment?
2. Is it theoretically possible to use optical activity as a means of measuring lipase-catalyzed hydrolysis of neutral fats? In what other way(s) could you measure this reaction?
3. A triglyceride has a saponification number of 189. A 100-g sample was dispersed in 1 liter of a solution of pancreatic lipase. When the reaction had gone for 17 hours at 37°C, there was no longer any increase in titratable fatty acid concentration with time. At this time a 100-ml sample required 21.4 ml of 1.13*N* KOH to titrate the fatty acids. Calculate the apparent equilibrium constant for this reaction.

### References

Mattson, F. H., and W. L. Beck, *J. Biol. Chem.*, **219,** 735 (1956).
Savary, P., and P. Desnuelle, *Biochim. et Biophys. Acta.*, **21,** 349 (1956).
Tattrie, N. H., R. A. Baily, and M. Kates, *Arch. Biochem. Biophys.*, **78,** 319 (1958).
Borgström, B., *Arch. Biochem. Biophys.*, **49,** 268 (1954).

EXPERIMENT

# 15. Complex Lipids: Extraction and Separation

## Theory of the Experiment

The chief constituent of adipose tissue is neutral fat. Other tissues rich in lipids, such as brain tissue, contain relatively small amounts of neutral fat. The chief lipid constituents of these tissues are complex lipids such as phospholipids and cholesterol.

The isolation of pure complex lipids is a

formidable task, owing to these facts: (1) groups of lipids often have similar chemical properties; (2) lipids are often unstable to ordinary laboratory manipulations (for example, many lipids undergo spontaneous oxidation in air); and (3) most lipids are soluble in nonaqueous solvent systems and are extremely insoluble in aqueous solvents. Furthermore, the solubility characteristics of one lipid are often markedly altered by the presence of another lipid. Thus there are few cases in which one general method may be used successfully to obtain in pure form the constituents of complex lipid mixtures.

The present experiment with complex lipids consists of the extraction and purification of complex lipids from brain tissue. This is accomplished by successive extractions with various solvents and subsequent fractionation of the extracts to obtain fractions rich in sterols, glycerophospholipids, and sphingolipids. This scheme is based on the fact that (1) unlike most complex lipids, sterols are soluble in acetone; (2) such glycerol phospholipids as the lecithins and phosphatidyl ethanolamine and phosphatidyl serine are soluble in ether and insoluble in acetone; and (3) sphingosine phosphatides and glycosides are soluble in hot alcohol and insoluble in ether and acetone.

It should be emphasized that the final products obtained in this experiment—with the possible exception of cholesterol in the sterol fraction—represent *purified*, not pure, materials. These constituents will be characterized in Experiment 16.

## Experimental Procedure

### MATERIALS

Calf brain
Diethyl ether
Blendor
Acetone
95% Ethanol
Chloroform:methanol (1:1)

**Caution:** Many of the solvents used in the following operations are highly flammable. Do *not* use a free flame. Use a steam bath for all heating operations. For evaporation of solvents, attach flasks with distillation heads to a water aspirator (evaporation is hastened by reduced pressure, and solvent vapors go down the sink—not into the air). Operate the blendor with care when using flammable solvents, since the sealing washer at the base of the blendor bowl may leak solvent into the motor housing and cause a fire.

### *Preparation of Fraction 1: Cholesterol*

Add 4 volumes of acetone (200 ml) to a 50-g sample of calf brain in a blendor. Operate the blendor for 1 min. Transfer the suspension to a beaker, and rinse the blendor bowl with an additional 25 ml of acetone. Allow the combined homogenate to stand for 5 min, with occasional stirring.

Filter the suspension through a Büchner funnel, using suction. Blend the residue for 1 min with an additional 100 ml of acetone in the blendor, and after allowing it to stand for 5 min, filter this suspension through the Büchner funnel. Rinse the blendor bowl with an additional 50 ml of acetone, and pour this rinse through the filter cake. The combined extracts and washings constitute fraction 1. Save the filter cake for the preparation of fractions 2 and 3.

Remove the acetone from fraction 1 under reduced pressure, using a temperature-regulated flash evaporator or a steam-warmed round-bottom flask fitted with a distillation head connected to the water aspirator. Then cool the solution with cold tap water, and collect the crude cholesterol on a Büchner funnel. Recrystallize the cholesterol by dissolving it in a *minimal* volume of hot ethanol, *filtering the solution while it is hot*, and then *allowing the filtrate to cool*. The filtration is best carried out by filtering the boiling suspension through a fluted filter paper placed in a funnel on an Erlenmeyer flask on the steam bath. Allow the filtrate in the flask to boil so that hot ethanol will condense and reflux on the exterior of the filter paper. Air-dry the crystalline cholesterol, weigh it, and save it for characterization in Experiment 16.

### *Preparation of Fraction 2: Lecithin and Cephalin*

Transfer the filter cake (the acetone-insoluble residue) to a 400-ml beaker, and extract it three times with ether, using 200 ml of ether per extraction. Perform each extraction by allowing the solid to stand in ether 4–5 min with occasional stirring; then vacuum filter the product. Save the residue for the isolation of fraction 3. Concentrate the ether extracts to 50 ml under reduced pressure as above. Pour the concentrated extract into 150–200 ml of acetone, and stir. Collect the precipitate (fraction 2) on a Büchner funnel. Discard the filtrate. Quickly place the precipitate in a tared 50-ml Erlenmeyer flask, and weigh it to the nearest milligram. Then dissolve the precipitate in chloroform:methanol (3:1), and stopper the flask, since air readily oxidizes fractions 2 and 3. Save the preparation for characterization in Experiment 16.

### *Preparation of Fraction 3: Sphingosine Phosphatides and Sphingosine Glycosides*

Extract the ether-insoluble residue obtained in the preparation of fraction 2 with 50 ml of boiling ethanol. Vacuum-filter the suspension while it is still hot. Discard the precipitate. Cool the filtrate, and collect the resulting precipitate (fraction 3) by vacuum filtration. Quickly place the precipitate in a tared 50-ml Erlenmeyer flask, and weigh it to the nearest milligram. Then dissolve the precipitate in chloroform:methanol (3:1), and stopper the flask. Save the preparation for characterization in Experiment 16.

## Report of Results

Express the dry weight of each fraction in terms of the percentage of the original wet weight of brain tissue.

Prepare a flow sheet showing the behavior of the various types of lipids as a result of the procedure of this experiment.

## Discussion

The complete separation of the various types of complex lipids in a given tissue on the basis of solubilities is hard to achieve. Although this experiment is rather effective in separating major groups of lipids, it does not, for example, completely separate sphingolipids from glycerophosphatides. Further, the complete separation of the various sphingolipids from each other, or the complete separation of lecithins from other glycerophosphatides on the basis of solubilities, is extremely difficult.

Recently developed chromatographic techniques have offered great promise of not only separating types of lipids but of yielding homogeneous materials (see Hirsch and Ahrens in References). No doubt, with an improvement in the methods available to separate mixtures of complex lipids, many new and interesting compounds will be discovered.

### Exercises

1. What is meant by the term adipose tissue?
2. Which of the fractions isolated in this experiment (if any) should contain phosphorus? Unsaturated double bonds? Nitrogen? Periodate-sensitive material?

### References

Colowick, S. P., and N. O. Kaplan (Editors), *Methods in Enzymology* (Vol. III), Academic, New York, 1957, p. 299.

Hirsch, J., and E. H. Ahrens., *J. Biol. Chem.*, **233**, 311 (1958).

EXPERIMENT

# 16. Complex Lipids: Partial Characterization

## Theory of the Experiment

In this experiment the complex lipid fractions obtained in Experiment 15 are examined in detail. Cholesterol will be subjected to the tests described below.

### *Determination of Melting Point*

The melting point of isolated fraction 1 is compared with that of cholesterol (149–151°C).

### *Salkowski Test*

When chloroform solutions of cholesterol are layered over an equal volume of concentrated $H_2SO_4$ and mixed gently, a characteristic color develops. This qualitative test is rather specific for cholesterol (see Salkowski in References). Although the exact mechanism of the reaction and the nature of the colored product(s) are uncertain, the color probably results from the formation of additional double bonds or from condensation of two cholesterol molecules to form bisteroids.

### *Liebermann-Burchard Test*

Acetic anhydride can condense with the 3-position OH groups of cholesterol and related sterols to yield the corresponding ester. If the original sterol also has a $\Delta^5$ unsaturation, a subsequent epimerization to the 3-$\alpha$ form and dehydration occur, with the formation of a characteristic color. Hence the reaction serves as a specific test for 3-hydroxysteroids with $\Delta^5$ unsaturation (see Brieskorn and Herrig in References).

### *Preparation of a Digitonide Derivative*

In nonpolar solvents the saponin, digitonin, will condense with equivalent amounts of 3-$\beta$ sterols (e.g., cholesterol) to yield an insoluble derivative, the digitonide. This reaction is very specific for 3-$\beta$ sterols and therefore serves as a qualitative test. Digitonides of $\Delta^5$ sterols will still yield a positive Liebermann-Burchard test.

The various materials of fractions 2 and 3 are detected by analysis of the products of acid hydrolysis of these fractions.

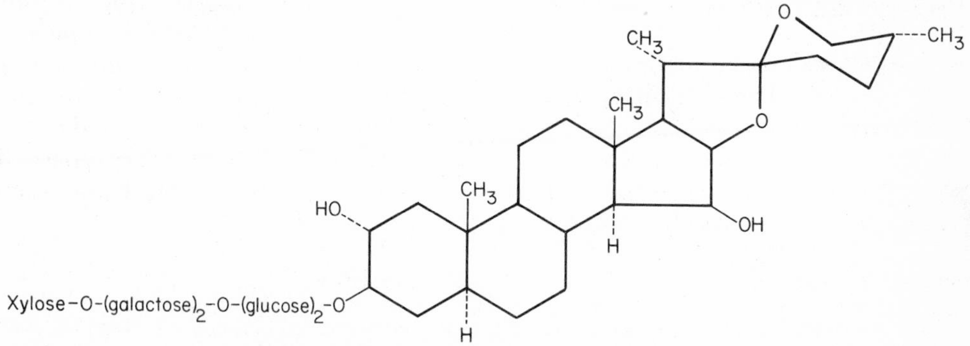

Digitonin ($C_{56}H_{92}O_{29}$), mol. wt. = 1229

## Experimental Procedure

### MATERIALS

Cholesterol
Chloroform
Methanol
Conc. $H_2SO_4$
Acetic anhydride
0.5% Digitonin in 50% ethanol
Acetone:absolute ethanol (1:1)
95% Ethanol
$2N$ HCl
1% Alcoholic phenolphthalein
$0.1N$ NaOH
10% Glycerol solution
2.5% Ammonium molybdate
$10N$ $H_2SO_4$
Phosphate standard (1 $\mu$mole/ml)
Reducing reagent
2% NaOCl (Clorox)
Acid molybdate reagent
5% Alcoholic $\alpha$-naphthol
$n$-Butanol:diethylene glycol:$H_2O$ (4:1:1)
0.2% Ninhydrin in ethanol
0.4% Stannous chloride in $3N$ HCl
2% Phosphomolybdic acid
Glass tray or pie plate
Carbohydrate chromatographic solvent
Serine solution (2 mg/ml)
Ethanolamine solution (2.2 mg/ml)
Choline chloride solution (10 mg/ml)
Sphingosine solution (10 mg/ml)
Aniline-acid-oxalate spray reagent
0.5% $NaIO_4$
0.5% Benzidine reagent
Whatman No. 1 paper
1% Solutions of glucose, galactose, glucosamine, and inositol
Cerebroside, sphingomyelin, stearic acid, $\alpha$-hydroxypalmitic acid
Chromatoplates (prepared with 100-mesh silicic acid and 2% plaster of Paris (calcined calcium sulfate)
0.2% Dichlorofluorescein in ethanol

### *Characterization of Fraction 1: Cholesterol*

#### DETERMINATION OF MELTING POINT

Determine the melting point of a small sample of the crystalline fraction 1, using a small capillary in an oil bath or some other appropriate device. Similarly, determine the melting point of authentic cholesterol and of a mixture of cholesterol and fraction 1.

#### SALKOWSKI TEST

Prepare 3 test tubes, the first containing 1 ml of chloroform; the second, 10 mg of fraction 1 in 1 ml of chloroform; and the third, 10 mg of cholesterol in 1 ml of chloroform. Add slowly 1 ml of concentrated $H_2SO_4$ to each, so as to form a lower phase, and observe any colors formed.

#### LIEBERMANN-BURCHARD TEST

Prepare 3 test tubes, the first containing 2 ml of chloroform; the second, 5–15 mg of fraction 1 in 2 ml of chloroform; and the third, 5–15 mg of cholesterol in 2 ml of chloroform (carefully avoid moisture). Add 1 ml of acetic anhydride to each, and mix well. Then add 2 ml of concentrated $H_2SO_4$ to each, and mix cautiously. Note the color changes after 30 min.

#### PREPARATION OF A DIGITONIDE DERIVATIVE

Separately dissolve weighed portions (100–500 mg) of fraction 1 and cholesterol in minimal volumes of acetone:absolute ethanol (1:1). Add 3 volumes of a 0.5% solution of digitonin in 50% aqueous ethanol to each; warm the mixtures for a few seconds on a steam bath, and allow the suspension to stand for an hour at 0–5°C. Collect the precipitates by centrifugation, and air-dry and weigh the products.

### *Thin-Layer Chromatography*

If time and availability of equipment permit, prepare two chromatoplates for thin-layer chromatography (TLC) (see Mangold in References) by coating 20 × 20 cm glass plates with a 0.25-mm layer of a slurry made by mixing 0.6 g of $CaSO_4 \cdot \frac{1}{2}H_2O$ and 29.4 g of silicic acid with 60 ml of $H_2O$. (A 0.25-mm layer can be obtained by using a specific applicator or by placing strips of masking tape on two edges of the glass plates, pouring the slurry into the area between the tape, and then sweeping off the excess slurry with a glass rod resting on both strips of tape.) Then air-dry and oven-dry (110°C, 1 hr) the plates before use. If necessary, store the chromatoplate at this stage in a covered container away from laboratory fumes and moisture and oven-dry again just before use.

To the chromatoplate apply 3-$\mu$liter samples of fractions 2 and 3, a 1% solution of fraction 1 in chloroform:methanol (3:1), and 1% solutions of standards (cerebroside, sphingomyelin, cholesterol, lecithin, phosphatidyl ethanolamine) all in chloroform:methanol (3:1). Arrange them in a row 2 cm from an edge and 2 cm apart.

Develop in each of two solvent systems:

(1) Hexane:ethyl ether:acetic acid (90:10:1),

(2) Chloroform:methanol:conc. $NH_4OH$ (25:75:4).

Development will be rapid (about 30 min). When the solvent front has moved well up the plate, remove and air-dry. Spray the plates with 0.2% dichlorofluorescein in ethanol, and view them under ultraviolet light, outlining the spots with a pin. Make a diagrammatic scale drawing of your results. What conclusions do you reach?

### *Characterization of Fractions 2 and 3: Lecithins, Phosphatidyl Ethanolamine, Phosphatidyl Serine, and Sphingolipids*

#### ACID HYDROLYSIS

Evaporate off the chloroform-methanol solution containing fractions 2 and 3 that you isolated in Experiment 15. Separately dissolve a 0.5-g sample of each in 6 ml of 95% ethanol by heating in a 50-ml Erlenmeyer flask on a steam bath. When dissolved, heat the solutions to the boiling point, and add 6 ml of 2*N* HCl to each. Further heat the mixtures on the steam bath until the alcohol has evaporated (usually 1 hr). Cool the aqueous solutions, and while they are still in the flasks extract each of the aqueous phases three times with 10-ml portions of ether (decanting the ether extracts into tared 125-ml Erlenmeyer flasks after each extraction). This procedure separates the ether-soluble fractions from the aqueous phase, or acid-soluble fraction.

Remove the ether under reduced pressure, and determine the weights of the dry residues. Add 10 ml of ether to the residues remaining after evaporation of the ether, and dry the solutions, thereby removing the last traces of hydrochloric acid. Determine the neutralization equivalents by suspending the solids and/or oils (obtained by hydrolysis and ether extraction of fractions 2 and 3) in separate 50-ml batches of warm ethanol and titrating to a phenolphthalein end point with 0.1*N* NaOH. (Gentle heating during the titrations will hasten solution of the solids; a solvent blank should be similarly titrated.)

Concentrate the ether-extracted aqueous solutions from fractions 2 and 3 almost to the point of dryness in a vacuum, using the steam bath. Do not overheat at this stage, for charring must be avoided. Add 5 ml of 95% ethanol to each fraction, and evaporate the mixtures to dryness (with the same precautions) to insure complete removal of hydrochloric acid. Dissolve the residues in 2–5 ml of water. These fractions will be analyzed in the sections that follow.

#### DETERMINATION OF TOTAL PHOSPHATE FROM ACID-SOLUBLE RESIDUE OF FRACTION 2

To determine the appropriate amount of sample for the digestion procedure below, assume that the sample is a phospholipid of a given structure. The analysis of several dilutions of this digest will enable you to find the proper range for analysis. The color values obtained for the diluted digests must be lower, of course, than the maximum on the standard curve (1 μmole phosphate).

Place an appropriate amount of sample in a test tube. Add 0.5 ml of 10*N* $H_2SO_4$ and heat the mixture in an oven or on a sand bath for 30–60 min at 130–160°C. until clear. Cool the tube in air, then *slowly* (**Caution:** concentrated $H_2SO_4$) add 2 ml of $H_2O$. Finally, heat this diluted solution for 10 min in a boiling water bath to hydrolyze any pyrophosphates formed during the digestion.

Prepare a water blank and phosphate standards (for example, 0.3, 0.6, and 1.0 ml of a 1 μmole/ml standard). Add 2 ml of water and 0.5 ml of 10*N* $H_2SO_4$ to the blank and the standards to bring the acid concentration up to that of the hydrolysed samples. Add 1 ml of 2.5% ammonium molybdate to *all* tubes (thus the acid molybdate concentration used in Experiment 10 is achieved by the separate addition of acid and molybdate). After shaking the tubes, add 1 ml of reducing reagent to each, and dilute the contents to 10 ml with water. After 20 min read the optical density of each tube at 660 mμ against the blank.

### DETECTION OF GLYCEROL IN AQUEOUS PHASE OF FRACTIONS 2 AND 3

Glycerol is detected in the aqueous phase remaining after acid hydrolysis and ether extraction by the oxidation of glycerol to glyceraldehyde and subsequent complexing of the glyceraldehyde with α-naphthol (see p. 000). Prepare three test tubes, the first containing 1 ml of the aqueous phase; the second, 1 ml of 10% glycerol solution; and the third, 1 ml of water. Add 1 ml of 2% NaOCl (Clorox) to each, and after 3 min add 3 drops of conc. HCl to each. Then boil the tubes in the hood for one minute to remove the chlorine. Finally, add 0.2 ml of 5% alcoholic α-naphthol and 4 ml of concentrated $H_2SO_4$, and shake the tubes cautiously. An emerald green color indicates the presence of glycerol.

### DETECTION OF NITROGENOUS BASES IN THE ACID-SOLUBLE RESIDUES OF FRACTIONS 2 AND 3

Using *p*H paper bring the acid-soluble components of fractions 2 and 3 to *p*H 1. Check the amounts of samples (knowns and unknowns) required for adequate detection by spotting on a separate paper and detecting the spots with the appropriate reagents (see below). The preferred range for separation and detection is 5–20 μg for ethanolamine and serine, 25–75 μg for choline, and 100 μg for sphingosine. Next prepare an ascending chromatogram containing standards and appropriate amounts of the acid soluble residues of fractions 2 and 3 (2 spots of each). After developing the chromatogram with *n*-butanol:diethylene glycol:water (4:1:1), air-dry the paper for at least an hour at room temperature, then for 30 min in an oven at 100°C. Cut off separate strips (perpendicular to the solvent front) to detect choline and other bases.

*Detection of Choline.* Immerse a paper strip in a 2% solution of phosphomolybdic acid for 1 min (in a glass tray). Next wash it first in *n*-butanol for 5 min (in another glass tray) and then in water for 5 min. Finally, slowly pass the paper through a freshly prepared 0.4% solution of stannous chloride in 3*N* HCl. Choline spots are dark blue on a light blue background. Choline forms an insoluble salt with phosphomolybdate, and phosphomolybdate is reduced (to give a blue color) by stannous chloride.

*Detection of Ethanolamine, Serine, and Sphingosine.* Spray the paper strips with 0.2% ninhydrin in ethanol, and heat them for 5 min in an oven maintained at 100°C. Compounds containing a primary amine (for example, ethanolamine, serine, and sphingosine) appear as purple spots (see p. 00 for nature of the reaction). The $R_f$ values for serine and ethanolamine are about 0.18 and 0.35, respectively, whereas that for sphingosine is usually near 1.0.

DETECTION OF CARBOHYDRATES FROM THE ACID-SOLUBLE RESIDUE OF FRACTION 3

The carbohydrate constituents present in the acid-soluble residue of fraction 3 (sphingolipids) are detected by chromatography. First, determine the amount of sample required to give a readily detectable colored spot on paper when sprayed with the aniline-acid-oxalate or periodate-benzidine spray systems of Experiment 3. Then apply the appropriate amounts of fraction 3 and 5 mm spots of 1% solutions of inositol, glucose, galactose, and glucosamine to a chromatogram, and develop it in the ascending manner with the carbohydrate chromatographic solvent provided. After drying the developed chromatogram, spray it with the desired reagent (inositol must be detected with the periodate-benzidine system), and record the $R_f$ values observed.

## Report of Results

Summarize the various structural features of the sterol(s) of fraction 1 revealed by the various tests performed. Further, calculate the purity of the "cholesterol" in fraction 1 by comparing the weight of the digitonide obtained from pure cholesterol with the quantity of digitonide obtained from an equal weight of fraction 1. Compare this calculation with the indication of purity expressed by the melting point of the recrystallized product.

Using the neutralization equivalents obtained from the ether-soluble extracts of the acid hydrolysates, draw representative lipids of fractions 2 and 3 containing acids whose molecular weights correspond with the neutralization equivalent found.

Tabulate the qualitative findings obtained from your examination of the aqueous residue remaining after ether extraction of the acid hydrolysates by listing the compounds identified, the fractions from which they were obtained, and the type of lipid indicated (for example, sphingomyelin, lecithin, cerebroside).

## Discussion

In this experiment, several fractions of complex lipids are characterized. In fraction 1 an apparently pure compound, cholesterol, is isolated. In other fractions, the analysis indicates the presence of one or more complex lipids.

It should be noted that in this experiment, fatty acid esters are hydrolyzed with acid rather than with alkali as in Experiment 13. Both methods of lipid hydrolysis have proved very effective in leading to the elucidation of the structure of simple and complex lipids.

**Exercises**

1. Why is it necessary to assure the removal of the HCl from the ether extract of the acid hydrolysate before determining the neutralization equivalent?
2. Devise a series of chemical or physical tests (no chromatography or melting point determinations) to distinguish between each of the following pairs of pure compounds: lecithin and cephalin; $\alpha$-glycerophosphoric acid and $\beta$-glycerophosphoric acid; $\alpha,\alpha'$-dipalmitin and $\alpha,\beta$-dipalmitin; $\alpha,\alpha'$-dipalmitin and tripalmitin; sphingomyelin and cerebroside (both containing lignoceric acid); cholesterol and trilaurin.
3. One gram of a white oily solid, compound *A*, is refluxed in 10 ml of 1.00*N* alcoholic KOH for several hours and then neutralized with 0.50*N* HCl. After removal of the alcohol by drying, this sample is

suspended in water, further acidified to *p*H 1, and extracted with ether. The residue, after the removal of the ether, contains an oil, compound *B* (N.E. = 256). Heating of the acidified aqueous phase remaining after ether extraction, followed by cooling, titration to *p*H 12, ether extraction of the basic solution, and drying of the ether and aqueous solutions gives two compounds, *C* and *D*. Compound *C* (from ether extract) reacts with ninhydrin and $IO_4^-$, and takes up 1 mole $I_2$ per mole *C*. Compound *D* (from aqueous phase) yields insoluble crystals when heated with conc. $HNO_3$ and then diluted with water. Propose a structure for each compound, and diagram the reactions mentioned. How many millititers of 0.05*N* HCl would be required to back-titrate the KOH remaining after complete reaction with the one gram of compound *A* (above)? Why is the heating of the acid-soluble residue necessary before adjusting to *p*H 12 and obtaining compounds *C* and *D*?

**References**

Salkowski, E., *Z. Physiol.*, **57,** 523 (1908).
Brieskorn, C. H., and H. Herrig, *Arch. Pharm.*, **292,** 485 (1959).
Mangold, H. K., *J. Am. Oil Chemists' Soc.*, **38,** 708 (1961).
Rapport, M. M., and B. Lerner, *J. Biol. Chem.*, **232,** 63 (1958).
Carter, H. E. et al., *J. Biol. Chem.*, **169,** 77 (1947).
Dawson, R. M. C., *Biochim. et Biophys. Acta*, **14,** 374 (1954).

EXPERIMENT

# 17. Production of Fatty Livers with Ethionine

## Theory of the Experiment

The accumulation of lipids, usually neutral fat, in the liver of animals is a condition referred to as "fatty liver." It can be caused by any of a number of pathological and nutritional conditions and by certain liver poisons, but the best known and most characteristic type of fatty liver is caused by a deficiency of dietary choline or methionine.

$$HO{-}CH_2{-}CH_2{-}{}^{+}N(CH_3)_3$$

Choline

$$CH_3{-}S{-}CH_2{-}CH(NH_3^+){-}C(=O){-}O^-$$

Methionine

$$CH_3{-}CH_2{-}S{-}CH_2{-}CH_2{-}CH(NH_3^+){-}C(=O){-}O^-$$

Ethionine

Young animals are particularly susceptible to this deficiency, which also produces hemorrhaging in the kidneys and permanent damage to the heart.

In older animals a "methionine or choline fatty liver" can be produced by injecting ethionine, a homolog of methionine. As will be noted from the structure, ethionine differs from methionine only in that it is an ethyl, rather than a methyl, thioether. Ethionine appears to act as a competitive antagonist of methionine, since administration of the latter can negate the effectiveness of ethionine in producing fatty livers.

In this experiment female rats are used because the increase in liver fat is much more pronounced in the female than in the male rat. Liver fat is measured as fatty acids released by saponification and made chloroform-soluble by acidification.

## Experimental Procedure

### MATERIALS

Female rats
Isotonic saline (0.9% NaCl)
95% Ethanol
Cotton
0.1% Bromcresol green
DL-Ethionine in isotonic saline
Saturated KOH
Chloroform
Aluminum foil
6*N* HCl

Obtain two female rats, one to be injected with ethionine and the other to be injected with isotonic saline to serve as a control. Inject 3 ml of a saline solution of DL-ethionine (16.7 mg of DL-ethionine/ml) intraperitoneally into a female rat. Inject the control female rat intraperitoneally with 3 ml of isotonic saline. Inject the same quantities of solution $2\frac{1}{2}$, 5, and 7 hr after the first injection. Withhold food but not water after the first injection.

Twenty-four hours after the first injection, stun the animals with a blow on the head, and then decapitate them; remove the livers, and weigh each separately on a triple-beam balance. Examine the livers of both animals for size, color, texture, and elasticity. Record any observed differences between the two livers.

Cut the livers into small pieces, and place the pieces of each in separate 125-ml Erlenmeyer flasks. Add 10 ml of saturated KOH and 15 ml of 95% ethanol to each. Heat the flasks on a steam bath for 3 hr. (If time dictates, the mixtures may be stored in a stoppered flask at this stage.) Add first a few drops of bromcresol green and then 6*N* HCl from a pipette until the color of the solution changes from blue to yellow. Add exactly 50 ml of chloroform, and stopper each flask with a rubber stopper covered with aluminum foil. Shake the tightly stoppered flasks vigorously for 3–5 min. At the end of the shaking period, remove the stoppers carefully. Allow the flasks to stand an additional 15 min (or until the chloroform layer has cleared) before pipetting 10-ml aliquots of the chloroform layer into accurately weighed (analytical balance) 125-ml Erlenmeyer flasks. Evaporate the chloroform in a vacuum, heating gently on a steam bath.

When the chloroform has evaporated, place the flasks in an oven for 30 min at 65°C. Then remove the flasks from the oven, and allow them to cool to room temperature. Wash the outside of the flasks with distilled water, and dry them well before weighing the fatty acid. Calculate the percentage of fatty acid in each liver on a wet-weight basis.

## Report of Results

Note the effect of the methionine analog both grossly and quantitatively. Record the total liver weights and the color and general appearance (that is, size, texture, and so on) of each liver in a tabular fashion. Include also the quantitative data on total fatty acids.

## Discussion

Experimental work has demonstrated that the effects of ethionine can be prevented by the simultaneous administration of methionine. Methionine itself is a lipotropic agent. Moreover, animals placed upon diets deficient in methionine or choline will develop fatty livers, and the administration of methionine or choline restores the livers of these animals to normalcy. Since methionine has been shown to donate its methyl groups in the synthesis of choline, it is assumed that these observations relate to transmethylation and methyl donors. On the other hand, the production of fatty livers by ethionine is not prevented by the simultaneous administration of choline or methyl donors other than methionine. It is of interest that the injection of large amounts of carbohydrate (for example, glucose) prevents the development of the ethionine fatty liver; only the administration of methionine, however, will prevent death from a toxic dose of ethionine.

Ethionine-induced changes in the lipid content of the liver may be only an indirect effect of the action of ethionine. This possibility is suggested by the fact that ethionine apparently inhibits protein synthesis in the liver—perhaps by competing with methionine, since ethionine is found in the proteins formed. Another fact which must be considered is the sex difference in susceptibility to ethionine. Castration of male rats causes these animals to become susceptible to ethionine, and resistance is restored by the administration of testosterone. Furthermore, the effects of ethionine are not limited to the liver. Ethionine causes destruction of a portion of the pancreas. It is clear that the action of ethionine is not fully understood at the present time. Further study of its metabolism may lead to new knowledge of the function of both ethionine and methionine in the animal body.

**Exercises**

1. Suggest a possible reason for fasting the animals during the 24-hr period following the first injection.
2. Lecithins are believed to be involved in the metabolism and transport of lipids. The ineffectiveness of choline in countering ethionine toxicity would appear to rule out interference of ethionine with lecithin metabolism. What experiments might be done to estimate the effect, if any, of ethionine upon lecithin metabolism?
3. Suggest an experimental approach to the ethionine problem, using techniques developed in this course. What other experiments would you do?

**References**

Farber, F., M. V. Simpson, and H. Tarver, *J. Biol. Chem.*, **182,** 91 (1950).

Jensen, D., I. L. Chaikoff, and H. Tarver, *J. Biol. Chem.*, **192,** 395 (1951).

Baruch, H., and I. L. Chaikoff, *Proc. Soc. Exp. Biol. Med.*, **86,** 97 (1954).

Gross, D., and H. Tarver, *J. Biol. Chem.*, **217,** 169 (1955).

# PART THREE **PROTEINS AND AMINO ACIDS**

## Introduction

There are many different proteins in nature. The chemical differences among certain of them are obvious. For example, the bulk proteins of muscle tissue—the specific proteins or enzymes which catalyze a variety of reactions—and the proteins of small molecular weight, which act as specific hormones, differ sharply from each other. Among other proteins the chemical differences are small or subtle; yet these small differences in structure may be associated with major differences in physiological function (for example, the various forms of human hemoglobin). Such differences in structure, whether major or minor, are the result of (a) differences in the total number and chemical nature of amino acids in a molecule of protein and (b) the sequence in which these amino acids are arranged. To understand these distinctions, it is necessary to study the structure of the amino acids and proteins.

### Protein Structure

Three elements of protein structure have been recognized. The so-called primary, secondary, and tertiary structures. The primary structure is the result of the major covalent bond, the amide or peptide linkage (—CO—NH—), which holds the individual amino acids together to form the peptide chain(s). These polymer chains vary in length from a few amino acid residues (oligopeptides) to molecules containing 2000 or more amino acids. One or more residues of each of approximately 20 natural amino acids may be present in each molecule of protein. The primary amide bonds can be partly or totally ruptured by exposure to strong acid or base at elevated temperatures, or by specific enzymes at ordinary temperatures, to yield the constituent amino acids or peptides.

Classification of the secondary and tertiary structure of proteins is based on forces which impart a specific three-dimensional structure to the polypeptide chain. X-ray crystallographic evidence obtained from certain fibrous proteins (for example, hair, muscle, fibrin) [1] and, more recently, other physical evidence from globular or soluble proteins [2] indicate that many proteins contain helically coiled polypeptide chains in which the various amino acid residues of each coil or helix form internal hydrogen bonds with residues in adjacent coils (Fig. 4). This helical structure, although probably not characteristic of the entire protein molecule, does impart stability to a large part of the molecule, owing to the internal hydrogen bonds. The helical form most commonly found in proteins, the $\alpha$-helix, has 3.7 amino acid residues per turn. Conformations other than the $\alpha$-helix have been found in certain fibrous proteins and in modified (stretched or contracted) proteins, but these forms also involve specific arrangements of polypeptide chains caused by hydrogen bonding. Such structural forms (helix or otherwise) are called secondary structures of proteins.

The tertiary structure of proteins—that is, the spatial arrangement of the helical coil(s) or chain(s) of a particular protein—is largely a function of the uncombined functional groups of the constituent amino acids and the frequency with which proline or hydroxyproline appears in the molecule. The occurrence of one or more of the following constituents constitutes a contribution

FIG. 4. *Secondary structure of proteins* ($\alpha$ *helix*).

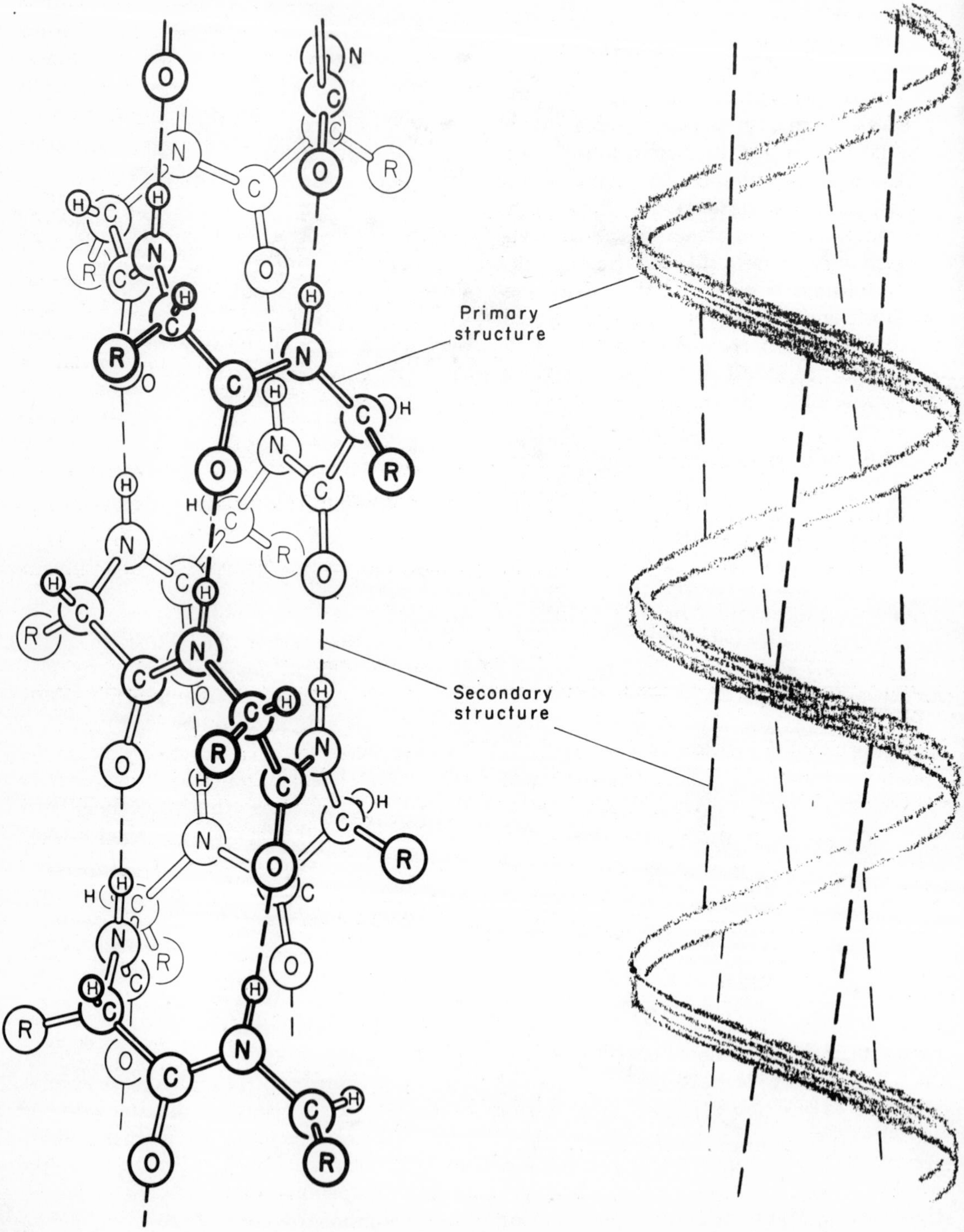

to the overall tertiary structure of a protein.

1. Disulfide bonds. The sulfhydryl groups of the amino acid cysteine are often involved in both intra- and/or interchain disulfide bonds, and thus impart a further, fixed, three-dimensional character to protein molecules. Although it may be logical to consider the disulfide bonds to be part of the primary covalent linkages of the protein molecule, it seems advisable to remember that they contribute to chain folding.
2. Hydrogen bonds. The phenolic group of tyrosine is frequently involved in hydrogen bonds with the carboxyl groups of glutamic acid and aspartic acid and with the carboxamide group of glutamine and asparagine.
3. Salt bonds. Ionizable functional groups of the constituent amino acids impart structural features to specific proteins. For example, an interchain salt bridge between two amino acids appears to be an essential component of the insulin molecule.
4. van der Waals forces. The hydrophobic alkyl groups of aliphatic amino acids and other substituents no doubt associate by weak van der Waals forces. The extent and importance of these forces in proteins is difficult to determine.
5. Proline and hydroxyproline residues. Since proline and hydroxyproline are secondary amines, and not primary amines like the other amino acids, they do not have a peptide-bond hydrogen to contribute to hydrogen bonds. Moreover, the conformation of proline and hydroxyproline does not fit into an $\alpha$-helix. Accordingly, a helical structure must be interrupted at each proline or hydroxyproline residue in the protein.

Thus a complete three-dimensional representation of the "backbone" of a typical

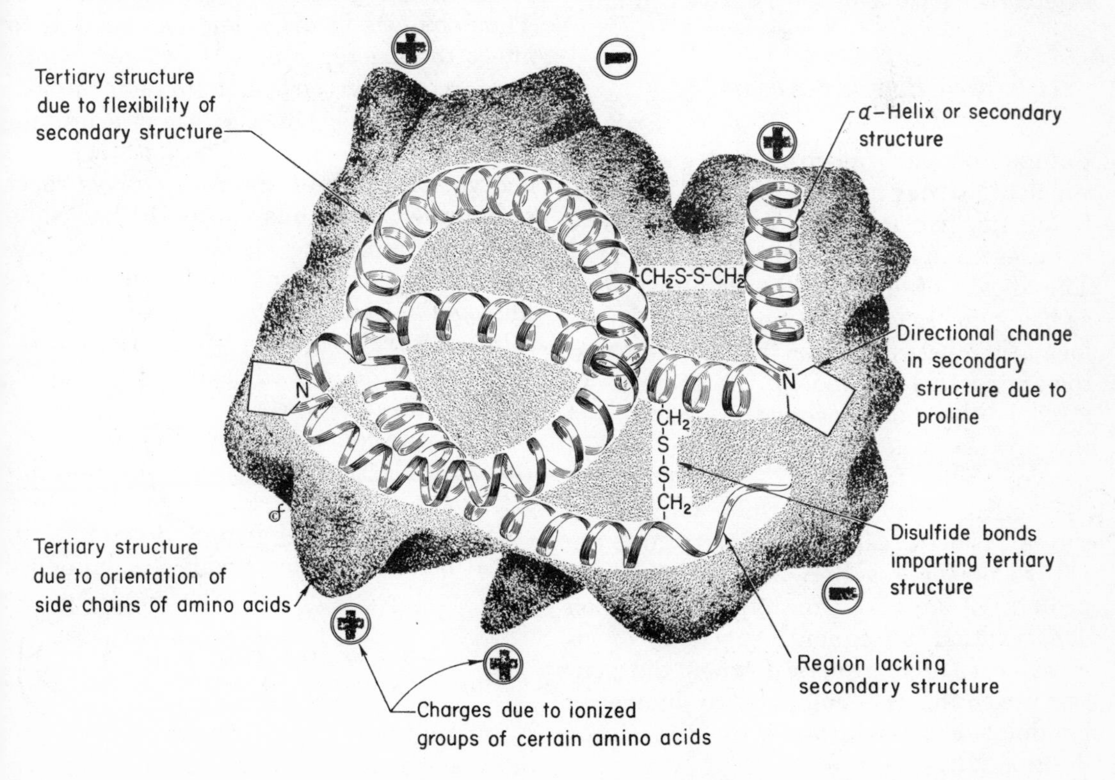


FIG. 5. *Schematic protein molecule.*

protein would describe a long polypeptide chain which is partially coiled in an $\alpha$-helix whose spring-like tubes are folded and intertwined upon themselves to form a specific shape. This backbone serves as the base for various different side chains of the constituent amino acids. The work of Perutz [3], Kendrew [4] and co-workers has led to an X-ray crystallographic confirmation of this three-dimensional structure in certain soluble proteins (myoglobin and hemoglobin). It is hoped that increased knowledge of the exact structure of specific proteins will permit a correlation to be made between structure and catalytic activity [5].

## Protein Denaturation

Proteins in their natural form are called native proteins. Any change in the structure of a protein from its native state is called denaturation. It is presumed that these changes occur in the secondary and tertiary structures of proteins.

The following are only a few of the many means of denaturing a protein. At any one time and under any one set of conditions, the extent of denaturation is undoubtedly a function of several, if not all, of these. Thus all of these parameters must be considered when working with native proteins.

### *Hydrogen Bonding Agents*

Compounds such as urea and guanidine, which are strong hydrogen bonding agents, frequently cause denaturation when they are present in high concentrations (4–6$M$). This means of denaturation is interpreted as being due to competitive hydrogen bond formation between such compounds and the amino acid residues of the protein. As a result, the coiled secondary structure of the protein is destroyed when the helical peptide chains form hydrogen bonds with the added agent rather than with themselves. The increase in viscosity and the nature of the reaction of urea with the protein [6] support this hypothesis. Such denaturation is frequently partially or almost completely reversed when the concentration of the competitive hydrogen bonding agent is lowered by dialysis or dilution [7].

### *Heat*

With few exceptions, proteins are denatured when heated in solution to higher than 50°C. Such harsh heat treatment is thought to alter the secondary and tertiary structures of proteins. The major cause of precipitation is probably formation of the random aggregates which result from destruction of the secondary structure.

### *Acid or Alkali*

Proteins are amphoteric polyelectrolytes. Thus changes in *p*H would be expected to affect the existence of salt bridges, which reinforce the tertiary structure of proteins. Further, when localized areas of a protein acquire a large net positive or negative charge, the ionizable groups involved repel each other and thus place the molecule under strain. The denaturation of the proteolytic enzyme, pepsin, occurs at alkaline *p*H's. This may be attributed to internal strain arising from mutual repulsions of the ionized carboxyls of aspartic and glutamic acids.

Moderate concentrations of certain acids (for example, trichloroacetic, phosphotungstic, or perchloric acid) usually produce complete denaturation and precipitation. Treatment with these acids is commonly used to stop enzymatic reactions.

### *Oxidation or Reduction of Sulfur Groups*

Many, but not all, proteins are sensitive to alterations in the oxidation-reduction potential of their environment. The effect is probably caused in part by oxidation of sulfhydryl groups or reduction of disulfide bonds. Because of differences in individual structure, proteins are not equally sensitive to such alterations. Frequently this characteristic is of great importance. The purification or assay of some proteins can be accomplished only be providing a reducing atmosphere (SH-glutathione, free cysteine, or mercaptoethanol).

### *Enzymatic Action*

Proteinases (cathepsins) present in crude or unpurified protein preparations often catalyze the breakdown of proteins by hydrolyzing the peptide bonds of proteins. Since these enzymes act more slowly at low temperatures, crude protein solutions are frequently kept cold (0–2°C) during the early stages of purification (see Heat, above).

### *Chemical Attack On Specific Sites*

Information about amino acids on or near the "active sites" of enzymes can be obtained by observing catalytic activity after treatment with chemical agents which alter specific amino acids. For example, iodoacetate or *p*-chloromercuribenzoate react with the free SH-groups of proteins, whereas diisopropyl phosphorofluoridate and related compounds add alkyl phosphates to serine residues of certain enzymes. Other specific reactions (for example, the photooxidation of histidine, bromacetic acid attack on histidine, and addition of organic residues to lysine) are frequently used.

## Protein Purification

There is no single or simple way to purify all proteins. Procedures useful in the purification of one protein may result in the denaturation of another. Further, slight chemical modifications in a protein may greatly alter its structure and thus affect its behavior during purification. Nevertheless, there are certain fundamental principles of protein purification upon which most fractionation procedures are based. Application of one or more of these procedures can lead to highly purified or crystalline proteins. Several, but not all of the factors of importance in protein purification are listed below.

### *Solubility Properties*

The solubility of most proteins in aqueous solutions can be attributed to the hydrophilic interaction between the polar molecules of water and the ionized groups of the protein molecules. Reagents which change the dielectric constant or the ionic strength of an aqueous solution would, therefore, be expected to influence the solubility of proteins. The addition of ethanol or acetone or certain salts such as ammonium sulfate, usually at low temperatures to avoid denaturation, frequently results in precipitation of proteins. Stepwise application of such techniques to heterogeneous protein solutions often results in fractionation (purification) because of the varying degrees of solubility among the proteins in the solution.

The precipitations noted above usually result when the attractive forces among protein molecules exceed those between the protein molecules and water. Proteins are polyelectrolytes with different numbers and types of ionizable groups, with the result that the net charge on any given protein molecule and, hence, its attraction to another protein molecule are functions of the *p*H of the medium. Therefore, any treatment designed to separate proteins by making use of relative solubility must take into consideration the *p*H of the medium. Often alteration of only *p*H under a given set of conditions is sufficient to effect an "isoelectric" precipitation. Choosing the correct condition of *p*H, ionic strength, or dielectric constant to carry out a given fractionation is somewhat arbitrary and is usually determined by trial and error.

### *Adsorptive Properties*

Under certain conditions of *p*H and at low ionic strength, certain proteins will be adsorbed by various substances. Calcium phosphate gel and alumina $C_\Gamma$ gel for example are frequently used to adsorb specific proteins from heterogeneous mixtures [8]. The adsorbed proteins can frequently be released from the isolated gels either by altering the *p*H or increasing the ionic strength. Thus fractions can be obtained by removing either the extraneous proteins or the desired protein with the gel, followed by elution.

Recently, the use of certain resins and modified celluloses as ion exchange components for the fractionation of proteins on columns has proved successful in protein purification [9]. Such materials as DEAE (diethylaminoethyl cellulose), CMC (carboxymethyl cellulose), IRC-50 (a carboxyl resin) and other resins are effective in the fractionation of heterogeneous protein solutions when elution is carried out with increasing concentrations of salts (gradient or stepwise).

### *Heavy-metal Salts of Proteins*

The addition of heavy-metal salts (for example, those of Hg or Pb) to protein solutions often results in precipitation of protein. Such precipitates may be denatured, but under carefully controlled conditions native proteins can be precipitated as their heavy-metal salts. Such protein-heavy-metal salts have proven extremely useful in X-ray crystallographic studies of proteins.

### *Heat Denaturation*

Since all proteins are not equally unstable when heated in aqueous solution, fractionation can often be achieved by controlled heating. The presence of a substrate will often stabilize a protein against heat denaturation.

### *Protein-Nucleic Acid Complexes*

Occasionally during protein purification it becomes necessary to remove basic proteins or contaminating nucleic acid. The controlled addition of sparingly soluble material of opposite charge (for example, addition of a basic protein such as protamine for RNA removal) results in the coprecipitation of the added and unwanted material.

### *Electrophoretic Separation (Electrophoresis)*

The net charge on a particular protein varies with the *p*H of the medium in which it is dissolved. Accordingly, application of an electric field to a buffered, heterogeneous protein solution will often result in a differential migration of proteins. Thus fractionations can be obtained either in free solution or in heterogenous systems with an "inert" supporting material, such as starch, when a potential is applied.

### *Criteria of Protein Purity*

It should be emphasized that there are no tests for protein purity, only for impurities. Crystallinity, the absence of contaminating enzymes, and homogeneous behavior in centrifugal fields (ultracentrifugation), in electric fields (electrophoresis), in ion exchange chromatography, or in immunological reactions may all be consistent with purity but do not prove it. One attempts to prove inhomogeneity with as many tests as are available. Practically, if a protein preparation behaves as a single entity in careful studies involving the above criteria, it may be assumed to have a high degree of purity.

## Quantitative Determination of Proteins

In general, there is no completely satisfactory or best way to determine the concentration of protein in a given sample. The choice of the method depends on the nature of the protein under test, the nature of the other components in the protein sample, and the desired speed, accuracy, and sensitivity of assay. Several of the methods commonly used for protein determination are discussed below.

### *Biuret Test*

Compounds containing two or more peptide bonds (for example, proteins) give a characteristic purple color when treated with dilute copper sulfate in alkaline solution. The name of the test [10, 11] comes from the compound biuret, which gives a typically positive reaction.

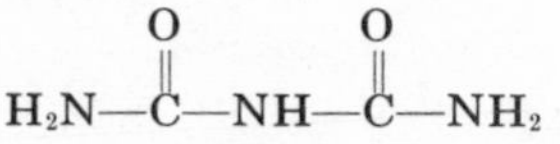

$$H_2N-\overset{\overset{\displaystyle O}{\|}}{C}-NH-\overset{\overset{\displaystyle O}{\|}}{C}-NH_2$$

Biuret

The color is apparently due to the coordination complex of the copper atom and four nitrogen atoms, two from each of two peptide chains (see figure below).

The biuret test is fairly reproducible for any given protein, but requires relatively large amounts of protein (1–20 mg) for color formation.

### *Folin-Ciocalteau Test*

The quantitative Folin-Ciocalteau test [12, 13] has the advantage of applicability to dried material as well as to solutions. In addition, the method is sensitive; samples containing as little as 5 μg of protein can be readily analyzed. The color formed by the Folin-Ciocalteau reagent is due to the reaction of protein with the alkaline copper in the reagent (as in the biuret test) and to the reduction of the phosphomolybdate-phosphotungstate salts in the reagent by the tyrosine and tryptophan of proteins.

### *Nitrogen Analysis: Kjeldahl Method*

Since proteins contain 12–18% nitrogen by weight, nitrogen determinations are frequently made on proteins and protein-like materials after the samples have been digested (oxidized) to $CO_2$ and $NH_4^+$ by boiling with concentrated sulfuric acid. The ammonium ion concentration can then be determined by the following two tests.

#### KJELDAHL TITRATION

The addition of excess alkali to acid solutions containing ammonium ion causes the formation of ammonia. Steam distillation of the ammonia from a closed system into standard acid solution is followed by subsequent titration of the unused acid with standard alkali [14].

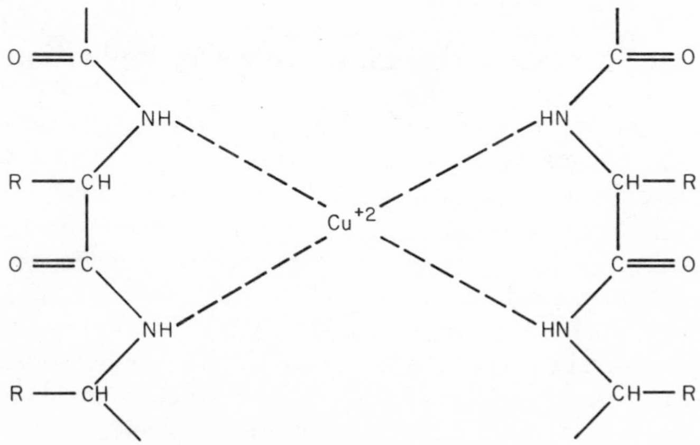

Biuret color complex

NESSLER'S TEST

In alkaline solutions, ammonia reacts with mercuric potassium iodide (Nessler's reagent) to yield a red-orange colored complex [15]. Addition of the Nessler's reagent and excess alkali to a solution containing ammonium ion produces a color which may be quantitatively measured at 440 mμ.

A handicap of Kjeldahl's and Nessler's tests is that they measure "total nitrogen"; therefore many materials that contain nitrogen (for example, nucleic acids or free amino acids) yield positive results. Accordingly, when using these methods for protein determination, the material under test should be free of other nitrogenous compounds.

### *Turbidimetric Determination*

When proteins are mixed with low concentrations of trichloroacetic acid, sulfosalicylic acid, or potassium ferrocyanide in acetic acid, turbidity results. Under controlled conditions of temperature, concentration, and time of exposure to precipitating agents, the resultant optical density at 600 mμ (turbidity) offers a quick, semiquantitative method of protein determination. This method has severe drawbacks: first, not all proteins yield similar precipitates; second, the acid-precipitatable nonprotein, materials, such as nucleic acids, interfere. Despite these disadvantages of the assay, its simplicity and rapidity favor its use [16, 17].

### *Spectrophotometric Assay of Protein*

The tyrosine and tryptophan residues of proteins exhibit an ultraviolet absorption at approximately 275 mμ and 280 mμ respectively. Since the concentration of these amino acids remains fairly constant in many proteins, the concentration of protein (in a pure solution) should be proportional to the absorption at 280 mμ [18, 19]. A number of pure proteins in solutions containing 1 mg of protein/ml exhibit an OD at 280 mμ of about 1.0 when viewed through a 1-cm light path. The advantages of this assay are its rapidity and the nondestruction of the enzyme.

Unfortunately, other compounds present in natural materials also exhibit absorption at 280 mμ. Specifically, nucleic acids with $E_{MAX}$ at 260 mμ still exhibit absorption at 280 mμ. One can make corrections for nucleic acid absorption at 280 mμ if the ratio of absorption at 280/260 mμ is known. Pure proteins have a 280 mμ/260 mμ ratio of approximately 1.75 (variations are due to differences in number and type of aromatic amino acids present), whereas nucleic acids have a 280 mμ/260 mμ ratio of 0.5. Intermediate ratios corresponding to various mixtures of protein and nucleic acid can be used to compute the concentration of each.

## Amino Acids: Identification and Quantitative Determination

Protein hydrolysates or supernatant phases from protein precipitates can be analyzed both qualitatively and quantitatively for total amino acids or for specific amino acids.

The most common assay for total amino acids is the ninhydrin method. Ninhydrin (triketohydrindene hydrate) reacts with amines (for example, primary amines and amino acids) to yield a purple product. Under controlled conditions the color development is quantitative [20, 21]. Thus the quantity of amino acid in a solution free of competing amines can be determined colorimetrically.

With amino acids, the products, other than the purple product, are $CO_2$ and aldehydes. Van Slyke [22] has developed a quantitative amino acid determination that measures manometrically the $CO_2$ evolved from amino acids in the presence of ninhydrin.

In addition to the ninhydrin reaction, there are several color reactions which are specific for individual amino acids. Some of the amino acids, and their procedures, are: tyrosine, Folin-Ciocalteau test [13] or Millon's test [23]; tryptophan, Hopkins-Cole-Benedict reaction [24]; arginine, Sakaguchi reaction [25]; and cysteine, nitroprusside reaction [26]. Such specific tests are often useful.

## Amino Acids: Ionic Properties

All amino acids contain ionizable groups that act as weak acids or bases, giving off or taking on protons when the *p*H is altered. As is true of all similar ionizations, these ionizations follow the Henderson-Hasselbalch equation:

$$pH = pKa + \log_{10} \frac{[\text{unprotonated form (base)}]}{[\text{protonated form (acid)}]}$$

The *p*Ka is the negative $\log_{10}$ of the ionization constant of the ionizable group. Examination of this equation leads to a further understanding of the term *p*Ka. When the ratio of the concentration of the unprotonated form to that of the protonated form equals 1, the entire $\log_{10}$ expression is canceled, hence *p*Ka can be defined as the *p*H at which the concentrations of unprotonated and protonated forms of a particular ionizable species are equal. The *p*Ka also equals the *p*H at which the ionizable group is at its best buffering capacity.

This useful equation is applied in the titration of glycine with acid and base. Glycine has two ionizable groups: a carboxyl group and an amino group, with *p*Ka values of 2.4 and 9.6 respectively. In water at *p*H 6, glycine exists as a dipolar ion, or zwitterion, in which the carboxyl group is unprotonated ($-COO^-$) and the amino group is protonated to give the substituted ammonium ion ($-NH_3^+$). (Verify this, using the Henderson-Hasselbalch equation, *p*H 6, and the respective *p*Ka values.) Addition of acid to the solution will lower the *p*H rapidly at first and then more slowly as the buffering action of the carboxyl is exerted (see Fig. 6). At *p*H 2.4 the

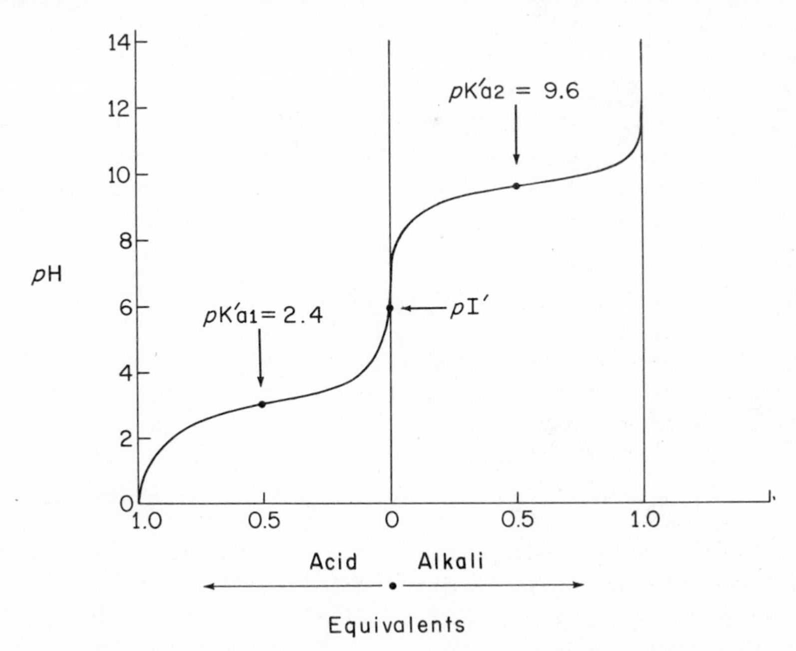


FIG. 6. *Titration curve of glycine.*

*p*Ka is reached, one half an equivalent of acid has been consumed, and the carboxyl group is half ionized and is most effective as a buffer. Further addition of acid results in titration of the remainder of the carboxylate ions in the solution (for example, at *p*H 1.4 only one carboxyl in eleven is ionized). Titration of the ammonium ion (amino) group with base follows a similar curve into the alkaline region. For each titration the concentration of the protonated and unprotonated species of each ion at any *p*H is calculated from the Henderson-Hasselbalch equation.

Most amino acids contain carboxyl and amino groups having *p*Ka values similar to those of glycine. In addition to these groups, many amino acids contain other ionizable groups, which introduce other "steps" or *p*Ka values into their titration curves. It is largely these other groups—those not tied up in peptide bonds—that account for the ionic properties of proteins. Table IV lists the ionizable groups of proteins and amino acids, the nature of their ionizations, and their approximate *p*Ka values. Notice that the *p*Ka values for groups on proteins are given as ranges rather than as single values. This is done because the ionization of various groups on proteins is modified by the presence of other ionizing groups in the vicinity.

**TABLE IV.** Dissociation of Ionizable Groups of Amino Acids and Proteins

| Group | Ionization | *p*Ka Range (on a protein) | *p*Ka (free amino acid) |
|---|---|---|---|
| Carboxyl | $\text{—COOH} \rightleftharpoons H^+ + \text{—COO}^-$ | 1.8–2.5 | 2.2 |
| Carboxyl (ω, of glu, asp) | $\text{—COOH} \rightleftharpoons H^+ + \text{—COO}^-$ | 3.0–4.7 | 3.9 |
| Ammonium | $\text{—NH}_3^+ \rightleftharpoons H^+ + \text{—NH}_2$ | 7.9–10.0 | 9.5 |
| *sec*-Ammonium | $\text{>NH}_2^+ \rightleftharpoons H^+ + \text{>NH}$ | 10.2–10.8 | 10.5 |
| Ammonium (ε, lysine) | $\text{—NH}_3^+ \rightleftharpoons H^+ + \text{—NH}_2$ | 9.4–10.6 | 9.7 |
| Phenolic hydroxyl (tyrosine) | $\text{—C}_6\text{H}_4\text{—OH} \rightleftharpoons H^+ + \text{—C}_6\text{H}_4\text{—O}^-$ | 9.8–10.4 | 10.1 |
| Imidazolium (histidine) | $\text{H—N}^+\text{(ring)N—H} \rightleftharpoons H^+ + \text{N(ring)N—H}$ | 5.6–7.0 | 6.1 |
| Guanidinium (arginine) | $\text{—NH—C(=}^+\text{NH}_2\text{)—NH}_2 \rightleftharpoons H^+ + \text{—NH—C(=NH)—NH}_2$ | 11.6–12.6 | 12.4 |
| Sulfhydryl (cysteine) | $\text{—SH} \rightleftharpoons H^+ + \text{—S}^-$ | 9.4–10.8 | 10.3 |
| *pri*-Phosphate (phosphoserine) | $\text{—O—P(=O)(OH)—OH} \rightleftharpoons H^+ + \text{—O—P(=O)(OH)—O}^-$ | 0.8–1.2 | — |
| *sec*-Phosphate (phosphoserine) | $\text{—O—P(=O)(OH)—O}^- \rightleftharpoons H^+ + \text{—O—P(=O)(O}^-\text{)—O}^-$ | 5.6–6.4 | — |

# Enzymology

## *Theory of Enzyme Action*

Enzymes are proteins which catalyze biological reactions. As is true of other catalysts, enzymes influence the rate at which equilibrium is obtained, but do not affect the overall equilibrium of the reaction. The reaction is helped along by providing a reaction route having a lower free energy of activation for the transition of substrate to product(s) than the uncatalyzed process. These two fundamental points are illustrated by the free energy profiles of hypothetical enzyme catalyzed and nonenzyme catalyzed reactions depicted in Fig. 7. The free energy levels of substrates and products are naturally the same in these systems, therefore the net free energy change in the overall reaction, $\Delta F$, is identical for both processes. The equilibrium constant $K_{eq}$ is directly related to $\Delta F^{\circ}$ as follows:

$$-\Delta F^{\circ} = RT \ln K_{eq} \tag{1}$$

Where $\Delta F^{\circ}$ is $\Delta F$ at standard states (one molal) of reactants and products, $R$ = gas constant, and $T$ = absolute temperature. It follows that the equilibrium constants for both enzymatic and nonenzymatic processes are identical as long as much less than stoichiometric amounts of catalyst are used.

The rate of a process is determined by the free energy levels of the rate-limiting transition state in the reaction pathway (the higher the free energy barriers the slower the rate). This relationship is defined by the absolute reaction rate theory as follows:

$$k_{vel} = \frac{k_b T}{h} e^{-\Delta F^{\ddagger}/RT} \tag{2}$$

where $k_{vel}$ = reaction velocity constant, $k_b$ = Boltzmann's constant, $h$ = Planck's constant, $R$ = gas constant, and $T$ = absolute temperature. The free energy differences between intermediates and transition states are indicated by $\Delta F^{\ddagger}$; the superscript ‡ emphasizes that the thermodynamic parameter refers to transitory complexes determining the rate of the reaction.

It was early recognized that the rate of *enzyme catalyzed* reactions increases with increasing substrate until a concentration is reached beyond which further additions give no increase in velocity. This phenome-

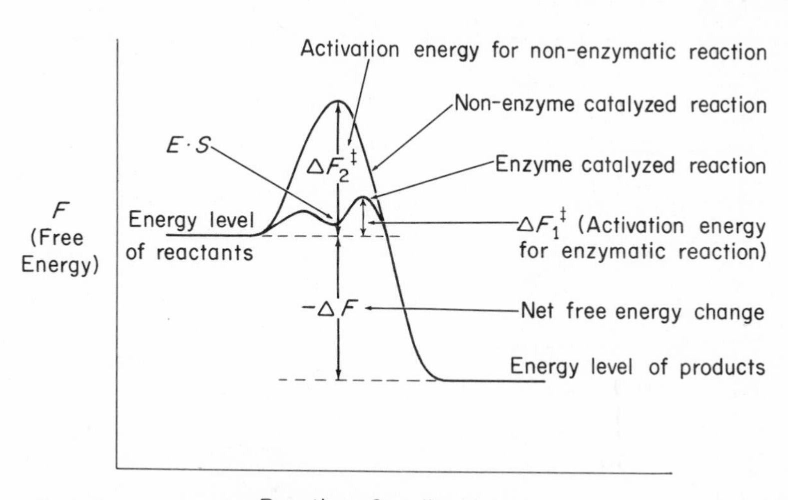


FIG. 7. *Energy profiles of enzymatic and nonenzymatic catalysis of a reaction.*

non, illustrated in Fig. 8 has been most commonly explained on the basis of catalytically active sites on the enzyme which react with the substrate. When all sites are occupied a maximum velocity is reached. The reaction pathway of an enzyme catalyzed reaction is generally depicted in terms of enzyme-substrate complexes as follows:

$$E_f + S \underset{k_2}{\overset{k_1}{\rightleftharpoons}} ES \underset{k_4}{\overset{k_3}{\rightleftharpoons}} EP \underset{k_6}{\overset{k_5}{\rightleftharpoons}} E_f + P \tag{3}$$

where $E_f$ = free-enzyme, $S$ = substrate, $P$ = product, $ES$ = enzyme-substrate complex, and $EP$ = enzyme-product complex. The initial rate of the reaction (product formation) is dependent upon the concentration of the enzyme-substrate complex, $[ES]$, as stated in equation (4), if events within the enzyme-substrate, enzyme-product complex are rate limiting (that is, if $k_3 \ll k_1$ or $k_2$ and $k_4 \ll k_5$ or $k_6$):

$$\text{Initial rate} = v_0 = k_3\,[ES] \tag{4}$$

A steady-state analysis of the reaction pathway (equation *3*) yields the following relationship if only the *initial* reaction is considered (no back reaction; $E_T$ = total enzyme):

$$\frac{d[ES]}{dt} = 0 = k_1\,[E_f][S] - (k_2 + k_3)[ES]$$

$$0 = k_1\,([E_T] - [ES])[S] - (k_2 + k_3)[ES]$$

$$0 = k_1\,[E_T][S] - k_1\,[ES][S] - (k_2 + k_3)[ES]$$

$$[ES] = \frac{[E_T][S]}{[S] + \left(\frac{k_2 + k_3}{k_1}\right)} \tag{5}$$

If $K_M$ (the Michaelis constant) is defined as

$$K_M = \frac{k_2 + k_3}{k_1} \tag{6}$$

then from (*4*), and multiplying both sides of (*5*) by $k_3$, we have

$$v_0 = k_3\,[E_T]\left(\frac{[S]}{[S] + K_M}\right)$$

$$v_0 = k_3\,[E_T]\left(\frac{1}{\frac{K_M}{[S]} + 1}\right) \tag{7}$$

Equation (*7*) states that the initial velocity is proportional to $[E_T]$ at all concentrations of $S$. The velocity of the reaction is dependent upon $[S]$ in relationship to $K_M$; thus, when $[S]$ is small relative to $K_M$

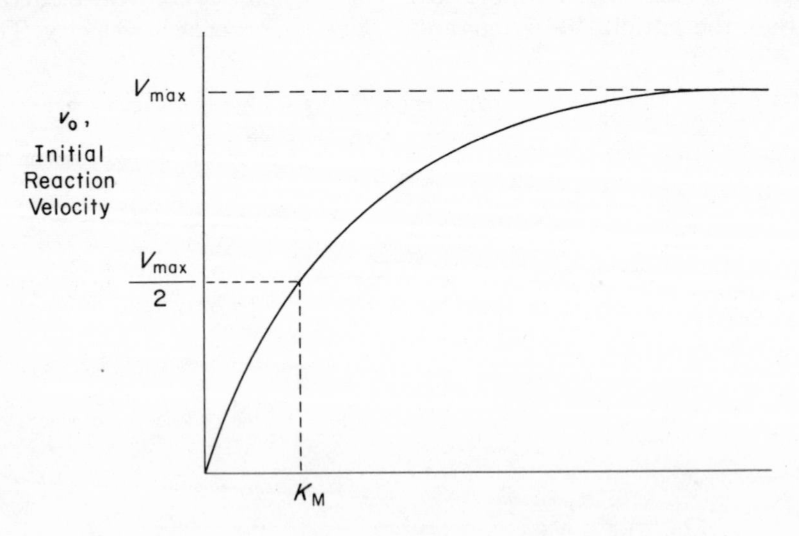


FIG. 8. *Relationship of substrate concentration and reaction velocity in an enzymatic reaction.*

($K_M/[S]$ is large), then the velocity will be a function of $[S]$; but when $[S]$ is large relative to $K_M$ ($K_M/[S]$ is small), the velocity will be largely dependent upon $[E_T]$. Thus when $[S] \gg K_M$,

$$v_0 = V_{MAX} = k_3 [E_T] \tag{8}$$

hence if we substitute (8) into (7) we obtain the usual form of the Michaelis-Menten equation:

$$v_0 = \frac{V_{MAX} [S]}{[S] + K_M} \tag{9}$$

If we rearrange (9) to give (10),

$$K_M = [S] \left(\frac{V_{MAX}}{v_0} - 1\right) \tag{10}$$

we can readily see that $K_M$ is numerically equal to a substrate concentration; $K_M$ therefore has the dimensions of moles/liter. The substrate concentration that yields a velocity equal to one-half the maximum velocity is equivalent to $K_M$.

The value of the $K_M$ reflects the stability of the enzyme-substrate interaction and is of great practical value. $K_M$ is not, however, the true dissociation constant of the enzyme-substrate complex. This dissociation constant is commonly termed $K_s$. Only when $k_2 \gg k_3$ does

$$K_M = K_s = \frac{k_2}{k_1} = \frac{[E_f] [S]}{[ES]} \tag{11}$$

The reciprocal of $K_s$ is the affinity constant of enzyme for the substrate. That is, the greater the affinity of the enzyme for the substrate, the smaller the $K_s$. Note that it is dangerous to consider $K_s$ and $K_M$ to be equivalent without evidence. (See Dixon and Webb in References.)

Although $K_M$ may be determined from equation (10), using data similar to that presented in Fig. 8, it is far more accurate and convenient to use one of the linear forms of the Michaelis-Menten equation. Lineweaver and Burk first pointed out that equation (12) can be obtained by inversion of equation (9):

$$\frac{1}{v_0} = \frac{K_M}{V_{MAX}} \frac{1}{[S]} + \frac{1}{V_{MAX}} \tag{12}$$

Velocities at different substrate concentrations may be plotted as reciprocals, as shown in Fig. 9.

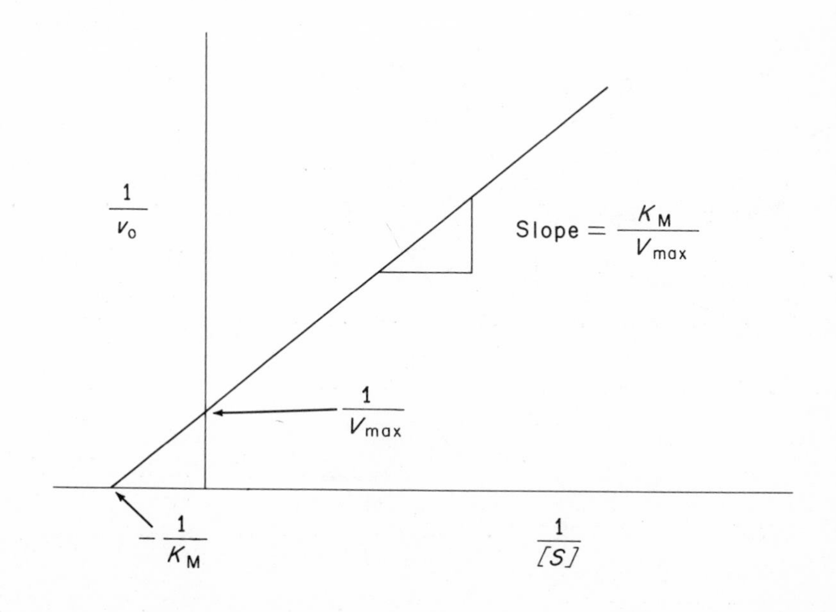


FIG. 9. *Lineweaver-Burk plot.*

Values for $K_M$ and $V_{MAX}$ are obtained by extrapolation of the line, as indicated. Other formulations may also be used to obtain the same information:

$$\frac{[S]}{v_0} = \frac{1}{V_{MAX}}[S] + \frac{K_M}{V_{MAX}} \tag{13}$$

$$v_0 = -K_M \frac{v_0}{[S]} + V_{MAX} \tag{14}$$

As implied in equation (*9*) the velocity of a reaction can be described in terms of $V_{MAX}$, $K_M$, and *[S]*. It follows, then, that changes in the reaction velocity which are not directly related to *[S]* must be due to alterations in either $K_M$, $V_{MAX}$, or both. A change in $K_M$ but not in $V_{MAX}$ would reflect an alteration in the reaction of enzyme with substrate ($k_2$ or $k_1$), whereas an alteration in $V_{MAX}$ would presumably reflect a change in the catalytic process ($k_3$), and not simply in the binding of substrate to enzyme. Note that an increase in $k_3$ also increases $K_M$ unless $k_2 \gg k_3$.

The action of certain compounds with the components of the enzyme catalyzed reactions will modify the kinetics of the system in a way that can be predicted and defined in terms of specific models. The following cases of inhibition are included as examples.

COMPETITIVE INHIBITION

$$E_f + S \rightleftarrows ES \longrightarrow\longrightarrow E_f + P$$
$$E_f + I \rightleftarrows EI$$

This model emphasizes that the inhibitor, *I*, only combines with free enzyme; it cannot combine with enzyme when the substrate is attached, hence the term *competitive inhibitor*. If we use assumptions similar to those employed in deriving (*9*), our model leads to the expression

$$v_0 = \frac{V_{MAX}}{1 + \frac{K_M}{[S]}\left(1 + \frac{[I]}{K_I}\right)} \tag{15}$$

where $[I]$ = inhibitor concentration, and where $K_I = [E_f][I]/[EI]$ = dissociation constant of *EI* complex, or, in the reciprocal formulation of Lineweaver-Burk:

$$\frac{1}{v_0} = \left[\frac{K_M}{V_{MAX}}\left(1 + \frac{[I]}{K_I}\right)\right] \cdot \frac{1}{[S]} + \frac{1}{V_{MAX}} \tag{16}$$

A series of velocity measurements at various substrate levels and at known inhibitor concentrations allows calculation of all the constants, as depicted in Fig. 10.

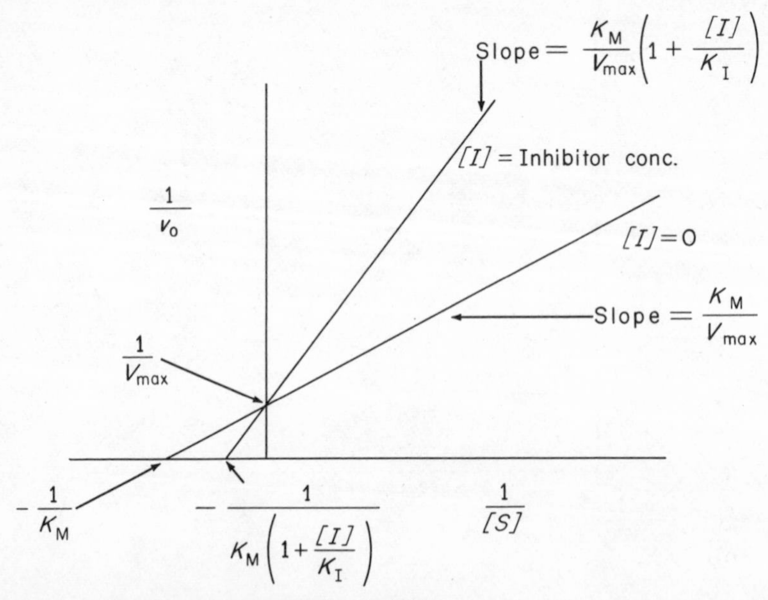


FIG. 10. *Lineweaver-Burk plot: competitive inhibition.*

NONCOMPETITIVE INHIBITION

In this model the reaction of inhibitor and enzyme is indifferent to the presence of substrate,

$$E_f + S \rightleftharpoons ES \longrightarrow\!\!\!\rightarrow E_f + P$$
$$E_f + I \rightleftharpoons EI$$
$$ES + I \rightleftharpoons ESI$$

and the complex $ESI$ is catalytically inactive.

This model leads to the following expression of the Lineweaver-Burk type:

$$\frac{1}{v_0} = \left(1 + \frac{[I]}{K_I}\right)\left(\frac{1}{V_{\text{MAX}}} + \frac{K_M}{V_{\text{MAX}}[S]}\right) \tag{17}$$

In this case

$$K_I = \frac{[E_f][I]}{[EI]} = \frac{[ES][I]}{[ESI]}$$

Lineweaver-Burk plots can yield a numerical value for the constants, as illustrated in Fig. 11.

Although other models can be imagined, the data obtained with many inhibitors have been shown to fit one of the above models (see Dixon and Webb (pp. 171–181) in References). Mention is occasionally made of so-called uncompetitive inhibition, in which the inhibitor is postulated to combine only with the $ES$ complex. This formulation leads to a double reciprocal plot where the uninhibited and inhibited systems show the same slope but the intercept term includes the multiplier, $[1 + ([I]/K_I)]$. Undoubtedly there are many "mixed" cases which require a more sophisticated treatment.

Studies of the action of specific inhibitors, activators, variation in the pH, temperature, and so on are always more informative when the effects are related to changes in $V_{\text{MAX}}$ and $K_M$. Such studies provide information about the specific groups on the enzyme that are involved both in binding the substrate and in the catalytic process. They also provide a specific basis for understanding the activity of the enzyme in vivo and the means by which the activity may be modified.

### *Enzyme Assay*

During the isolation of an enzyme from any source, it is necessary to determine the amount of enzyme present in the prepara-

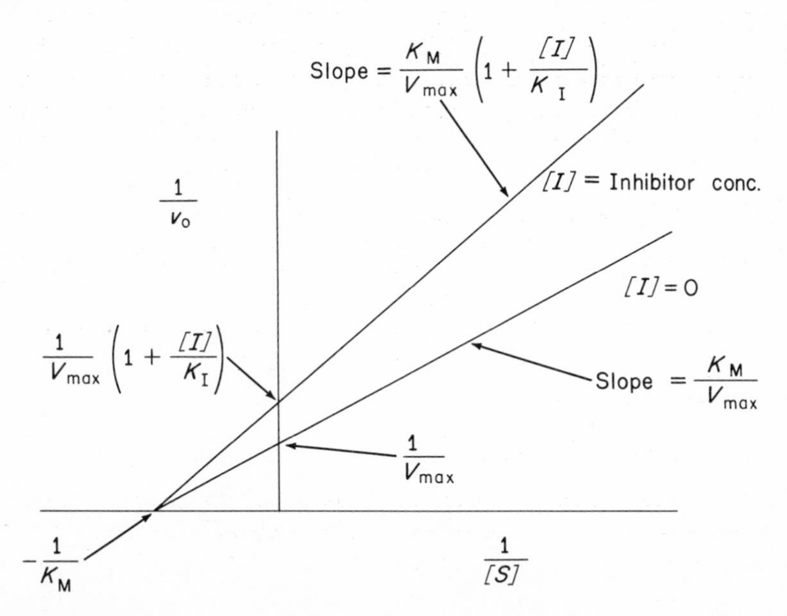


FIG. 11. *Lineweaver-Burk plot: noncompetitive inhibition.*

tion(s). This is usually accomplished indirectly, by measuring the catalytic activity of the enzyme under some fixed conditions of $p$H, temperature, and added ions. Examination of equation (*9*) tells us this is straightforward, provided the value of $[S]$ is defined ($k_3$ and $K_M$ are fundamental for the system). The initial velocity, $v_0$, corresponds to the velocity at initial substrate concentration $[S_0]$, as defined in (*18*):

$$(18) \quad v_0 = \frac{k_3\,[E_T][S_0]}{[S_0] + K_M} = k_3\,[E_T]\left(\frac{1}{\frac{K_M}{[S_0]} + 1}\right)$$

The initial velocity is directly proportional to $[E_T]$, hence careful measurement of $v_0$ values would allow an estimation of the enzyme concentration. The major problem in enzyme assays, therefore, is to establish conditions that allow accurate determination of the initial velocity, $v_0$.

The simplest procedure in obtaining $v_0$ is to use such high concentrations of substrate that a negligible loss occurs during the period of assay: $[S_0] - [S_T] \cong [S_0]$. In such cases the velocity after 5 min, or even 30 min, may be the same as the initial velocity; eventually, of course, the substrate will become limiting. If a low $[S]$ is used (that is, $[S] \cong K_M$) the velocity will continually decrease as substrate concentration decreases, and it will be difficult to estimate $v_0$ (see Fig. 8).

Various states of relative enzyme and substrate concentration are possible. One can have a high $[S]$ and so little enzyme that the velocity cannot be accurately determined (see assay 1, Fig. 12). Similarly, too much enzyme makes it difficult to determine an extremely high velocity (see assays 3 and 4, Fig. 12).

A so-called kinetic assay in which the reaction rate is followed continuously is advantageous, since it is then possible to witness the linearity or nonlinearity of the response with time. Many enzyme assays, however, are based on a single measurement at a defined time, a so-called fixed-time assay. It is usually not possible to predict the appropriate amount of enzyme in either kinetic or fixed-time assays to obtain an optimum velocity like that of assay 2 in Fig. 12. This may be empirically determined by a dilution experiment in two stages. At first, constant volumes of serial 10-fold dilutions of enzyme are assayed to find the range of dilution in which the calculated activity is maximal and constant (see Fig. 13).

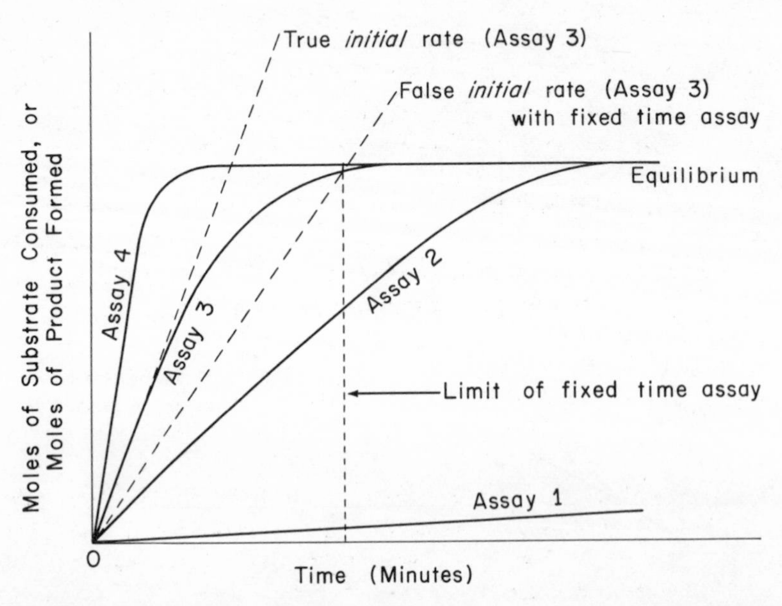


FIG. 12. *Time course of enzyme catalyzed reactions.*

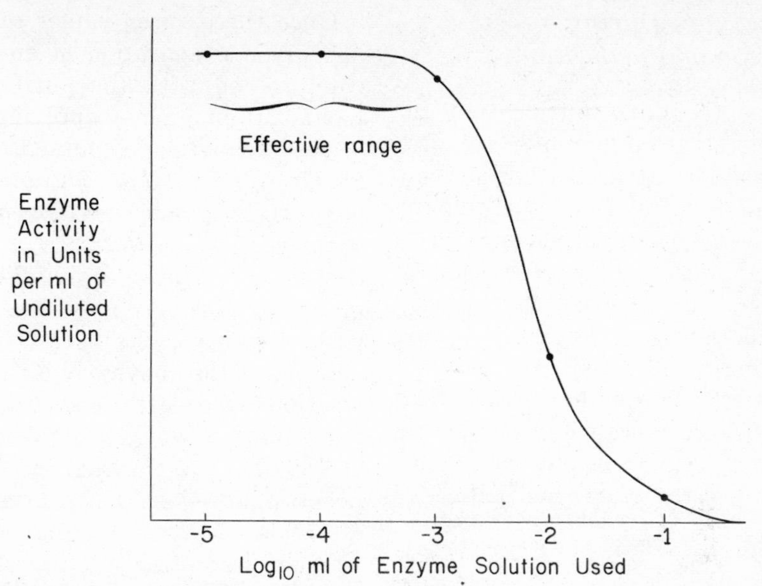


FIG. 13. *Effect of enzyme dilution on apparent assay.*

Such "range finding" may frequently straddle the optimal enzyme concentration for assay. Thus a second series of enzyme dilutions is run to pinpoint the enzyme concentration over a narrower range. Usually 4 or 5 dilutions over a 10- to 20-fold concentration range are adequate.

The prime test of $v_0$ assays is a graphical test to see that the rates observed are a linear function of enzyme concentration, as illustrated in Fig. 14.

This test should be applied to all assays where reaction rates are used to measure enzyme concentrations.

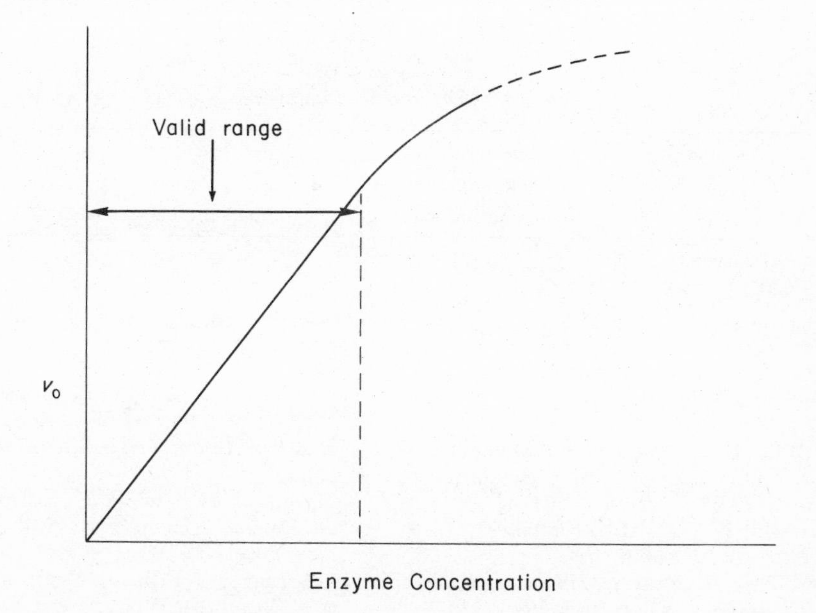


FIG. 14. *Relationship of enzyme activity to enzyme concentration.*

### *Assay Procedure During Fractionation or Purification*

Since enzyme (protein) purification involves the selective removal of other proteins, it is necessary to assess the amount of enzymatic activity relative to the amount of protein present. A measure of enzymatic activity per milligram of protein can be employed to indicate the degree of purity of the enzyme in the various fractions obtained during purification.

This quantity, the specific activity, can be obtained from a measure of enzymatic activity (true $v_0$) and a protein determination (for example, the biuret method).

$$\text{Specific activity} = \frac{\text{units}}{\text{mg protein}}$$

$$= \frac{\mu\text{moles substrate used/min}}{\text{mg protein}}$$

or

$$= \frac{\mu\text{moles product formed/min}}{\text{mg protein}}$$

Other values of importance are

$$\text{Total activity} = (\text{specific activity}) \times (\text{total mg protein in preparation})$$

$$\%\ \text{Yield} = \frac{\text{total activity of given preparation}}{\text{total activity of the starting material}} \times 100$$

Once these three values are known for any given preparation of an enzyme, one can proceed with the purification of the particular enzyme by applying one or more of the techniques discussed on pages 73–74. *It is essential to calculate all three of these values for every fraction obtained during a purification. An increase in specific activity indicates a purification.* The usefulness of a particular step may then be evaluated with reference to the increase in the specific activity of the enzyme and the yield in the fractions of greatest enrichment. An ideal fractionation would provide complete enrichment (pure enzyme) in 100% yield; in practice, however, procedures are not usually found which are so selective, and various, less ideal fractionation steps are combined. Of course, if enrichment is great, a lower yield may be allowed in a particular step. The table below shows a convenient way to present the results obtained during a purification procedure.

Results of Enzyme Fractionation

| Enzyme Fractionation Step | Specific Activity $\left(\frac{\mu\text{moles product/min}}{\text{mg protein}}\right)$ | Total Activity $\left(\frac{\mu\text{moles product/min}}{\text{total protein}}\right)$ | Yield |
|---|---|---|---|
| Original enzyme | 1 | 1000 | 100 |
| First $(NH_4)_2SO_4$ ppt. | 7 | 800 | 80 |
| Eluant from $Ca_3(PO_4)_2$ gel | 28 | 600 | 60 |

## General References


Boyer, P. D., H. Lardy, and K. Myrbäck, *The Enzymes* (Vol I) (2nd ed.), Academic, New York, 1959.

Dixon, M., and E. C. Webb, *Enzymes*, Academic, New York, 1958.

Fox, S. W., and J. Foster, *Introduction to Protein Chemistry*, Wiley, New York, 1957.

Laidler, K. J., *The Chemical Kinetics of Enzyme Action*, Oxford, Clarendon Press, 1958.

Neilands, J. B., and P. K. Stumpf, *Outlines of Enzyme Chemistry* (2nd ed.), Wiley, New York, 1958.

Neuberger, A. (Editor), *Symposium on Protein Structure*, Wiley, New York, 1958.

**Specific References**


1. Pauling, L. et al., *Proc. Nat. Acad. Sci.*, **37,** 205 (1951).
2. Doty, P., and R. D. Lundberg, *Proc. Nat. Acad. Sci.*, **43,** 213 (1957).
3. Perutz, M. F. et al., *Nature*, **185,** 416 (1960).
4. Kendrew, J. C. et al., *Nature*, **185,** 422 (1960).
5. Anfinsen, C. B., *in* A. Neuberger (Editor), *Symposium on Protein Structure*, Wiley, New York, 1958.
6. Kauzmann, W., *in* W. D. McElroy, and B. Glass (Editors), *Mechanism of Enzyme Action*, Johns Hopkins Press, Baltimore, 1954, p. 70.
7. Swenson, A. D., and P. D. Boyer, *J. Am. Chem. Soc.*, **79,** 2174 (1957).
8. Singer, T. P., and E. B. Kearney, *Arch. Biochem.*, **29,** 190 (1950).
9. Sober, H. A. et al., *J. Am. Chem. Soc.*, **78,** 756 (1956).
10. Gornall, A. G. et al., *J. Biol. Chem.*, **177,** 751 (1949).
11. Robinson, H. W., and C. G. Hogden, *J. Biol. Chem.*, **135,** 727 (1940).
12. Lowry, O. H. et al., *J. Biol. Chem.*, **193,** 265 (1951).
13. Folin, O., and V. Ciocalteau, *J. Biol. Chem.*, **73,** 627 (1927).
14. Hawk, Philip B. et al., *Practical Physiological Chemistry* (12th ed.), McGraw-Hill, New York, 1951, p. 814.
15. *Ibid.*, p. 818.
16. Kunitz, M., *J. Gen. Physiol.*, **35,** 423 (1952).
17. Heppe, F. et al., *Hoppe-Seyler's Z. Physiol. Chem.*, **286,** 207 (1951).
18. Warburg, O., and W. Christian, *Biochem. Z.*, **310,** 384 (1941).
19. Kalckar, H. M., *J. Biol. Chem.*, **167,** 461 (1947).
20. Yemm, E. W., and E. C. Cocking, *Analyst*, **80,** 209 (1955).
21. Moore, S., and W. H. Stein, *J. Biol. Chem.*, **176,** 367 (1948).
22. Van Slyke, D. D. et al., *J. Biol. Chem.*, **141,** 622 (1941).
23. Hawk, *op. cit.* (note 14), p. 154.
24. Shaw, J. D. D., and N. D. McFarlane, *Can J. Research*, **16B,** 361 (1938).
25. Albanese, A., and J. E. Frankston, *J. Biol. Chem.*, **159,** 185 (1954).
26. Grunert, R. R., and P. H. Phillips, *Arch. Biochem. Biophys.*, **30,** 217 (1951).

EXPERIMENT

# 18. Dietary Proteins and Essential Amino Acids

## Theory of the Experiment

The growth of various living forms can occur only when the dietary requirements are met. The law of "constancy of composition" says that an organism will not form abnormal tissues or cells and that if an essential component which cannot be made by the organism is not provided, growth will cease. Although several minor exceptions to this rule have been described, it is still a useful concept and is the basis of bioassay procedures which have been described for most of the nutrients.

It has been shown that for optimal growth and body maintenance in higher animals, certain amino acids must be present in the diet. Since proteins constitute the major source of the amino acids, as a result of their hydrolysis to amino acids during digestion, it follows that the amount and amino acid composition of the ingested protein will determine the amino acid adequacy of the diet. In the young rat (see Rose, 1938) at least ten amino acids must be present in the diet for optimal growth. (These are L-arginine, L-histidine, L-isoleucine, L-leucine, L-lysine, L-methionine, L-phenylalanine, L-threonine, L-tryptophan, and L-valine). In other higher animals the essential amino acids may be different (for example, chicks require glycine in addition to those given above). In all higher animals the requirement for an amino acid is indicative of a need for the amino acid in certain metabolic processes, such as protein synthesis, and of the inability of the animal to biosynthesize the required amino acid from other materials in the diet. On the other hand, plants and certain bacteria can synthesize all of the amino acids required.

Factors other than amino acid composition also affect the dietary usefulness of a given protein preparation. The abundance of any one amino acid relative to others

Possible Diets (in grams of material to make 1 kilo)

| Constituent | 1<br>Control | 2<br>Casein | 3<br>Wheat gluten | 4<br>Wheat gluten-lysine-threonine | 5<br>½ casein | 6<br>Soybean meal | 7<br>Auto-claved soybean meal* |
|---|---|---|---|---|---|---|---|
| Protein | 1000 g food pellets | 200 g casein | 200 g wheat gluten | 190 g casein gluten | 100 g casein | 200 g soybean meal | 200 g auto-claved soybean meal* |
| L-Lysine · HCl | —— | —— | —— | 6 | —— | —— | —— |
| DL-Threonine | —— | —— | —— | 4 | —— | —— | —— |
| Sucrose | —— | 690 | 690 | 690 | 790 | 690 | 690 |
| Corn oil | —— | 50 | 50 | 50 | 50 | 50 | 50 |
| Salt mixture | —— | 40 | 40 | 40 | 40 | 40 | 40 |
| Vitamin mixture | —— | 20 | 20 | 20 | 20 | 20 | 20 |
| TOTALS | 1000 g | 1000 g | 1000 g | 1000 g | 1000 g | 1000 g | 1000 g |

*Autoclave at 15 lb pressure for 30 min.

affects growth rates in young animals. In cases of major inbalance of dietary amino acids, growth rates are reduced. Certain proteins act as inhibitors of the proteolytic enzymes of the digestive tract and thereby limit the availability of all amino acids present in the dietary proteins.

In this experiment wheat gluten (a cereal protein with low L-lysine and L-threonine content) is compared with casein (a fairly complete milk protein) as a source of amino acids for the growth of the young rat. The experiment further contrasts the growth rates of rats on the above diets with the growth rates of rats on diets fortified with added amino acids (lysine and threonine). Additional diets, including a protein-deficient diet, a soybean meal diet (containing an inhibitor of the digestive enzyme, trypsin), and a soybean meal diet in which the trypsin inhibitor has been denatured, are also studied.

## Experimental Procedure

### MATERIALS

Casein
Wheat gluten
Soybean meal
Autoclaved soybean meal
L-Lysine·HCl
DL-Threonine
Scale or animal balance
Salt mixture
Vitamin mixture
Food pellets
Supplemented Haliver oil
Corn oil
Sucrose

*Diet preparation*

Select the diets to be tested, and group yourselves into appropriate sized groups, each of which will be responsible for one dietary treatment. Prepare 1 kilo of each diet by thoroughly mixing the components together so as to yield homogeneous diets.

*Assay*

Select weanling rats weighing 40–50 g each, and randomly distribute them into a number of groups (5 rats/group) equal to the number of diets to be studied. Weigh each rat, record its weight, and place it in a separate cage; or, if necessary, 2 to 5 rats on the same diet may be placed in the same cage. Provide an adequate supply of water and the particular diet to be studied.

Maintain the rats on the selected diets for 20–25 days, replenishing the food and water supply each day and recording the weight of each rat at least every 3 days (or alternately recording the average weight of all the rats on a single diet). Once each week force-feed 2 drops of Supplemented Haliver oil to each rat by gently placing 2 drops of oil on the tongue of each rat.

If time permits after the 3- to 4-week test period, animals in groups which grew poorly can be fed a more complete diet to demonstrate their recovery from the deficiency state.

## Report of Results

Prepare a large wall chart showing the average weight of the rats in each group (ordinate) versus time in days (abscissa). After each weighing, record the average weight of the rats in your group on the chart. Record the growth data, and prepare growth curves. Then record your interpretations of the data.

## Discussion

This experiment illustrates some aspects of the nutritional approach which over a period of many years has led to the discovery of many new compounds and facts of biochemical importance. For example, the vitamins, compounds needed in very small quantities for many living forms, were discovered largely by experimentation of this type. The amino acid threonine was discovered by a comparison of casein hydrolysate with a mixture of the amino acids thought to be in casein. Facts about overall energy requirements of animals have been obtained by studies of overall caloric ingestion.

**Exercises**

1. What factors influence the growth rate of young rats? Might the differences among groups of rats be due to differences in palatability of the diets? How would you test for this possibility?
2. List the required nutrients for the rat.
3. Would you expect the tissues of rats that grew at a submaximal rate to have the same amino acid and enzymatic content as those which grew at the maximum rate?

**References**

Flodin, N. W., *Agri. and Food Chem.*, **1,** 222 (1953).

Henderson, L. M. et al., *J. Biol. Chem.*, **201,** 698 (1953).

Phillips, P. H., and E. C. Hart., *J. Biol. Chem.*, **109,** 657 (1935).

Rose, W. C., *Physiol. Rev.*, **18,** 109 (1938).

EXPERIMENT

# 19. Amino Acids and Proteins: Ionic Properties

## Theory of the Experiment

All amino acids are ampholytes; that is, they contain at least one acidic (carboxyl) group and one basic ($\alpha$-amino) functional group. Some amino acids contain other readily ionizable groups: amino, carboxyl, *p*-hydroxyphenyl, sulfhydryl, guanidium, and imidazole groups. Since the basic structure of the polypeptide chain involves the linkage of the carboxyl group of one amino acid to the amino group of the adjacent amino acid, the ionic character of the polypeptide is due largely to these additional ionizable groups together with the terminal amino and carboxyl groups.

In this experiment the reaction of typical amino acids with hydrogen ions is studied. The change of the *p*H is observed as an acid or a base is added to solutions of amino acids. After allowances for the dilution of the acid or base during titration, it is found that in all cases the results may be predicted on the basis of the general Henderson-Hasselbalch equation.

## Experimental Procedure

MATERIALS

Amino acids in powder form
$2N$ NaOH
$2N$ $H_2SO_4$
*p*H meter

### *Titration of Amino Acid*

Using a sample of standard buffer, familiarize yourself with the operation of the *p*H meter. Then weigh about 400 mg of a neutral (monoamino, monocarboxylic) amino acid such as glycine to the nearest milligram, and dissolve it in 20 ml of $H_2O$. Using a burette and a *p*H meter, titrate the dissolved amino acid and another 20 ml of water with $2N$ $H_2SO_4$: 1 drop at a time for the first 10 drops, 2 drops at a time for the next 10 drops, and finally 4 drops at a time until the solution reaches *p*H 1.2. After the addition of each increment of acid, mix the solutions, and record the *p*H and the volume (from burette) of acid consumed.

Dissolve a second 400 mg of the sample (weight known to 1 mg) in 20 ml of $H_2O$, and titrate the amino acid and another 20 ml of $H_2O$ with 2*N* NaOH in the same manner as for the acid titration until the solution reaches *p*H 12.

If time permits, repeat these operations with a trifunctional amino acid (for example, histidine, glutamic acid, or lysine).

## Report of Results

To determine the true titration curve of any substance, it is necessary to determine how much acid or base is consumed in titrating the solvent (water) to each *p*H and then subtract this amount from the total amount of acid or base consumed in reaching that *p*H.

The following example with the acid side of the titration of an amino acid illustrates one of several available methods of correcting for such acid or base dilution.

1. For both the sample and the water blank, plot the volume of acid added versus the *p*H reached (see Fig. 15).
2. From the graph or the original data, prepare a table like the one shown on p. 92, listing the amount of acid required to reach each *p*H. Then subtract the volume of acid required to bring the water blank to any given *p*H from the volume of acid required to bring the sample to the same *p*H. This difference represents the amount of acid consumed in the titration of the sample.
3. Using the data from your table, plot the *p*H versus the number of equivalents of acid needed to titrate the amino acid sample to a given *p*H (see Fig. 16).

Similar methods of correcting for dilution can also be applied to the base side of

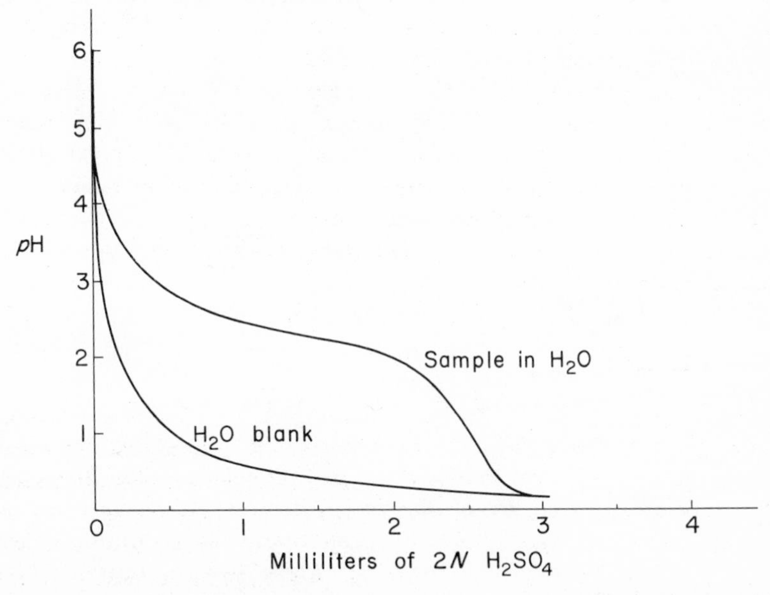


FIG. 15. *Uncorrected titration curves of water blank and sample.*

*Amounts of Acid to Titrate Sample and Water Blank in Solution*

| | Volume (ml) of acid (2N $H_2SO_4$) | | |
|---|---|---|---|
| pH | Water plus Sample | Water Blank | Difference |
| 3.5 | 0.103 | 0.003 | 0.100 |
| 3.0 | 0.335 | 0.020 | 0.315 |
| 2.5 | 0.667 | 0.032 | 0.635 |
| 2.2 | 1.063 | 0.063 | 1.000 |
| 2.0 | 1.425 | 0.200 | 1.225 |

the curve. Prepare complete, corrected titration curves for the amino acids tested.

The number of equivalents of acid or base consumed in passing through one full inflection of a curve (in the zone of a single ionization, as in Fig. 16) represents the quantity of acid required to titrate one ionizable group in the quantity of amino acid you have assayed. Using this relationship and noting the number of full inflections on your curve, calculate an equivalent weight for your amino acid. Compare this with the known molecular weight of the compound. Further, compare the observed *p*Ka values with the known values for the amino acid in question.

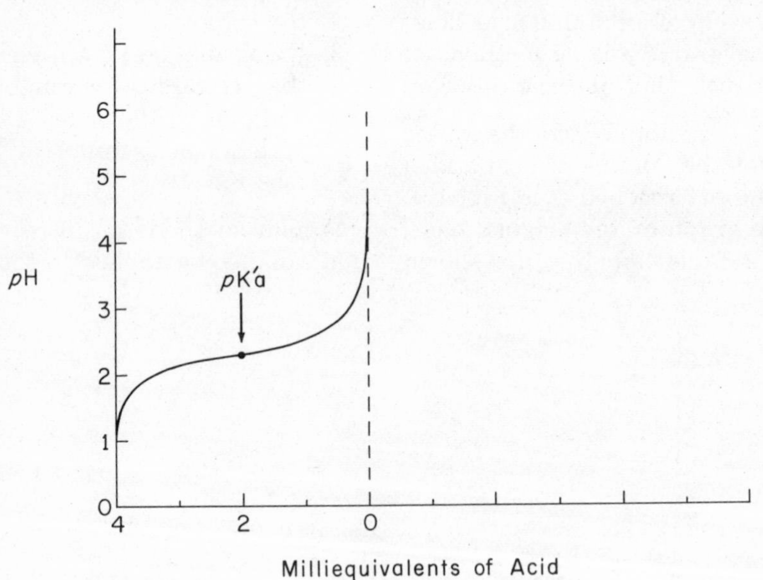


FIG. 16. *Corrected titration curve of sample.*

## Discussion

In proteins, the $\alpha$-amino and $\alpha$-carboxyl groups of all but the terminal amino acids are tied up in peptide bonds. Even so, proteins and polypeptides are usually good polyelectrolytes. This is because of the dissociation of the terminal groups and the free ionizable groups of the trifunctional amino acids (for example, lysine, histidine, glutamic acid). Accordingly, proteins have well-defined titration curves, isoelectric points and electrophoretic behavior (that is, movement in a buffer solution under an applied voltage).

### Exercises

1. How do the true titration curves of aspartic acid and lysine differ from that of glycine? Explain your answer.
2. Is the true titration curve of leucyl valine roughly similar to that of glycine? Explain your answer.
3. The quantity 432 mg of a monoamino monocarboxylic amino acid is observed to consume 7.39 milliequivalents of acid and base in the titration range $p$H 0.8–12.0. What is the name of the amino acid?
4. Give an example of a protein structure which is not a polyelectrolyte.
5. Calculate the isionic point for the following polyfunctional amino acids: (a) histidine, (b) lysine, and (c) aspartic acid. (See the table of $p$Ka values on p. 78.)
6. Derive the Henderson-Hasselbalch equation in terms of $HA$, $A^-$, and $H^+$ for the acid $HA$. Further, show how this equation can be converted into the more generalized equation

$$pH = pK + \log \frac{\alpha}{1 - \alpha}$$

involving only the term

$$\alpha = \frac{[A^-]}{[A^-] + [HA]}$$

7. You are given three bottles, the first containing $K_2HPO_4$; the second, $Na_2HPO_4$; and the third, $H_3PO_4$. How many moles of each would you use to make 1 liter of buffer solution ($p$H 5.8) which is *1 molar* with respect to sodium and *1 molar* with respect to potassium? The three $p$Ka values of phosphoric acid are:

$p\text{Ka}_1 = 1.96 \qquad p\text{Ka}_2 = 6.8 \qquad p\text{Ka}_3 = 12.3$

### References

Cohn, E. J., and J. T. Edsall, *Proteins, Amino Acids, and Peptides*, Reinhold, New York, 1943.

Fruton, J. S., and S. Simmonds, *General Biochemistry* (2nd ed.), Wiley, New York, 1958.

Christensen, H. N., *pH and Dissociation*, Saunders, Philadelphia, 1963.

EXPERIMENT

# 20. Amino Acid Composition of Proteins

## Theory of the Experiment

Hydrolysis followed by examination of the products is a good method of obtaining information about the composition of proteins. Hydrolysis can be carried out by treating the protein with acid, alkali, or proteolytic enzymes. Acid hydrolysis is accomplished with hydrochloric or sulfuric acid at a high temperature (100°C). Sulfuric acid is sometimes preferable because sulfate ions can be readily removed by the addition of barium ions. Barium hydroxide is often used for alkaline hydrolysis because barium, too, is easy to remove (as barium sulfate) after hydrolysis. Enzymatic hydrolysis of proteins is carried out by incubating the particular protein with catalytic quantities of one or more proteolytic enzymes and enzyme cofactors. Under controlled conditions all three methods of hydrolysis—acid, alkali, and enzyme—yield free amino acids at the completion of hydrolysis.

Each method has certain disadvantages. During acid hydrolysis a number of amino acids, especially tryptophan and to a lesser extent serine and threonine, are destroyed by prolonged treatment. Moreover, when carbohydrates are present, other amino acids are destroyed, and the hydrolysates contain black material (humin) which hinders subsequent separations and analyses. Alkaline hydrolysis results in the partial or complete destruction of arginine, cystine, cysteine, serine, and threonine; it also causes racemization of the remaining amino acids. For these reasons, alkaline hydrolysis is rarely used. Enzymatic hydrolysis, inherently the most satisfactory, has severe disadvantages: proteolytic enzymes catalyze the hydrolysis of specific peptide bonds. Thus a mixture of proteolytic enzymes must be used in order to hydrolyze a protein completely to its constituent amino acids. Individual proteins may present a special problem in this respect. For example, some proteins are completely resistant to enzymatic attack and must be treated chemically before enzymatic hydrolysis is possible. Another difficulty is that proteolytic enzymes, through self-hydrolysis, alter the results of an amino acid analysis of a protein.

In this experiment protein samples are hydrolyzed by treatment with both acid and alkali. These procedures are followed by means of the biuret and ninhydrin reactions (see pp. 75 and 76). Qualitative analysis of the constituents of the hydrolysates is carried out by two-dimensional paper chromatography (see Appendix II, p. 198). In this section of the experiment, the neutralized protein hydrolysates are chromatographed on paper, first in phenol saturated with $H_2O$ (run in an ammonia atmosphere) and then (at 90° to the phenol solvent) in *n*-butanol:formic acid:$H_2O$ (100:30:25). (*Note*: This second solvent must be made up just before use, since gradual esterification occurs in the mixture, altering its chromatographic usefulness.) The amino acids are located on the chromatograms by the ninhydrin reaction (ninhydrin spray).

## Experimental Procedure

### MATERIALS

Protein
8*N* $H_2SO_4$
$Ba(OH)_2 \cdot 8H_2O$
16*N* $H_2SO_4$
1*N* KOH
Biuret reagent
0.001*M* Glycine
0.3% $NH_4OH$
0.01*M* Amino acid standards
Diethyl ether
Glass tray or pie plate
Ninhydrin solution reagent
50% Aqueous *n*-propanol
Phenol saturated with water
*n*-Butanol
Formic acid
Ninhydrin spray reagent
Whatman No. 1 paper
Stapler
100 ml Saturated $Ba(OH)_2$

#### *Acid Hydrolysis*

To 0.5 g of protein in a 50-ml Erlenmeyer flask add 5.0 ml of 8*N* $H_2SO_4$. Label the flask using a tag, and plug the top with cotton. Autoclave the flask at 15 lb for 5 hr. Record the appearance of the sample after this treatment.

Neutralize the hydrolysate by warming the solution with solid barium hydroxide (before adding barium hydroxide, calculate approximately how much will be required). After each addition, wait for the barium hydroxide to dissolve before adding the next portion so that neutrality is not exceeded. Use *p*H paper as a *p*H indicator. Carry out the last stages in neutralization (*p*H 3–4 or higher) by adding saturated barium hydroxide solution. Remove the voluminous white precipitate by filtration or centrifugation, wash it twice with 5 ml of boiling $H_2O$, and discard it. Make up the combined filtrates to 25 ml, and store the solution in the cold room for further analysis.

#### *Alkaline Hydrolysis*

Place 0.5 g of protein, 10 ml of boiling water, and 6.36 g of $Ba(OH)_2 \cdot 8H_2O$ in a 50-ml Erlenmeyer flask. Plug the flask with cotton, and warm gently while mixing to dissolve most of the $Ba(OH)_2$. (Avoid excessive mixing to minimize $BaCO_3$ formation.) Label the flask with a tag, and autoclave at 15 lb for 5 hr. Record the appearance of the sample after autoclaving. Titrate the solution by adding about 2.5 ml of 16$N$ $H_2SO_4$. (Check the $p$H with $p$H paper after adding each increment of acid.) When the $p$H has dropped to 10, use less-concentrated $H_2SO_4$ until $p$H 7 is reached. Either centrifuge or filter off the precipitate, and wash it twice with 5 ml portions of boiling water. Make the combined filtrates up to 25 ml with $H_2O$, and store the hydrolysate in the cold room for further analysis.

### *Chemical Tests*

Dissolve 50 mg of protein in 5 ml of $H_2O$ (add a few drops of 1$N$ KOH to hasten solution if necessary). Then perform biuret and ninhydrin assays on this protein solution and on the hydrolyzed protein samples as follows.

#### BIURET ASSAY

Prepare a 2-ml water blank and three assay tubes containing 4, 10, and 16 mg of protein (0.4, 1.0, and 1.6 ml of the above protein solution); add $H_2O$ to bring each assay tube up to 2 ml. In addition, prepare two tubes containing 0.5 and 1.0 ml of the acid hydrolysate and two tubes containing 0.5 and 1.0 ml of the alkaline hydrolysate; add $H_2O$ to bring volumes to 2 ml. (*Note*: 1.0 ml of a hydrolysate represents 20 mg of original protein.) Add 8 ml of the Biuret reagent to all tubes, and mix the solutions. After 30 min measure the optical density of the tubes at 550 m$\mu$ against the water blank (see Appendix II).

#### NINHYDRIN ASSAY

Prepare a 2-ml water blank, two assay tubes containing 0.01 and 0.005 ml of the unhydrolyzed protein solution (0.1 and 0.05 ml of a 1:10 dilution, representing 0.2 and 0.1 mg of protein), and two more tubes containing 0.002 and 0.001 ml of the hydrolyzed samples (0.2 ml and 0.01 ml of a 1:100 dilution of the hydrolysates, representing 0.04 and 0.02 mg of original protein); add $H_2O$ to bring volumes to 0.5 ml. To all of these tubes, and to a tube containing 0.5 ml of 0.001$M$ glycine (a standard), add 1.5 ml of ninhydrin solution (do not confuse this with the ninhydrin spray reagent). Mix the contents of each tube, and place in a boiling water bath for 20 min. After the heating period, cool the tubes in a water bath, and then add 8 ml of 50% aqueous $n$-propanol to each tube (thorough mixing is essential at this step, therefore we suggest blowing the 8 ml of 50% aqueous propanol from a 10-ml serological pipette). Further stir the contents of the tubes by shaking. Allow the mixed tubes to stand for 10 minutes to develop full color, then read the optical densities at 570 m$\mu$ against the blank.

### *Two-dimensional Paper Chromatography of Protein Hydrolysates*

Make a single spot of the neutralized acid hydrolysate and a single spot of the neutralized alkaline hydrolysate on separate sheets of 21 x 21-cm Whatman No. 1 paper on the left-hand corner 3 cm from each edge (spots should be no larger than 4 mm in diameter). Using a pencil, mark one paper "acid hydrolysate" and the other "alkaline hydrolysate," and circle the spots. Staple the papers in the form of a cylinder (avoid overlapping the edges of the papers), and place the cylinders (spot down) into chromatogram jars containing a 1-cm layer of phenol saturated with water and 100 ml beakers containing 20 ml of 0.3% $NH_4OH$. (**Caution:** Phenol can cause harmful burns; wash all phenol off of skin and clothing with excess ethanol.) Close the chromatograph jar, and allow the solvent front to come within a few centimeters of the top (5–10 hr). Then remove the chromatograms, mark the solvent fronts with a pencil, and allow the chromatograms to air-dry overnight in a hood to remove the phenol. When the phenol odor on the paper has nearly disappeared, dip the chromatograms in a trough of diethyl ether (in the hood) to remove the residual phenol. Then air-dry the papers to remove the ether. When the chromatograms are completely dry and free of phenol, reroll the papers in a cylinder at 90° to the first cylinder, and staple as before. Mix 200 ml of *n*-butanol with 60 ml HCOOH and 50 ml $H_2O$; add this solution to two clean chromatography jars; and after placing the papers in the jars with the original spots down, develop the chromatograms until the solvent fronts are a few centimeters from the edges of the papers.

Using amino acid standards, prepare two chromatograms (containing samples of each standard) on Whatman No. 1 paper for single-dimension chromatography, using the phenol solvent for one chromatogram and freshly prepared *n*-butanol: HCOOH:$H_2O$ (100:30:25) for the other. Develop these in one dimension until the solvent fronts are a few centimeters from the edges of the paper, then air-dry the papers overnight in a hood.

Locate the amino acids on any or all chromatograms by spraying them lightly with the ninhydrin spray reagent and then heating the papers at 100°C in an oven for 5 min. The amino acids appear as blue or purple spots (except proline and hydroxyproline, which appear as a yellow spot in this ninhydrin system). Circle all the spots with pencil, and save the chromatograms for further study.

## Report of Results

Contrast the appearances of the original acid and alkaline hydrolysates. Explain the reason for these gross differences.

Using the known specificities of the biuret and ninhydrin reactions, draw both qualitative and quantitative conclusions from your data as to the nature of the material present before and after hydrolysis (for example, mg protein/ml hydrolysate and $\mu$moles amino acid/ml hydrolysate, before and after hydrolysis).

Determine the $R_f$ value of each amino acid in each solvent. Then, using these $R_f$ values and the slight differences between the ninhydrin colors, identify each spot on the two-dimensional chromatograms. Comment on any abnormalties or deficiencies in either chromatogram.

## Discussion

The biuret assay of this experiment is linear within the range 1–20 mg of protein and is rather specific for proteins, since few other materials found in biological systems contain the sequence of atoms necessary for biuret color formation. Familiarize yourself with this assay and the other protein determination procedures, for they will be used extensively in Experiments 22–24.

### Exercises

1. Why are amino acids not extracted from a chromatogram by an ether wash?
2. Using your ninhydrin reaction data ($\mu$moles amino acid/ml hydrolysate), and knowing the amount of protein originally hydrolyzed, calculate an average molecular weight for the amino acids of the hydrolyzed protein. What sources of error are there in this method of average molecular weight determination?

### References

Gornall et al., *J. Biol. Chem.*, **177**, 751 (1949).

Moore, S., and E. H. Stein, *J. Biol. Chem.*, **176**, 367 (1948).

# 21. Sequence Determination on a Dipeptide

## Theory of the Experiment

'he compound 1-fluoro-2,4-dinitrobenzene FDNB) will react with free amino, imidzole, and phenolic groups at neutral to lkaline *p*H to yield the corresponding, olored dinitrophenyl (DNP) compounds. 'hus the N-terminal and ϵ-amino (lysine) roups, the phenolic OH (tyrosine) groups, nd the imidazole (histidine) groups of roteins, peptides, and amino acids all re-ct to form DNP derivatives.

$$O_2N-C_6H_3(NO_2)-F + H_2N-CH(CH_3)-C(=O)-NH-CH(CH_2C_6H_5)-C(=O)-O^- \rightarrow O_2N-C_6H_3(NO_2)-NH-CH(CH_3)-C(=O)-NH-CH(CH_2C_6H_5)-C(=O)-O^- + HF$$

DNP-Alanylphenylalanine

f this derivative is acid-hydrolyzed and, hile acidified, extracted with ether, the onpolar, N-terminal DNP amino acids re removed with the ether phase (exceptions: arginine and cysteic acid), whereas ll other amino acids (including ϵ-DNP ysine, imidazole-DNP-histidine, *bis*-DNP-histidine, and O-DNP-tyrosine) will remain in the aqueous phase, owing to the charged α-amino group.

Chromatographic identification of the N-terminal DNP derivative(s) and of other DNP-substituted and free amino acids yields useful data on the sequence of the amino acids in the material examined.

In this 2-period experiment, milligram amounts of an unknown dipeptide will be examined by the above procedure. In the first period, we suggest that you start the FDNB treatment (Part B) first, and then start the total hydrolysis of the dipeptide (Part A) once the FDNB treatment is underway. Then on the second day, complete Part A before continuing with Part B.

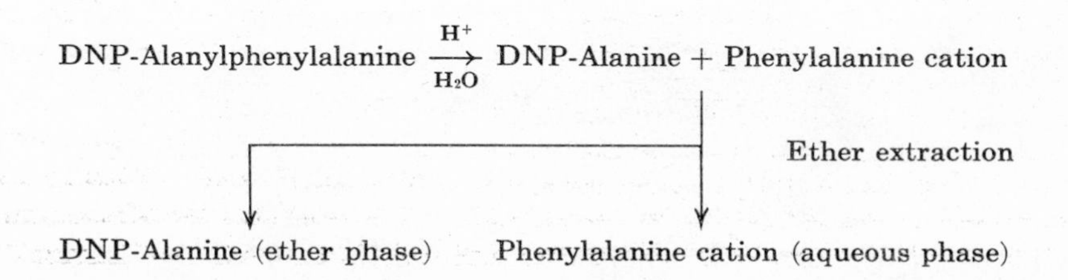

## Experimental Procedure

### MATERIALS

Dipeptides
DNP Amino acid standards
Amino acid standards
Peroxide-free ether
*t*-Amyl alcohol saturated with phthalate buffer
Heat lamp
*n*-Butanol
Whatman No. 1 paper
Formic acid
Hydrolysis vials
6*N* HCl
FDNB solution
4.2% $NaHCO_3$
Phthalate-buffered paper
Pasteur disposable pipettes
Plastic planchets or porcelain spot plate
Phenol saturated with $H_2O$
0.3% $NH_4OH$
Ninhydrin spray reagent

### *Part A. Total Hydrolysis of the Dipeptide*

Draw out a piece of common glass tubing in the Bunsen burner (wing top), and cut off a 10-cm piece (approx.) having an internal diameter of 1.0–1.5 mm. Seal one end by drawing out in a microburner. Weigh approximately 1 mg of the unknown dipeptide onto a plastic planchet. Add 30 μl of 6*N* HCl to the dipeptide by transferring the HCl from a 0.1-ml pipette. When mixed, transfer the 30 μl containing the dipeptide to the capillary tube with a disposable Pasteur pipette or a drawn-out piece of glass tubing such as that shown in Fig. 17. Seal the tube by drawing out, label it, an place it in the oven overnight at 110°(

Scratch the capillary tube with a shar file, open it, and, with the same techniqu used in filling the tube, remove the cor tents to a labeled plastic planchet. Evapc rate the contents under the heat lamp, ad 20 μl of water, and evaporate again (don let the sample char!). Add 50 μl of water t the sample, and chromatograph one 5-, and one 10-μl sample on Whatman No. paper, as described in Experiment 2( Identify the spots by comparison wit amino acid standards as in Experiment 2(

### *Part B. Preparation of DNP Derivative*

Place 2 mg of the dipeptide in a 12-n conical centrifuge tube and then add 0.2 n of water, 0.05 ml of 4.2% $NaHCO_3$, an 0.4 ml of FDNB solution. (*Note: Do n* pipette this material by mouth. Use pipett and propipette provided by instructor. Be come familiar with the operation befor pipetting the FDNB solution.) Shake th resultant suspension frequently for a pe riod of one hour. Maintain the *p*H aroun 8–9 with additional 4.2% $NaHCO_3$ (us *p*H paper). If a large amount of precipitat develops, the *p*H is too low. After 1 hou add 1 ml of water and 0.05 ml of NaHCO Extract the suspension three times (sti ring with a blunt stirring rod to assur good mixing) with equal volumes of pe

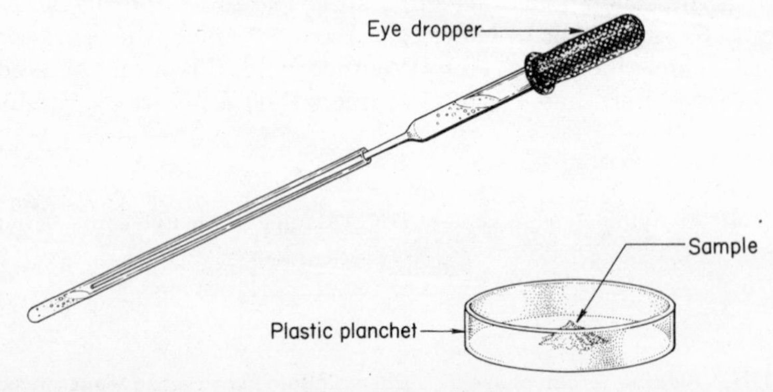


FIG. 17. *Tools for transfer of dipeptide to capillary tube.*

ɔxide-free ether to remove the unreacted FDNB. Centrifuge (if necessary) to sepa-rate the layers, then remove the ether with a Pasteur disposable pipette fitted with a rubber bulb. Adjust to approximately $p$H 1 with 6$N$ HCl using $p$H paper (approxi-mately 0.1 ml of acid). Extract with 2 ml of ether three times, and carefully evapo-rate the ether extracts in a test tube by placing the tube in a beaker of warm water and directing a stream of air over the surface (Fig. 18).

To the dried DNP dipeptide add 0.2 ml of acetone, and transfer to a hydrolysis vial. Several transfers may be necessary. (Alternate Method: The acidified ether ex-tract may be dried directly in the hydroly-sis vial. All ether must be removed before hydrolysis commences.) Remove the ace-tone as described for the ether, and add 0.5 ml of 6$N$ HCl. Label the vial, and give it to the assistant. The labeled vial will be sealed by the assistant and placed in an oven overnight at 100°C.

Open the vial, and transfer the contents to a test tube with a disposable pipette. Add 2 ml of water, and extract with 2 ml of ether three times. Concentrate the com-bined ether extracts in a test tube as before. Dissolve the DNP amino acid in 0.5 ml of acetone, and chromatograph duplicate 10-$\mu$l and 20-$\mu$l samples and standards on phthalate-buffered paper. Develop the as-cending chromatogram with $t$-amyl alcohol saturated with phthalate buffer.

Evaporate the aqueous phase of the hy-drolysate with a heat lamp. Add 0.10 ml of water, and evaporate again to remove the last of the HCl. Add 50 $\mu$l of water to the sample, and chromatograph one 5-$\mu$l and one 10-$\mu$l sample, plus appropriate standards on Whatmann No. 1 paper using the phenol solvent of Experiment 20. Save the remainder for use if necessary.

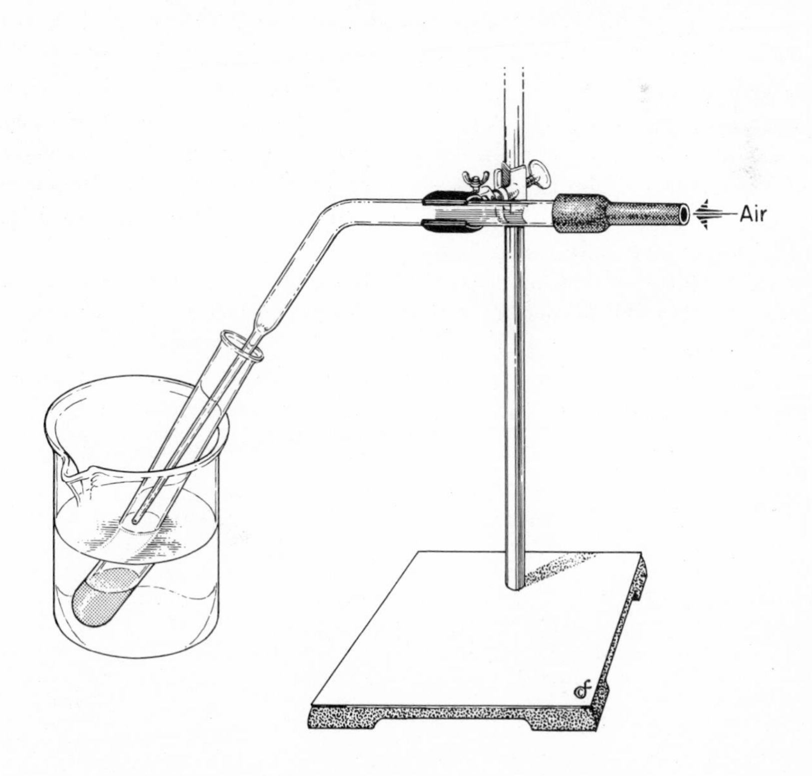


FIG. 18. *Setup for ether extract evaporation.*

## Report of Results

Determine the amino acid residues present by comparing the $R_f$ values observed with those of standard amino acids. In the same way ascertain what amino acid is N-terminal. Usually, two yellow spots will be observed. One is dinitrophenol; the other is the DNP amino acid. Dinitrophenol is colorless below *p*H 4. Either spray the chromatogram with dilute HCl or add a drop of dilute HCl to each yellow spot to check its identity. (If the DNP amino acid spot is hard to find, examination under UV light may prove useful.) From these data suggest a sequence for your dipeptide.

## Discussion

The compound FDNB has proven extremely effective in sequence determination on proteins and polypeptides. In this work, peptide pieces, originating from the intact protein as a result of acid or enzymatic hydrolysis, are first separated by electrophoresis and/or chromatography. Treatment of these peptides with FDNB, followed by hydrolysis and chromatography, yields the N-terminal amino acid in each peptide. These data—in combination with knowledge of (1) the total amino acid composition of each peptide; (2) the specificities of any enzymes used in peptide formation; and (3) the C-terminal amino acid (obtained, for example, by short-time carboxypeptidase treatment) of each peptide—often allow the determination of the original sequence. Such an approach was first used by Sanger (see References) to elucidate the structure of the hormone insulin. Subsequent work has permitted sequence determinations on enzymes.

### Exercises

1. Why did the DNP peptide remain in the aqueous layer during the first ether extraction but enter the ether phase in the second series of extractions?
2. What is the advantage of Edman's phenylisothiocyanate method over the present method of determining amino acid sequences?
3. A peptide was isolated and found to contain arg, val, tyr, glu, lys, ala, and gly (equimolar).
   (a) On trypsin treatment, the following peptides and amino acids were obtained:
      (i) arg,
      (ii) ala-lys,
      (iii) peptide containing glu, tyr, gly, val. Treatment of (iii) with chymotrypsin gave val-tyr and glu-gly.
   (b) On short-time treatment with carboxypeptidase, free glycine was found.
   (c) On chymotrypsin treatment two peptides were isolated, glu-gly, and a peptide with lys, val, ala, tyr, arg.
   What is the sequence of the peptide?

### References

Frankel-Conrat, H. et al., *in* D. Glick (Editor), *Methods of Biochemical Analysis*, Interscience, New York, 1955.
Blackburn, S., and A. G. Lowther, *Biochem. J.*, **48,** 126 (1951).
Sanger, F., and H. Tuppy, *Biochem. J.*, **49,** 463 (1951).
Sanger, F., and E. O. P. Thompson, *Biochem. J.*, **53,** 366 (1953).

EXPERIMENT

# 22. Purification and Catalytic Properties of Yeast Invertase ($\beta$-fructofuranosidase)

## Theory of the Experiment

Among the earliest known and most widely studied of enzymes are those that hydrolyze sucrose. Berthelot, in 1860, working on the purification of a sucrose-hydrolyzing enzyme from yeast, named the enzyme *ferment inversif* because of the inversion (+ to −) in the optical rotation during the hydrolysis of sucrose. Subsequently, the names invertase, sucrase, saccharase, and more recently, $\beta$-fructofuranosidase have been used with reference to the enzymatic hydrolysis of sucrose.

Invertase activity has been demonstrated in yeasts, molds, many bacteria, plants, and higher animals. The basic reaction for yeast invertase is that of a $\beta$-fructofuranosidase. Many $\beta$-fructoturanosides are hydrolyzed by the enzyme, but the activity of the enzyme towards sucrose far exceeds that towards other natural substrates; for example, raffinose or stachyose.

The hydrolytic reaction is most conveniently assayed by observing the change in optical rotation (a kinetic assay) or in reducing power (a fixed-time assay) upon incubation of the enzyme with sucrose. This experiment uses the fixed-time-assay procedure, which consists in stopping the reaction with alkali and subsequently measuring the reducing power by means of the 3,5-dinitrosalicylate method (see p. 26).

In the present experiment, dried yeast

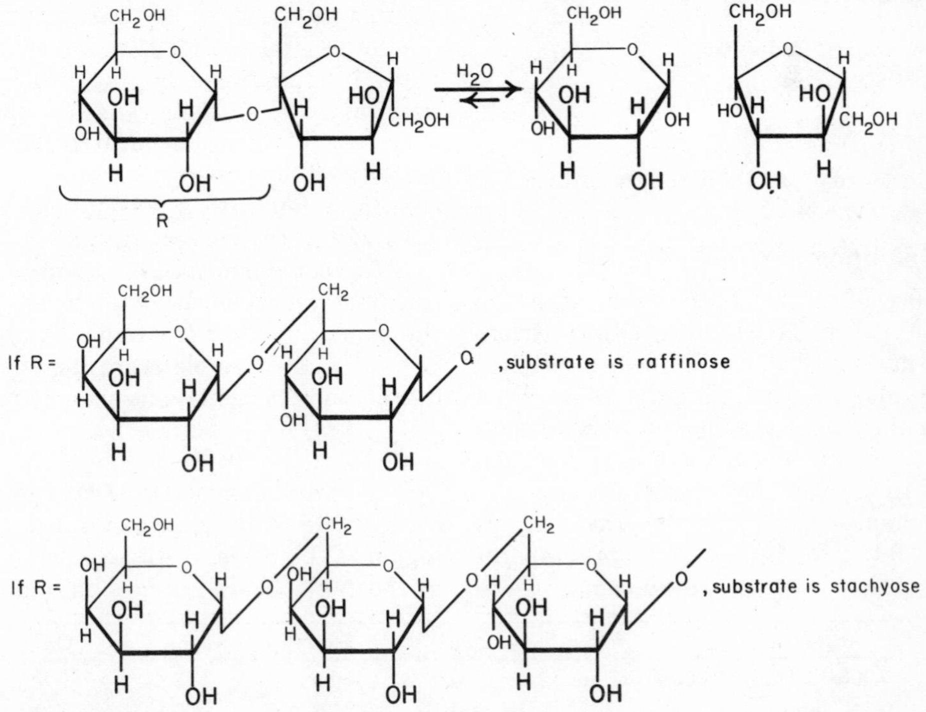

CH₂OH CH₂OH H₂O CH₂OH CH₂OH
R
If R = ..., substrate is raffinose
If R = ..., substrate is stachyose

Invertase-catalyzed hydrolysis of sucrose

cells are disrupted by autolysis, and after removal of the cell debris the specific activity and total activity of the crude autolysate are determined. The crude enzyme is then purified by removal of contaminating materials (proteins and other molecules) as their picrate salts, followed by acetone fractionation of the remaining preparation. The specific activity, total activity, and percent yield are calculated for each step in the fractionation. Finally, certain aspects of the catalytic behavior of the purified invertase are studied ($K_M$, activation energy, *p*H optimum, action of inhibitors).

The principles considered and the operations performed in this experiment are generally representative of those discussed earlier. It is necessary to study appropriate sections before undertaking this experiment.

# Experimental Procedure

## MATERIALS

Standard brands 20:40 dried yeast
0.1*M* $NaHCO_3$
Constant-temperature water bath
High-speed centrifuge
0.05*M* Sodium acetate buffer (*p*H 4.7)
0.3*M* Sucrose
3,5-Dinitrosalicylate reagent
0.005*M* Glucose, 0.005*M* fructose
Cotton
Picric acid solution
Acetone (R.G.)
0.1% $Na_2EDTA$ (*p*H 7.2)
Salt
Dialysis supplies and setup
Biuret reagent
Protein standard (10 mg/ml)
$1.5 \times 10^{-4}M$ PMB in $2 \times 10^{-3}M$ glycyl glycine (*p*H 8.0)
0.2*M* Citrate, 0.01*M* EDTA buffers (*p*H 2.5, 3.5, 4.5, 5.5, 6.5, and 7.5

### *Preparation of Crude Enzyme*

Mix 50 g of 20:40 Dried Yeast with 150 ml of 0.1*M* $NaHCO_3$ in a 1-liter Erlenmeyer flask until all the lumps are softened by the solution. Then plug the flask with a cotton plug, place the flask in a water bath maintained at 40–45°C, and allow the preparation to autolyze for 24 hours.

After autolysis, centrifuge the mixture at 15,000 × g for 15 min at room temperature or below in a high-speed centrifuge to sediment cellular debris. Decant the clear, amber-colored supernatant into a graduate; record the volume; and then pour the solution into a beaker maintained in an ice bath to cool it to 0–2°C.

This solution constitutes the original or crude enzyme, fraction 1. You must accurately determine the specific activity and total activity of this preparation from the assay results on given dilutions of enzyme, the protein concentration, and the volume of the enzyme. Assay a range of aliquots (1 ml to $1 \times 10^{-4}$ ml) of this enzyme by the range finding procedure to determine the suitable range for the fixed-time assay. Then accurately determine velocities that are proportional to enzyme concentration with a second series of assays within the correct range.

If further purification is planned, try to continue the purification of the enzyme on the same day that fraction 1 is obtained. If this is not possible store the enzyme at 0–2°C, and fractionate as soon as possible.

### *Assay Procedure*

Before every enzymatic assay, prepare a series of 18 × 150-mm test tubes consisting of a blank and three standards as in the table below (values in milliliters).

| Substance | Blank | $Std_1$ | $Std_2$ | $Std_3$ |
|---|---|---|---|---|
| 0.005*M* Glucose, 0.005*M* fructose | —— | 0.4 | 0.8 | 1.2 |
| 0.3*M* Sucrose | 1.0 | 1.0 | 1.0 | 1.0 |
| $H_2O$ | 2.0 | 1.6 | 1.2 | 0.8 |

*Note:* The concentration of reducing sugar in the most concentrated standard is 12 μmoles of reducing sugar. This is the approximate limit of linearity in the assay; therefore, only OD values equal to or less than the OD of the standard containing 12 μmoles of reducing sugar are usable.

Place the appropriate aliquots of enzyme in 18 × 150-mm test tubes, and make each solution up to 1.5 ml with water. Add 0.5 ml of 0.05*M* sodium acetate buffer (*p*H 4.7) to each tube, and allow each tube to equilibrate to 25°C for at least 2 min, placing the tubes in a water bath if necessary. Then add 1.0 ml of 0.3*M* sucrose, also previously equilibrated to 25°C, to each tube, and allow the assay mixtures to incubate at 25°C for exactly 5 min. (*Note:* If necessary, start and stop the individual assays on 15- or 30-sec intervals to assure assays of exactly 5-min duration.) Stop each incubation by adding 2.0 ml of 3,5-dinitrosalicylate reagent and shaking the contents. Then add 2.0 ml of 3,5-dinitrosalicylate to the standards and blank.

Develop the color in all the tubes—assays, standards, and blank—by placing the tubes in a boiling water bath for 5 min. Then cool, dilute to 20 ml with $H_2O$, and determine the OD of each tube at 540 mμ against the blank. (*Note:* Some turbidity may occur when using high concentrations of enzyme in assays. This should be removed by centrifugation or filtration before attempting OD determinations.) Using the standards, plot the results as required, and express the results in terms of velocity—that is, in units of μmoles of sucrose hydrolyzed per min.

### *Protein Determination by the Biuret Method*

In practice, any one of the protein determination methods discussed in the Introduction to Proteins and Amino Acids can be used in this experiment. On the other hand, the ready availability of protein, the tyrosine content of the crude autolysate and the picrate content of the purified fractions lead one to favor the biuret assay over results obtained by 280:260 measurements or the Folin-Ciocalteau method. The biuret method of Gornall et al., used in Experiment 20 (that is, making up a blank, appropriate dilutions of enzyme and standards containing 0–20 mg protein to 2.0 ml with $H_2O$, adding 8 ml of biuret reagent, mixing, and after 30 min, reading the $OD_{550}$ against the blank), works well in this experiment as long as dilution yields a protein assay in the range 1–20 mg. Record the results in terms of milligrams of protein per milliliter.

### *Purification of Crude Enzyme*

All of these operations should be performed so as to keep the temperature of the enzymes in the range 0–2°C (cold room if possible).

To the cold (0–2°C) crude homogenate (fraction 1) in an iced beaker add 0.35 volumes of ice-cold picric acid solution, and gently stir once or twice to assure even distribution. Allow the mixture to remain at 0°C for 1 hr; then remove the precipitate by centrifugation in the cold at 10,000 × g for 10 min. Discard the precipitate. Measure the volume of the yellow supernatant solution in a precooled graduate, then remove 3 ml of this solution (fraction 2), and store it at 0°C for future assay. If this experiment is a group project, some students should start the assays and others continue the fractionation.

Without further delay, immerse a beaker containing fraction 2 in a salted ice bath, and cool the solution until it is at 0°C or below. Then pour in three volumes of freezer-cooled (−15°C) acetone over a 3-min period, adding the acetone as a thin stream, while gently stirring the enzyme mixture. Add further salt to the ice bath, and allow the mixture to sit in the salted bath for 15 min. Next, decant as much of the clear supernatant solution as possible from the gummy precipitate, and centrifuge the remaining mixture at 10,000 × g for 10 min at −4 to +1°C. (*Note:* Use centrifuge tubes made of acetone-insoluble materials such as polyethylene.)

Decant off the remaining supernatant solution, and redissolve the precipitate in a minimum of ice-cold 0.1% $Na_2EDTA$ solution, *p*H 7.2. Dialyze this solution at 0–2°C against several changes of large volumes of precooled $H_2O$ over a 24- 48-hr period until the solution in the dialysis reservoir appears colorless. Then remove the enzyme (fraction 3), and determine its specific and total activity as before. Store fraction 3 at 0–2°C as required.

### *Determination of* $K_M$

Prepare a series of assay tubes containing 1.5-ml volumes of a constant dilution of fraction 3 known to yield velocities directly proportional to enzyme concentration when assayed in the usual manner with 1 ml of 0.3*M* sucrose. Add 0.5 ml of 0.05*M* acetate buffer (*p*H 4.7) to each tube. After temperature equilibration (25°C), add 1 ml of the following sucrose solutions to the assay tubes at timed intervals 0.3*M*, 0.03*M*, 0.015*M*, 0.003*M*, 0.0015*M*. Allow each assay to incubate for exactly 5 min before adding 2 ml of 3,5-dinitrosalicylate reagent. Then add 2 ml of 3,5-dinitrosalicylate reagent to the usual blank and standards, and develop the color in all tubes as before. Evaluate the assay tubes in terms of velocities—that is, μmoles of sucrose hydrolyzed per min.

### *Inhibition of Reaction with p-Mercuribenzoate (PMB)*

Prepare 10 enzyme dilutions, each 1.3 ml in volume and containing a quantity of enzyme known to yield velocities directly proportional to enzyme concentration when assayed in the usual manner with 1.0 ml of 0.3*M* sucrose. Add 0.2 ml of $1.5 \times 10^{-4}M$ PMB to 5 of the tubes, and mark them as having a final PMB concentration of $1 \times 10^{-5}M$; add 0.2 ml of $7.5 \times 10^{-5}M$ PMB to the other 5 tubes, and denote these as $5 \times 10^{-6}M$ PMB tubes. Allow the tubes to incubate for at least 2 min at room temperature, then proceed with an assay similar to the $K_M$ assay, first adding to all tubes 0.5 ml of acetate buffer (*p*H 4.7) and then starting the reactions with 1-ml volumes of various sucrose concentrations (0.3*M*, 0.03*M*, 0.015*M*, 0.003*M*, 0.0015*M*), testing each of the PMB treatments with a complete spread of sucrose concentrations.

After color development and OD determination, express the results in terms of μmoles of sucrose hydrolyzed per min.

### *pH Optimum of Enzymatic Hydrolysis*

Prepare the usual blank and standards and a series of 6 assay tubes, each containing 1 ml of 0.3*M* sucrose. Next add 0.5-ml of each of the 0.2*M* citrate EDTA buffers to the separate assay tubes, one buffer to each tube. After noting or recording the distribution of buffers in the six assay tubes, add to each assay tube at timed intervals 1.5-ml volumes of a constant enzyme dilution known to yield assays proportional to enzyme concentration when assayed with 1 ml of 0.3*M* sucrose. Stop each incubation after 5 min with 2 ml of 3,5-dinitrosalicylate, and determine for each assay the velocity or μmoles of sucrose hydrolyzed per min.

### *Effect of Temperature on the Invertase Reaction*

Prepare the usual blank and standards and a series of six assay tubes containing 1.5 ml of a constant dilution of enzyme known to yield a reasonably high level of sucrose hydrolysis (at 25°C) that is still directly proportional to enzyme concentration. Then add 0.5-ml volumes of 0.05*M* acetate buffer (*p*H 4.7) to each assay tube. Next, prepare a series of six test tubes, each containing 1.5 ml of 0.3*M* sucrose. At 15- or 20-sec intervals place pairs of tubes (one assay tube and one sucrose tube) in separate constant temperature baths (or in large beakers of $H_2O$) operated at various temperatures over a range of 10–70°C (6 baths in all). At timed intervals (15 or 20 sec)—as soon as each tube has received 2 min of temperature equilibration (do not exceed this duration, particularly at the higher temperatures, as heat denaturation will occur)—transfer 1 ml of 0.3*M* sucrose from the equilibrating tube of each bath to the assay tube of the same bath. Terminate all assays after 5 min duration by adding 2 ml of 3,5-dinitrosalicylate reagent. Also add 2 ml of 3,5-dinitrosalicylate reagent to the blank and the standards. Develop the colors in the usual manner. Determine the velocity ($\mu$moles of sucrose hydrolyzed per min) for each assay.

## Report of Results

Report your assay data of fractions 1, 2, and 3 by constructing two types of graphs. In the first set of graphs (one for each fraction) show the range-finding data, plotting $\mu$moles of sucrose hydrolyzed/min per milliliter of undiluted enzyme versus the $\log_{10}$ of the volume (ml) of enzyme used in the assay—(Fig. 13). In the second set of graphs (one for each fraction) show the results of the assays, once the correct range has been found ($\mu$moles of sucrose hydrolyzed/min versus milliliters of undiluted enzyme used in the assay). (See Fig. 14.)

In order to summarize the results of the purification, construct a table showing for each fraction the volume of the enzyme solution in milliliters, the protein concentration (mg protein/milliliter), the specific activity, the total activity, and the percent yield.

Prepare a Lineweaver-Burk plot of the data obtained in the $K_M$ determination. Further, on the same figure, plot the data obtained from the assays in the presence of PMB. Then determine the $K_M$ and the $K_I$ of the enzyme towards sucrose and PMB, respectively. Compare the $K_M$ determination from the Lineweaver-Burk plot with a $K_M$ determined from a plot of $V$ versus $[S]$.

From the data obtained from assays run at different *p*H, prepare a *p*H optimum curve for the invertase reaction.

Prepare a graph for the temperatures studied, listing velocity ($\mu$moles of sucrose hydrolyzed/min) versus temperature of the assay. Explain the reasons for the shape of the curve. In addition, prepare an Arrhenius plot ($\log_{10}$ velocity versus 1/°K, and calculate the activation energy ($-\Delta E$) from the slope of the line in the appropriate region.

## Discussion

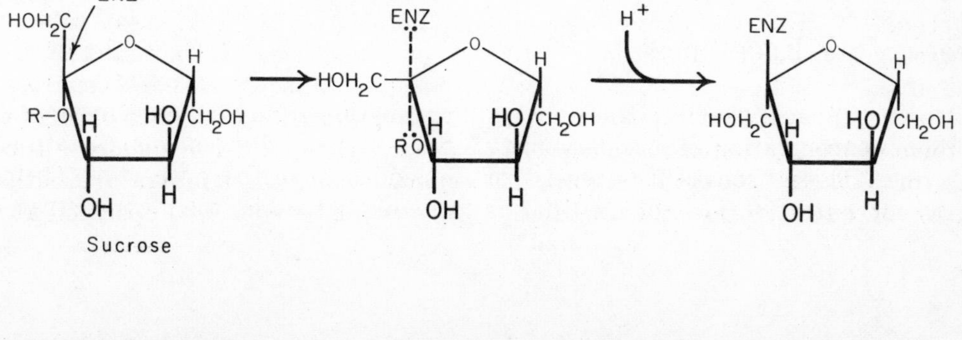

When an invertase-catalyzed hydrolysis of sucrose is run in $H_2O^{18}$, it is observed that the glucose moiety is released without the uptake of $0^{18}$ from the $H_2O^{18}$. Therefore the cleavage of the glucose-fructose bond must be between the bridge oxygen and the fructose residue. Further, it is known that substrate specificity is determined by the presence of a terminal $\beta$-fructofuranoside residue. Thus we may envision the mechanism of formation of the *ES* complex as a displacement reaction, and the subsequent release of free fructose as a second displacement (see below).

Our belief in the existence of such an enzyme-fructose intermediate is strengthened by the fact that incomplete hydrolysis of sucrose by yeast invertase yields, in addition to fructose and glucose, a series of oligosaccharides such as glucosyl,1-2,-fructosyl,6-2,fructose. Thus during the invertase-catalyzed hydrolysis of sucrose, and before final equilibrium is achieved, organic residues occasionally replace $OH^-$ in the above equation; accordingly, invertase acts as a transfructosidase.

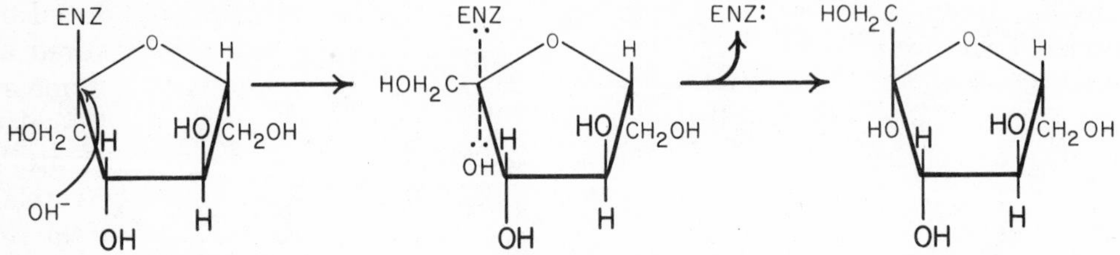


### Exercises

1. Glucose is known to inhibit invertase-catalyzed hydrolysis of sucrose. Devise an experimental procedure to allow study of this inhibition and determination of its type.
2. What factor (or factors) determines the lower limit of sucrose concentration to be used in $K_M$ determinations?
3. Devise an experiment to contrast the activation energy ($-\Delta E$) of the enzyme catalyzed hydrolysis of sucrose with that of a nonenzyme catalyzed hydrolysis of sucrose (for example, dilute acid catalysis).
4. In a study of the invertase reaction as measured by the formation of reducing sugar, it is observed that at a fixed enzyme concentration, sucrose is hydrolyzed at a certain maximum velocity. Further, it is observed that the tetrasaccharide stachyose, at $V_{MAX}$ concentration in the presence of the same level of enzyme, is hydrolyzed at a rate equal to 5% of that of sucrose. Predict the appearance of a Lineweaver-Burk plot obtained with a constant enzyme level and varied sucrose concentrations in the presence of and absence of a constant level of stachyose.

### References

Neuberg, C., and I. Mandel, *in* Summer, J.B., and K. Myrbäck, *The Enzymes* (1st ed.), Chapter 14, Academic, New York, 1950.

Koshland, D. E., Jr., and S. S. Stein, *J. Biol. Chem.*, **208,** 139 (1954).

Myrbäck, K., *in* Boyer, P. D., H. Lardy, and K. Myrbäck, *The Enzymes* (2nd ed.), Academic, New York, 1961.

EXPERIMENT

# 23. Study of the Properties of β-Galactosidase

## Theory of the Experiment

Cells of organisms usually have a fixed minimum concentration of a wide variety of enzymes. These "constitutive levels" of enzyme concentration are not depleted as a function of available substrates. In a wide variety of organisms, certain of these enzymes have the property of being "inducible"; that is, their concentration will increase if their particular "inducing agent" —often their particular substrate—is present. Thus organisms with inducible enzyme systems have the ability to alter their enzymatic constitution according to the various molecular species available to them.

The enzyme β-galactosidase of *E. coli* strain $K_{12}$ is an example of an inducible enzyme. This enzyme catalyzes the hydrolysis of a wide variety of β-galactosides. When *E. coli* $K_{12}$ is grown in the absence of lactose or other β-galactosides (for example, with glucose as the major carbon source) β-galactosidase is present at a very low level in the cells (approximately one enzyme molecule per cell). Alternatively, when *E. coli* $K_{12}$ is grown in the presence of a β-galactoside (for example, with lactose as the major carbon source), the level of β-galactosidase is greatly increased. Therefore, cell extracts from *E. coli* $K_{12}$ grown in the presence of lactose have a higher specific activity (μmoles of substrate hydrolyzed/min per milligram of protein) towards β-galactosides than cell extracts from *E. coli* $K_{12}$ grown in the absence of a β-galactoside.

In this experiment, the relative levels of β-galactosidase activity in induced and non-induced *E. coli* will be measured and the properties of the induced enzyme studied. The compound *o*-nitrophenyl-β-galactoside (ONPG) is used as a substrate. Under alkaline conditions, this substrate allows a continuous or kinetic assay of the reaction: as *o*-nitrophenol (ONP) is released into alkali as a result of the enzyme catalyzed hydrolysis of ONPG, the ONP is instantly converted to a yellow chromogen which absorbs at 420 mμ.

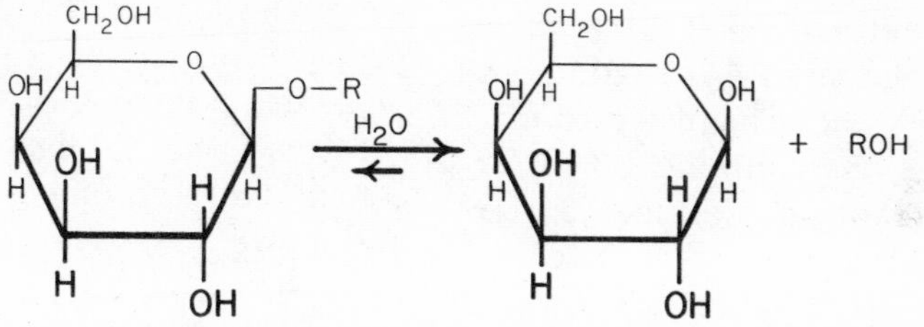

$CH_2OH$ ... O—R $\xrightarrow[\leftarrow]{H_2O}$ ... + ROH

When R 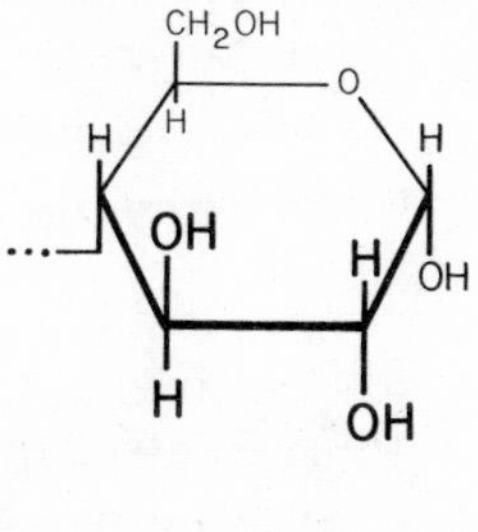 = ..., substrate is lactose.

When R = ..., substrate is o-nitrophenyl-β-galactoside

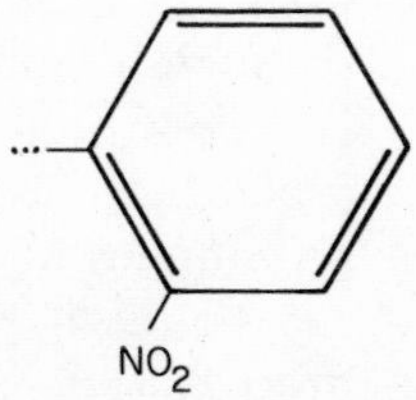

## Experimental Procedure

### MATERIALS

Materials for AC medium
Materials for lactose medium
*E. coli* strain $K_{12}$
Autoclave
Shaker
Fermentation tank and carboys
Alumina or sonic oscillator
Constant-temperature baths
Vacuum desiccators
Folin-Ciocalteau reagent
Protein standard (1 mg/ml)
High-speed centrifuge
Continuous centrifuge
0.08*M* Sodium phosphate (*p*H 7.7)
0.0025*M* ONPG (freshly prepared)
1*M* $Na_2CO_3$
$P_2O_5$
2% $Na_2CO_3$ in 0.1*N* NaOH
0.5% $CuSO_4 \cdot 5H_2O$ in 1% sodium tartrate

### *Growth of Cells*

Prepare and autoclave four test tubes containing AC medium, four 2-liter flasks (two containing 500 ml of AC medium and two containing 500 ml of lactose medium), and two carboys (one containing 10 liters of AC medium and one containing 10 liters of lactose medium).

*AC Medium*

| | |
|---|---|
| Tryptone | 1.0% |
| Yeast extract | 1.0% |
| $K_2HPO_4$ | 0.5% |
| Glucose | 0.1% |

*Lactose Medium*

| | |
|---|---|
| $NH_4Cl$ | 0.2% |
| Lactose | 0.6% |
| Monosodium glutamate | 0.1% |
| Yeast extract | 0.1% |
| $KH_2PO_4$ | 0.15% |
| $Na_2HPO_4$ | 1.35% |
| $MgSO_4 \cdot 7H_2O$ | 0.02% |
| $CaCl_2$ | 0.001% |
| $FeSO_4 \cdot 7H_2O$ | 0.00005% |

Using sterile technique, inoculate the four test tubes with stock cultures of *E. coli* $K_{12}$. Incubate these tubes at 30°C until actively growing cultures are obtained; then use the contents of these tubes to inoculate the four flasks. Shake the flasks for 12–24 hr on a mechanical shaker until good growth is obtained. Then inoculate the carboy containing AC medium with the flasks containing AC medium; similarily inoculate the carboy containing lactose medium with the flasks containing lactose medium. Incubate the carboys aerobically (air bubbler, 2–4 liters/min) in a fermentation tank at 30°C. Observe the rates of growth in the two carboys by removing aliquots and determining their optical density at 660 mμ. When the peak of the log phase of growth is reached—as determined from a plot of $OD_{660}$ versus time—harvest the cells of the two carboys separately by centrifugation in a continuous centrifuge (Sharples). Resuspend the cells in 1 volume of cold (0–2°C) $H_2O$, and recentrifuge in a cooled high-speed centrifuge (15 min, 15,000 × g) in order to wash away the last of the growth media. (The preparation may be used directly, or frozen, or dried, as below.)

Spread the batches of washed cells in $\frac{1}{8}$-inch layers on one or more Petri dishes, and store in evacuated desiccators over $P_2O_5$ for several days until dry. Then scrape the dried cells from the dishes, and grind them into a fine powder. Respread the powder in Petri dishes, and further dry as before for an additional 24 hr.

### *Preparation of a Cell-free Extract*

Several methods for the preparation of cell-free extracts from bacterial cells are available; for example, grinding with alumina at low temperatures, homogenization with very small glass beads, extrusion through a small orifice in the frozen state (Hughes press), and disruption by sonic oscillation. In this experiment alumina grinding and sonic oscillation are presented. Either method yields satisfactory preparations for further assay, the choice between them being dictated by the availability of equipment. In both methods all of the operations with cells and subsequent cell-free extracts should be performed at the 0–4°C range unless otherwise stated.

ALUMINA GRINDING

This method is presented on a 0.5-g scale, but it can, of course, be expanded for larger amounts of cells. Add 0.5 g of dried cells (either induced or wild type) and 1.0 g of powdered alumina to a precooled (−5 to 0°C) mortar. Grind the mixture with a pestle for 3 min, occasionally adding aliquots of ice cold 0.008*M* sodium phosphate buffer (*p*H 7.7) until 2.5 ml of buffer is added. (*Note:* If wet or frozen cells are used, to 1.5 g of cells and 1.0 ml of buffer, add 1.5 g of alumina over a 3-min grinding period; then continue as follows.) Grind this sticky mixture for an additional 5 min, and then slowly stir in an additional 2.5 ml of precooled buffer.

Centrifuge the mixture at 15,000 × g for 15 min, and discard the precipitate containing unbroken cells, alumina, and cell debris. Store the viscous supernatant at 0–2°C until used.

SONIC OSCILLATION

Suspend 5 g of dried cells in 25 ml of ice cold 0.001*M* sodium phosphate buffer (*p*H 7.7), and pour the suspension into the precooled sonic oscillator cup. (*Note:* If washed wet or frozen cells are used directly instead of dried cells, add 15 g of cells and 15 ml of buffer to the cup.) Rupture the cells by sonic oscillation for 20 min in a 200-watt, 10 KC Raytheon oscillator or similar instrument. At the end of this time remove the cellular debris by centrifugation at 15,000 × g for 15 min. Store the supernatant at 0–2°C until needed.

*Kinetic Assay*

Add 1 ml of 0.0025*M* ONPG and 8 ml of 0.08*M* sodium phosphate buffer (*p*H 7.7) to a series of 18 x 150-mm colorimeter tubes. Add 1 ml of $H_2O$ to one of the tubes, and, after mixing by inversion, use it as a blank, and adjust the colorimeter to zero optical density (100% transmission) at 420 mμ. Next, noting the time to the second, add 1 ml of an appropriate dilution of enzyme to an assay tube, quickly mix by inversion, and place it in the colorimeter. Observe the $OD_{420}$ at 30-sec intervals over a 5-min period. Repeat this procedure for each assay.

Essentially, ONPG is completely hydrolyzed at equilibrium. Therefore, relate the observed $OD_{420}$ values during assay to the μmoles of ONPG hydrolyzed by assuming that the equilibrium $OD_{420}$ values represent complete hydrolysis of the known amount of ONPG in the system.

Plot the μmoles of ONPG hydrolyzed versus time, and calculate the initial velocity for each assay in terms of μmoles of ONPG hydrolyzed per min.

*Protein Determination*

The high nucleic acid content of the crude, unpurified enzyme tends to rule out the turbidometric, Kjeldahl, and 280:260 spectrophotometric protein determinations discussed in the Introduction to Proteins and Amino Acids (see p. 75). Accordingly, use either the biuret method of Experiment 20 or, preferably, the Folin-Ciocalteau method discussed below for protein determinations.

For protein determination by the Folin-Ciocalteau method, prepare a series of tubes including a blank (1.2 ml $H_2O$), standards (30–600 μg protein), and assay tubes containing 1.2-ml volumes of an appropriate dilution series to yield protein values in the 30–600 μg range. Add 6 ml of alkaline copper solution [freshly mixed 2% $Na_2CO_3$ in 0.1*N* NaOH: 0.5% $CuSO_4 \cdot 5H_2O$ in 1% sodium tartrate (50:1)] to all tubes, and mix each by shaking. After 10 min, add 0.6 ml of Folin-Ciocalteau reagent to each tube, and thoroughly mix the contents of each. After 30 min read the optical densities at 500 mμ against the blank. Calculate the protein concentration(s) of the original solution from $OD_{500}$ values within the range of the standards.

### *Determination of the Specific Activities of Induced and Noninduced Cell Extracts*

Using the above assay procedures, assay 1-ml dilutions of enzyme containing 0.5, 0.1, 0.01, 0.001, and 0.0001 ml of the original extracts of both the induced and noninduced cell extracts of *E. coli* $K_{12}$ to find the correct range of assay for each extract. Then, having determined the correct range of assay, run three dilutions of each extract in this range to establish enzyme dependency. Further, determine the protein concentration (milligrams of protein per milliliter) of both extracts. From these data, calculate the respective specific activities.

### *Determination of $K_M$*

Having determined a dilution of cell-free enzyme from the induced *E. coli* $K_{12}$ that yields a convenient assay when incubated with 1 ml of $2.5 \times 10^{-3}M$ ONPG, run a series of kinetic assays in 10-ml total volumes—that is, with 1 ml of the dilution of enzyme plus 8 ml of 0.08 sodium phosphate buffer (*p*H 7.7) but varying the quantity of ONPG present (1.0, 0.8, 0.6, 0.4, and 0.2 ml of $2.5 \times 10^{-3}M$ ONPG—make up the difference with $H_2O$). Plot the results in terms of $\mu$moles of ONPG hydrolyzed versus time. From the plot of the assays, measure the initial velocities, $v_0$ ($\mu$moles of ONPG hydrolyzed per min), and construct two plots, the first a plot of initial velocities versus ONPG concentration, and the second a Lineweaver-Burk plot. Calculate a $K_M$ for ONPG from both plots, and compare the values.

### *Determination of Activation Energy of Enzyme Catalyzed Hydrolysis*

It is necessary to use a fixed-time assay to evaluate the effect of temperature on reaction rate, since temperature control in the colorimeter is difficult. Accordingly, perform the following operations.

Prepare a series of colorimeter tubes containing 1 ml of $2.5 \times 10^{-3}M$ ONPG and 7.9 ml of $0.08M$ sodium phosphate buffer (*p*H 7.7). Place pairs of these tubes in a series of constant-temperature baths in the range 5–60°C. After adequate temperature equilibration, add to the two tubes in each bath, at timed intervals, 0.1-ml volumes of two different enzyme dilutions, one known to yield a linear assay for 4 min at room temperature when assayed with 1 ml of $2.5 \times 10^{-3}M$ ONPG, and the other, half that of the first. After each tube has incubated 4 min, add 1 ml of $1M$ $Na_2CO_3$ to each, and remove the tube from the bath. (*Note:* This makes the solution sufficiently alkaline to be beyond the range of *p*H where the enzyme is active.) Then read the $OD_{420}$ against a blank (1 ml of ONPG solution, 9 ml of buffer), and convert the results to $\mu$moles of ONPG hydrolyzed per min.

Prepare a graph, plotting velocities for both enzyme concentrations versus temperature. Note whether all the rates are directly proportional to enzyme concentration. Establish this point for the higher enzyme concentrations, and if so, use the data obtained at the higher enzyme concentration for further analysis; otherwise use the data obtained at the lower enzyme concentration.

From the ascending (left to right) region of the above graph, construct an Arrhenius plot ($\log_{10} v_0$ versus $1/T$ (°K) ), and calculate the activation energy of the enzyme catalyzed reaction.

### *Determination of the Activation Energy of Enzyme Denaturation*

Prepare twelve test tubes containing a constant 1-ml dilution of cell-free extract (induced cells) known to yield assays directly dependent upon enzyme concentration when examined in the kinetic assay at room temperature over a 4-min period.

Divide these samples into four groups, and place each group in separate constant-temperature baths maintained at 45°C, 48°C, 51°C, and 54°C. At 30 sec, 1 min, and 2 min after beginning heat treatment, remove the enzyme samples from each bath, and place them in an ice bath to cool. When cool, allow the tubes to equilibrate to room temperature, and add 7 ml of $0.08M$ sodium phosphate buffer ($p$H 7.7) to each tube. Finally, at timed intervals, add 1 ml of $2.5 \times 10^{-3}M$ ONPG to each tube. After each tube has incubated at room temperature for 4 min, add 1 ml of $1M$ $Na_2CO_3$ to each to stop the reaction (fixed-time assay). Determine the optical densities by using a blank containing 1 ml of $2.5 \times 10^{-3}M$ ONPG and 9 ml of $0.08M$ sodium phosphate ($p$H 7.7).

Determine the extinction coefficient from the $OD_{420}$ of a solution with a known concentration of completely hydrolyzed ONPG. Calculate the rates of hydrolysis shown by each enzyme preparation. Then prepare a graph showing data from each preincubated enzyme sample, plotting velocity in each assay versus duration of preincubation. Calculate the rates of denaturation from the slopes of the curves on this graph. Finally, prepare an Arrhenius plot—that is the $\log_{10}$ of the rates of denaturation versus $1/T$ (°K), and calculate $\Delta E$ for the heat of denaturation.

### *Purification of β-Galactosidase from Cell-free Extract of Induced Cells*

Several methods for purifying β-galactosidase from *E. coli*, varying in complexity and in time required, are available in the literature (see Rotman and Spiegelman or Kuby and Lardy in the References). We suggest that one or more of these methods be tried by groups in the class, depending upon class size and availability of equipment and time.

Remember that all determinations of specific activities must start with data from assays where response is directly proportional to enzyme concentration.

## Report of Results

Plot all the graphs, and perform all the calculations called for.

If you make a purification study, construct a table listing the volume, protein concentration, specific activity, total activity, and percent yield of each fraction obtained.

## Discussion

Enzyme induction appears to represent a net synthesis of new protein rather than a conversion of some pre-existing protein into an active form (see Rotman and Spiegelman or Hogness et al. in the References). Therefore, studies on enzyme induction are studies of protein synthesis and of the factors controlling the dynamic turnover of protein.

### Exercises

1. In the study of bacterial growth rates in this experiment, you used a plot of $OD_{660}$ versus time. Yet, it is known that OD represents a log term, $\log_{10} I_0/I$. Why is it possible to use a value representing a log function to measure bacterial growth rates?
2. Considering your value for the activation energy of enzyme denaturation, what factors other than the direct action of heat on β-galactosidase may contribute the observed value? How could these factors be controlled so as to yield a less ambiguous determination?

**References**

Rotman, B., and S. Spiegelman, *J. Bact.*, **68,** 419 (1954).

Kuby, S., and H. Lardy, *J. Am. Chem. Soc.*, **75,** 890 (1953).

Lederberg, J., *J. Bact.*, **60,** 381 (1950).

Wallenfels, K., and O. P. Malhotra, *In* Boyer, P. D., H. Lardy, and K. Myrbäck (Editors), *The Enzymes* (Vol. 4), Academic, New York, 1961.

Hogness, D. S. et al., *Biochim. et Biophys. Acta*, **16,** 99 (1955).

Lowry, O. H. et al., *J. Biol. Chem.*, **193,** 265 (1951).

EXPERIMENT

# 24. Purification and Catalytic Properties of Rabbit Muscle 3-Phosphoglyceraldehyde Dehydrogenase

## Theory of the Experiment

The enzyme 3-phosphoglyceraldehyde dehydrogenase (also called triose phosphate dehydrogenase, or TDH), one of the enzymes of the glycolytic scheme (see Experiment 30, p. 146), catalyzes the oxidation of 3-phosphoglyceraldehyde to 1,3-diphosphoglyceric acid (reaction *1*).

This enzyme usually constitutes a relatively high proportion of the total protein in many cells which have a prominent glycolytic system—approximately 5% and 10% of the total protein in rabbit muscle and yeast, respectively. Thus 3-phosphoglyceraldehyde dehydrogenase can be obtained in high yield and purity from a number of sources by applying but a few efficient fractionation procedures.

In this experiment crystalline triose phosphate dehydrogenase is obtained from rabbit muscle by a four-step procedure. First, TDH is efficiently extracted from minced muscle with 0.03*M* KOH. Second, contaminating proteins are then removed from this extract by two successive $(NH_4)_2SO_4$ precipitations to yield a clear $(NH_4)_2SO_4$-containing solution with most of the original enzymatic activity. Third, TDH is crystallized from this by changing the *p*H. Fourth, subsequent recrystallization from a Versene-$(NH_4)_2SO_4$ solution

$$(1)\quad \mathrm{H{-}C(=O){-}CH(OH){-}CH_2{-}O{-}P(=O)(O^-){-}O^-} + \mathrm{DPN^+} + \mathrm{HPO_4^{-2}} \rightleftharpoons \mathrm{O{=}C(O{-}P(=O)(O^-){-}O^-){-}CH(OH){-}CH_2{-}O{-}P(=O)(O^-){-}O^-} + \mathrm{DPNH} + \mathrm{H^+}$$

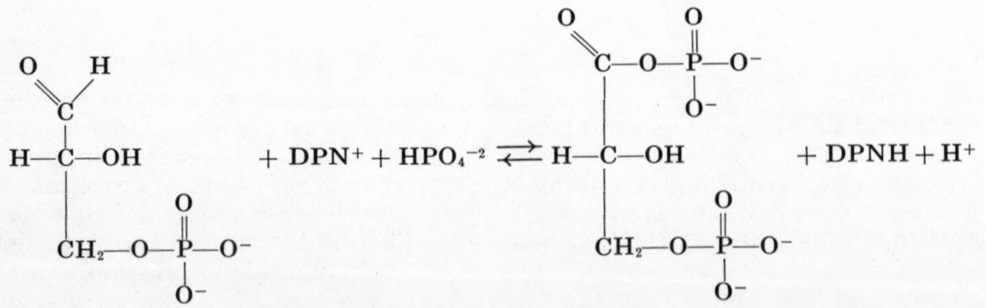

removes traces of heme proteins and yields an apparently pure protein which remains stable for a long time.

A spectrophotometric assay of the formation of DPNH, which absorbs at 340 mμ, is usually used for the assay of TDH. Since the equilibrium constant of reaction (1) at *p*H 7 (excluding $H^+$) is approximately 1, two convenient features of the reaction can be used to shift the reaction toward DPNH formation, improving the efficiency of the assay. First, when $HAsO_4^{-2}$ is used in place of $HPO_4^{-2}$, the arsenate replaces the phosphate as a substrate to form DPNH and 1-arseno-3-phosphoglyceric acid. 1-Arseno-3-phosphoglyceric acid is unstable at neutral or alkaline *p*H, and rapidly hydrolyzes to 3-phosphoglyceric acid and arsenate. Thus the reaction in the direction of DPNH is favored, since one of the products of the reaction in this direction is removed (reaction *2*). Second, since one of the products of the reaction is a proton, assays at neutral or alkaline *p*H favor DPNH formation.

Since 3-phosphoglyceraldehyde is both unstable and expensive, it is more convenient to generate this substrate through use of a second reaction, the aldolase reaction; that is, by adding fructose-1,6-diphosphate and the enzyme aldolase (reaction *3*).

After a recrystallized enzyme and the data concerning its purification have been obtained, you will study two features of TDH. First, you will examine the phosphate requirement and calculation of the equilibrium constant for the reaction. Second, you will study the role of sulfhydryl groups in the active site of the enzyme, using the sulfhydryl reagents, iodoacetate (IAA) and *p*-mercuribenzoate (PMB) (reaction *4*), which react with free SH groups on enzymes or substrates to yield the substituted thio derivative (see p. 114).

(2)

$$\begin{array}{c} O{=}C{-}O{-}As(O)(O^-){-}O^- \\ | \\ H{-}C{-}OH \\ | \\ CH_2{-}O{-}P(O)(O^-){-}O^- \end{array} \xrightarrow{H_2O} \begin{array}{c} O{=}C{-}O^- \\ | \\ H{-}C{-}OH \\ | \\ CH_2{-}O{-}P(O)(O^-){-}O^- \end{array} + HAsO_4^{-2} + H^+$$

(3)

$$\begin{array}{c} CH_2{-}O{-}P(O)(O^-){-}O^- \\ | \\ C{=}O \\ | \\ HO{-}C{-}H \\ | \\ H{-}C{-}OH \\ | \\ H{-}C{-}OH \\ | \\ CH_2{-}O{-}P(O)(O){-}O^- \end{array} \underset{\text{aldolase}}{\rightleftharpoons} \begin{array}{c} CH_2{-}O{-}P(O)(O^-){-}O^- \\ | \\ C{=}O \\ | \\ CH_2OH \end{array} \;+\; \begin{array}{c} O{=}C{-}H \\ | \\ H{-}C{-}OH \\ | \\ CH_2{-}O{-}P(O)(O^-){-}O^- \end{array} \underset{}{\overset{\text{TDH}}{\rightleftharpoons}}$$

$$\text{ENZ–SH} + \text{ICH}_2\text{–}\overset{\text{O}}{\overset{\|}{\text{C}}}\text{–O}^- \longrightarrow \text{ENZ–S–CH}_2\text{–}\overset{\text{O}}{\overset{\|}{\text{C}}}\text{–O}^- + \text{H}^+ + \text{I}^-$$

(4)

$$\text{ENZ–SH} + \text{Cl–Hg–C}_6\text{H}_4\text{–}\overset{\text{O}}{\overset{\|}{\text{C}}}\text{–O}^- \longrightarrow \text{ENZ–S–Hg–C}_6\text{H}_4\text{–}\overset{\text{O}}{\overset{\|}{\text{C}}}\text{–O}^- + \text{H}^+ + \text{Cl}^-$$

Thus if an enzyme is inhibited by either of these reagents, one tentatively assumes that an SH group from the amino acid cysteine is involved in or near the active site. Final proof requires knowledge of the three-dimensional structure.

## Experimental Procedure

MATERIALS

0.003*M* DPN$^+$
0.03*M* Sodium pyrophosphate (*p*H 8.4)
0.4*M* $Na_2HAsO_4$
0.6*M* Fructose-1,6-diphosphate; Na salt in 0.03*M* pyrophosphate buffer
Aldolase
Cysteine
Spectrophotometer for use at 340 m$\mu$
Cheesecloth
Saturated $(NH_4)_2SO_4$ (*p*H 7.5)
0.001*M* Versene (*p*H 7.5)
Filter paper
$3 \times 10^{-3}M$ Iodoacetate
0.4*M* Sodium phosphate (*p*H 8.3)
Rabbit
Dissecting tools
Cotton
5% Nembutal
Hypodermic syringe
Meat grinder
0.03*M* KOH
*p*H meter
Saturated $(NH_4)_2SO_4$ in 0.001*M* versene (*p*H 8.4)
High-speed centrifuge
7.5*N* $NH_4OH$
Folin Ciocalteau reagents

### *Preparation of Crystalline Enzyme*

Perform *all* the operations of this procedure, other than killing the rabbit, at 1–4°C. Since rapidity of operation and adequately cool temperatures are essential in the early stages, check to see that all materials are ready before starting the preparation (see Cori et al. in References).

Kill a rabbit either by a quick blow on the neck or by injection of 5 ml of 5% Nembutal into an ear vein dilated by rubbing with cotton soaked in 95% EtOH. Then *immediately* skin the hind legs and back of the rabbit; rapidly rip and cut out the large leg, back, and shoulder muscles; Place these in crushed ice as soon as they are obtained. When most of the muscle has been removed, pass the chilled muscle through a precooled meat grinder into a precooled beaker. Quickly stir the minced muscle in an equal volume of precooled (0–2°C) 0.03*M* KOH, and stir occasionally over a 10-min period. Pour the resultant mixture onto two layers of cheesecloth over a 1-liter beaker, then gather the edges of the cloth, and squeeze the mixture over the beaker for approximately 30 sec, kneading the mixture with the fingers to assure the removal of extract from the moist muscle mince. Repeat the KOH extraction operation with the same cheesecloth and a volume of cold 0.03*M* KOH equal to the first volume. Finally, extract the muscle mince with a volume of cold water equal to one-half that of each KOH volume. Then check the *p*H of an aliquot of the combined extract, and adjust to *p*H 6.8–7.2 with $H_2SO_4$ or $NH_4OH$ if necessary. Measure the volume of the extract, and remove a 0.5-ml volume for future determination of specific activity (fraction 1).

While maintaining the extract in an ice bath at 0°C, add to each 100 ml of extract 108 ml of "room-temperature saturated" $(NH_4)_2SO_4$ solution (*p*H 7.5) over a 5 min period, making the solution 52% saturated with $(NH_4)_2SO_4$. Gently stir the mixture to assure even distribution. Then allow the mixture to remain for 20 min at 0°C before starting to filter off the precipitate by means of several large, fluted filter papers (Whatman No. 1 or No. 12). Allow the filtration to proceed until complete (often 2–4 hr) before measuring the volume of the solution and removing a 0.5-ml volume for future assay (fraction 2).

Slowly add 13 g of solid $(NH_4)_2SO_4$ with gentle stirring (avoid vigorous stirring and bubbles, as these lead to denaturation) for each 100 ml of clear filtrate, and continue to stir occasionally until all the crystals are dissolved. Begin filtering off the resultant precipitate by gravity filtration with two or more Whatman No. 1 or No. 12 papers as soon as the crystals of $(NH_4)_2SO_4$ dissolve. Pour the filtrate back onto the same filter papers, and refilter if necessary in order to obtain a clear filtrate.

Next add dropwise small aliquots of 7.5*N* $NH_4OH$ with stirring, checking the *p*H on five-fold dilutions of the enzyme solution, until a *p*H of 8.4 is obtained. Transfer the solution to one or more large Erlenmeyer flasks; recheck the *p*H on a 1:5 dilution of the enzyme, and readjust to 8.4 if necessary; cork the mouth(s) of the flask(s), and allow the closed flasks to remain for 1–4 days at 0–2°C. Crystals, as evidenced by a sheen in the solution upon swirling, should appear after 12–15 hr and accumulate in quantity over the next several days. If no crystals appear after 24 hr, recheck the *p*H on a 1:5 dilution of the enzyme, and readjust to *p*H 8.4 if necessary with 7.5*N* $NH_4OH$.

After a satisfactory quantity of crystals has accumulated (usually after 3 days at *p*H 8.4, but 2 days is adequate if time is lacking), spin down the crystals by one or more successive centrifugations at 12,000 × g for 5 min. Discard the supernatant if adequate crystals are obtained.

Redissolve the crystals obtained from one rabbit in 40–50 ml of 0.001*M* Versene (*p*H 7.5). After clarification by centrifugation to remove undissolved materials and debris, remove a 0.25-ml aliquot for future assay (fraction 3). Then add two volumes of saturated $(NH_4)_2SO_4$ in 0.001*M* Versene (*p*H 8.4), and allow the solution to stand overnight in the cold. Crystals will appear within the first hour and accumulate over a 12- to 24-hr period. If time permits, repeat the recrystallization of the enzyme one or two more times by successive centrifugation, dissolving, and reprecipitation to remove the last of the heme proteins and other materials. During each recrystallization, remove a 0.25 ml aliquot for assay (fractions 4, 5, and so on). Store the final crystal suspension at 0–2°C until used.

Determine the specific activity of each of the aliquots of the various fractions as soon as time is available during the fractionation procedure, and then calculate the total activity of each fraction. Make all necessary enzyme dilutions with *freshly prepared* 0.004*M* cysteine in 0.03*M* sodium pyrophosphate (*p*H 8.4).

In order to determine the specific activity from the kinetic assay, two things are required: (a) the appropriate range of enzyme concentration for assay must be found ("range finding"), and (b) a closer examination must be made of enzyme concentrations within the correct range. See the section on Enzymology (pp. 79–86) for a discussion of the principles involved.

### Protein Determination

The Folin-Ciocalteau determination of protein described below is suggested for this assay. Other protein determinations may also be applied with success to the fractions obtained during this purification. Consult the Introduction to Proteins and Amino Acids for details (pp. 75–76).

For protein determination by the Folin-Ciocalteau method, prepare a series of tubes including a blank (1.2 ml of $H_2O$), standards (30–600 $\mu$g of protein in 1.2 ml of $H_2O$), and assay tubes containing 1.2-ml volumes of an appropriate dilution series to yield protein values in the 30–600 $\mu$g range. Add 6 ml of alkaline copper solution [freshly mixed 2% $Na_2CO_3$ in 0.1$N$ NaOH; 0.5% $CuSO_4 \cdot 5H_2O$ in 1% sodium tartrate (50:1)] to all tubes, and mix each by shaking. After 10 min add 0.6 ml of Folin-Ciocalteau reagent to each tube, and thoroughly mix the contents of each. After 30 min read the optical densities at 500 m$\mu$ against the blank. Calculate the protein concentration(s) of the original solution(s) from $OD_{500}$ values within the range of the standards.

### General Assay Procedure

The following description applies to a general assay for a 3-ml reaction volume. Only the volume of the TDH solution to be added is listed, thus appropriate dilutions of TDH that yield measurable reaction rates must be selected for meaningful assays.

Add the following reagents, in the order listed, to a cuvette capable of passing 340 m$\mu$ light (350 m$\mu$ if B & L Spectronic 20 is used).

| Reagent | Volume |
|---|---|
| 0.03$M$ sodium pyrophosphate, 0.004$M$ cysteine (pH 8.4) (freshly prepared) | 2.55 ml |
| Aldolase (1 mg/ml) | 0.10 ml |
| 0.003$M$ $DPN^+$ | 0.10 ml |
| 0.4$M$ sodium arsenate | 0.10 ml |
| TDH dilution (e.g., 1:40 dilution) | 0.05 ml |

Allow the reactants to equilibrate to room temperature for 5 min, then measure the $OD_{340}$ of the mixture against a water blank. Initiate the reaction by adding 0.10 ml of 0.06$M$ fructose-1,6-diphosphate (HDP) in 0.03$M$ sodium pyrophosphate buffer ($p$H 8.4). Measure the $OD_{340}$ at 15-sec intervals for 2–5 min.

### Phosphate Requirement and $K_{eq}$ Measurement

Omitting the 0.1 ml of 0.4$M$ arsenate of the general assay procedure, run an assay with a dilution of TDH known to yield measurable rates in the general assay. After 2 min add 0.1 ml of 0.4$M$ sodium phosphate ($p$H 8.3), and while taking readings at 15-sec intervals, allow the reaction to proceed to equilibrium. Finally, add 0.1 ml of 0.4$M$ arsenate, and allow the reaction to go to completion.

### Inhibition Studies with Iodoacetate ($IAA$)

Set up two assays having appropriate enzyme dilutions and lacking in HDP, as described in the general assay. To the first add 0.05 ml of $3 \times 10^{-3}M$ IAA, and mix; then initiate the reaction with 0.1 ml of 0.06$M$ HDP. Measure the $OD_{340}$ at 15-sec intervals.

To the second, add 0.1 ml of 0.06$M$ HDP to initiate the reaction, and observe the $OD_{340}$ at 15-sec intervals for 1 min. Then quickly add and mix in 0.05 ml of $3 \times 10^{-3}M$ IAA, and continue to measure the $OD_{340}$ at 15-sec intervals for several minutes.

## Report of Results

For all assays, subtract the zero-time $OD_{340}$ reading from the timed assays to obtain valid $\Delta OD_{340}$ readings. If a 1-cm light path was used, use a value of $6.22 \times 10^3$

for the molar extinction coefficient ($E_{1\,cm}^{1M}$) of DPNH at 340 m$\mu$ to convert each timed $\Delta OD_{340}$ value into units of $\mu$moles of DPNH formed. If a round colorimeter tube was used instead of a flat-sided, 1-cm-light-path cuvette, use an extinction coefficient determined by allowing a reaction to go to completion with a known amount of $DPN^+$ added in order to convert $\Delta OD_{340}$ values to $\mu$moles DPNH formed. Prepare a plot of $\mu$moles of DPNH formed versus time in minutes for each assay or set of assays. Then determine the initial velocities ($\mu$moles of DPNH formed/min) of each assay.

Using data in which the observed initial reaction rate is proportional to enzyme concentration, prepare a table showing the volume, milligrams of protein per milliliter, specific activity ($\mu$moles of DPNH formed/min per milligram of protein), total activity, and percent yield for each fraction assayed.

From the data illustrating the phosphate requirement, calculate the extent of $DPN^+$ reduction when phosphate is used as a substrate. Using this value and the known amounts of reactants in the cuvette, plus a $K_{eq}$ for aldolase at 25°C of

$$\frac{(\text{DHAP})(\text{triose P})}{(\text{HDP})} = 2 \times 10^{-3}$$

calculate an equilibrium constant, $K_{eq}$, where

$$K_{eq} = \frac{(\text{DPNH})(1{,}3\ \text{di PGA})(\text{H}^+)}{(\text{DPN}^+)(3\ \text{PGA})(\text{Pi})}$$

for the reaction at $p$H 8.3.

Compare this value with literature values (see Burton and Wilson in References). Explain the effect of the added arsenate in these studies. Using IAA, interpret the data from both studies, and note the order in which the reagents were added.

## Discussion

In actuality, the velocity determinations of this experiment only approximate the true velocities. This is because the reaction follows a more complicated reaction order than zero-order kinetics. Examination of the reaction reveals a requirement for several reactants. It is possible to control the quantities of these components so as to operate in a true second-order reaction (see Cori et al. in References). If time permits, allow an assay to proceed for several minutes, and then from a plot of the rate ($OD_{340}$ versus time) determine the reaction order. Consult any introductory physical chemistry textbook for the methods involved.

**Exercises**

1. Impure fractions of 3-phosphoglyceraldehyde dehydrogenase from rabbit muscle contain fair quantities of the enzymes triose isomerase and $\alpha$-glycerolphosphate dehydrogenase.
   Consider this fact in light of the possible effect on the specific activity determination of fraction 1 in this purification.

$$\text{OHC–CH(OH)–CH}_2\text{–O–PO}_3^{2-} \underset{\text{Triose isomerase}}{\rightleftharpoons} \text{HOCH}_2\text{–C(=O)–CH}_2\text{O–PO}_3^{2-} \underset{\alpha\text{-glycerophosphate dehydrogenase}}{\overset{\text{DPNH + H}^+}{\rightleftharpoons}} \text{HOCH}_2\text{–CH(OH)–CH}_2\text{–O–PO}_3^{2-} + \text{DPN}^+$$

2. What is the standard state free energy change at $p$H 7 ($\Delta F'$) for the reaction catalyzed by TDH?
3. The assays of invertase and $\beta$-galactosidase activity (Experiments 22 and 23) can be run in a manner such that the reaction is zero order with respect to substrate. How is this possible when both sugar and $H_2O$ are required for reaction?

### References

Cori, G. T. et al., *J. Biol. Chem.*, **173,** 605 (1948).
Taylor, J. R. et al., *J. Biol. Chem.*, **173,** 619 (1948).
Velick, S. F., *J. Biol. Chem.*, **203,** 563 (1953).
———, and J. E. Hayes, Jr., *J. Biol. Chem.*, **203,** 545 (1953).
——— et al., *J. Biol. Chem.*, **203,** 527 (1953).
Burton, K., and T. H. Wilson, *Biochem. J.*, **54,** 86 (1953).

# PART FOUR **NUCLEIC ACIDS**

## Primary Structure

Nucleic acids are components of all living cells. Most nucleic acids exist in the form of nucleoproteins, a complex of basic proteins (for example, protamines and histones) and acidic nucleic acids. Since Miescher first isolated nucleoprotein in 1869 [1], several ways of separating nucleic acids from their associated proteins have been developed. Early in this work it was found that not all nucleic acids were identical. Tissues having large cell nuclei (for example, thymus) contain predominantly one type of nucleic acid (originally called thymus nucleic acid), whereas others, having relatively small nuclei (for example, yeast), contained predominantly a second type (yeast nucleic acid). On the basis of their chemical differences, these two types of nucleic acids are now named deoxyribonucleic acid (DNA) and ribonucleic acid (RNA), respectively.

DNA is composed largely of phosphate, the purine bases adenine and guanine, the pyrimidine bases cytosine and thymine, and the sugar 2-deoxy-D-ribose.

RNA differs from DNA in that it contains the pyrimidine base uracil, instead of thymine, and the sugar D-ribose, instead of 2-deoxy-D-ribose.

The available evidence indicates that the basic chemical formula, common to both types of nucleic acid, consists of a chain or backbone of alternating sugar (furanose) and phosphate groups, joined in regular 3′,5′-phospho-diester linkages (Fig. 19). The purine, or pyrimidine, bases are attached to this sugar-phosphate chain by N-glycosyl β-linkage to the 1′-position of the ribose or deoxyribose (see p. 120).

The subunit of the overall structure, composed of a nitrogenous base and a sugar, is called a nucleoside. Nucleosides

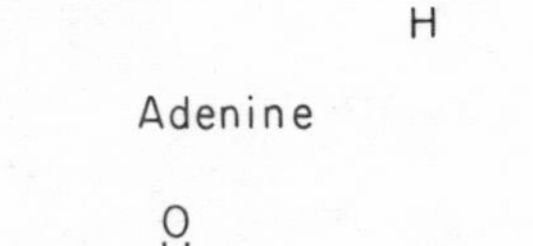

Adenine

Guanine

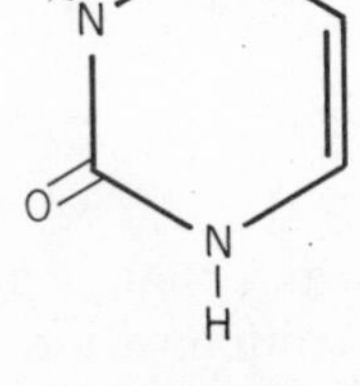

Uracil

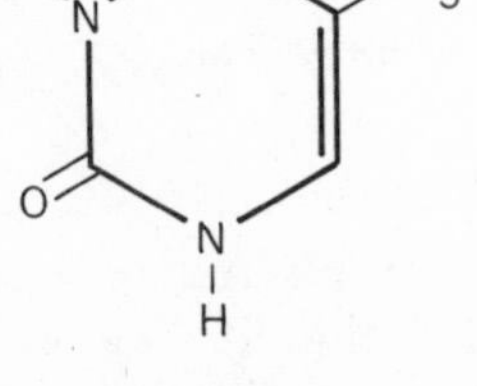

Thymine

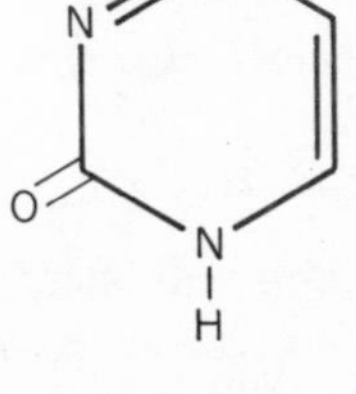

Cytosine

containing a phosphate in either the 2′, 3′, or 5′ position are called nucleotides (a number with a prime denotes position on the sugar; a number without a prime denotes position on the base). Compounds formed by the association of two or more nucleotides, making a small molecule, are called oligonucleotides.

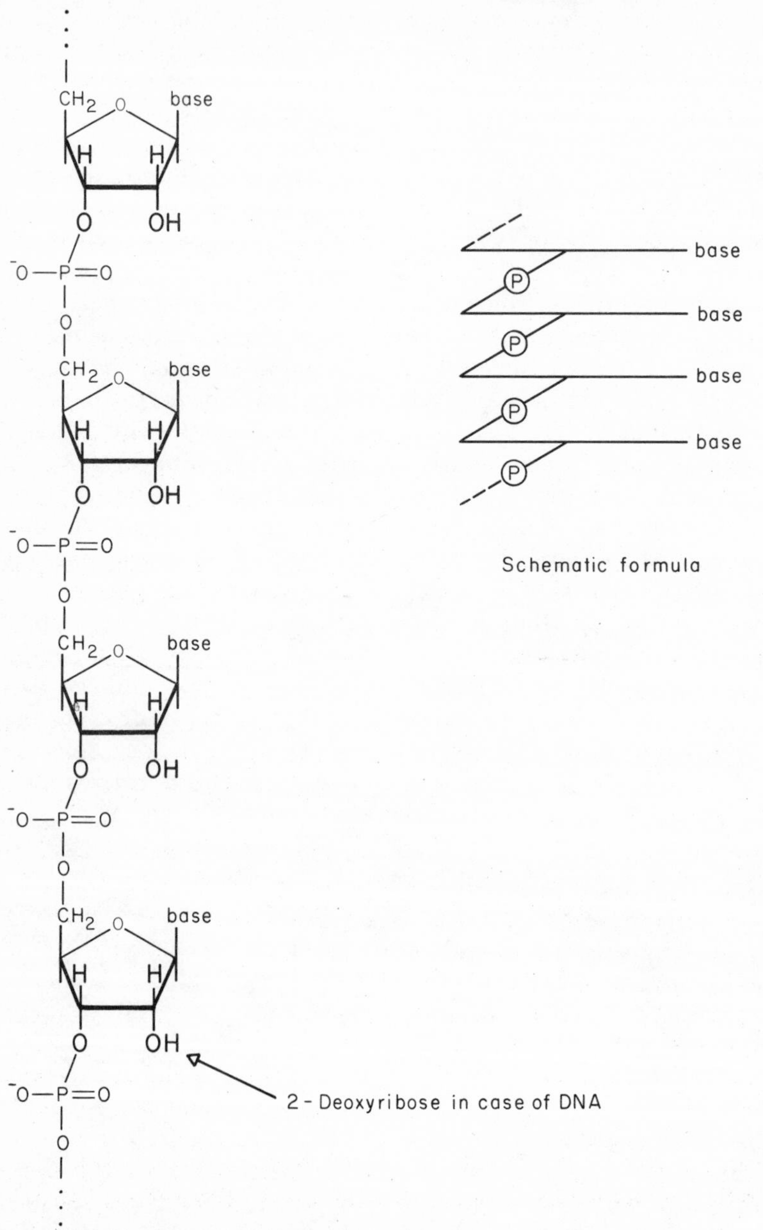

FIG. 19. *Basic chemical structure of nucleic acids.*

## Secondary Structure

Various physiochemical studies on DNA in aqueous solution have revealed that it is a polymer of high molecular weight ($>5 \times 10^6$) and that it is rather stiff and extended in configuration. One consequence of the high molecular weight and rigid form of DNA molecules is the extremely high viscosity of aqueous DNA solutions; for example, a solution containing $10^{-4}$ g of DNA/ml is twice as viscous as water. Another consequence is the extreme ease with which shearing forces (such as those produced by high-speed stirring, pipetting, or squirting of DNA solutions through a syringe) cause a reduction of the molecular weight of DNA by producing chain scissions. Because it is nearly impossible to extract DNA from living material without some shearing, the molecular weights observed for isolated DNA ($5\text{–}10 \times 10^6$) may have little bearing on the size of DNA molecules in vivo.

Isolated fibrous DNA gives a sharp X-ray diffraction pattern. This, by itself, is de facto evidence for a highly regular secondary structure in DNA. In 1953, Watson and Crick [2, 3] proposed a secondary structure for DNA derived from these X-ray patterns and in agreement with the findings of Chargaff [4], which showed that cetain pairs of DNA bases exist in equimolar quantities. In this Watson-Crick structure, two chains of DNA twist helically about a common central axis, yielding a double-stranded molecule about 20 Å in diameter (Fig. 20). The sugar-phosphate chains are on the outside of the right-handed helixes, whereas the purine and pyrimidine bases are hydrogen bonded on the inside, with the planes of their rings perpendicular to the central axis. Only base pairings between adenine and thymine or between guanine and cytosine are allowed. Furthermore, the sequence of atoms in the sugar-phosphate backbones is reversed in one chain as compared to the other.

Although the DNA of a given organism is relatively homogenous as regards (1) base composition, (2) molecular weight, and (3) intracellular localization, there exist at least two classes of RNA molecules which are distinguishable in these respects. One type of RNA is a low-molecular-weight material in the range 20,000–40,000. This RNA is named soluble RNA (or transfer RNA) because it remains in the supernatant fraction when cell homogenates are centrifuged at $100{,}000 \times g$ for a few hours [5]. The second type of RNA represents the major part of the RNA present in most organisms. This RNA has a higher molecular weight ($5 \times 10^5\text{–}2 \times 10^6$, or 1500–6000 nucleotides) and is found as nucleoprotein. Much of this RNA is located in specific subcellular particles such as microsomes (ribosomes in bacteria).

Although no detailed model of the secondary structure of RNA has yet been established, certain deductions can be made. Most striking is the dissimilarity between the solution properties of RNA and those of DNA. For example, when viscosities of DNA and RNA of the same molecular weight are compared, the viscosity of increment due to DNA is twenty times greater than that of the corresponding RNA solution [6]. This would indicate a randomly coiled polymer structure for RNA. Thus a double-stranded helical structure must be ruled out. X-ray diffraction patterns of most RNA fibers contain fuzzy patches rather than sharp spots, indicating the absence of any regular, repeating, three-dimensional structure. The absence of apparent structural regularity in isolated RNA does not necessarily imply a similar disorder in the naturally occurring nucleoproteins from which RNA was extracted. A regular packing of RNA has been demonstrated by X-ray diffraction for two nucleoproteins: the tobacco mosaic virus particle, and the turnip yellow mosaic virus particle.

## Possible Function of Nucleic Acids

Since the base pairing in DNA dictates that the two strands are complementary, separation of the two strands into single strands and assembly of new, complementary, hydrogen bonded strands from monomer units would yield two double-stranded

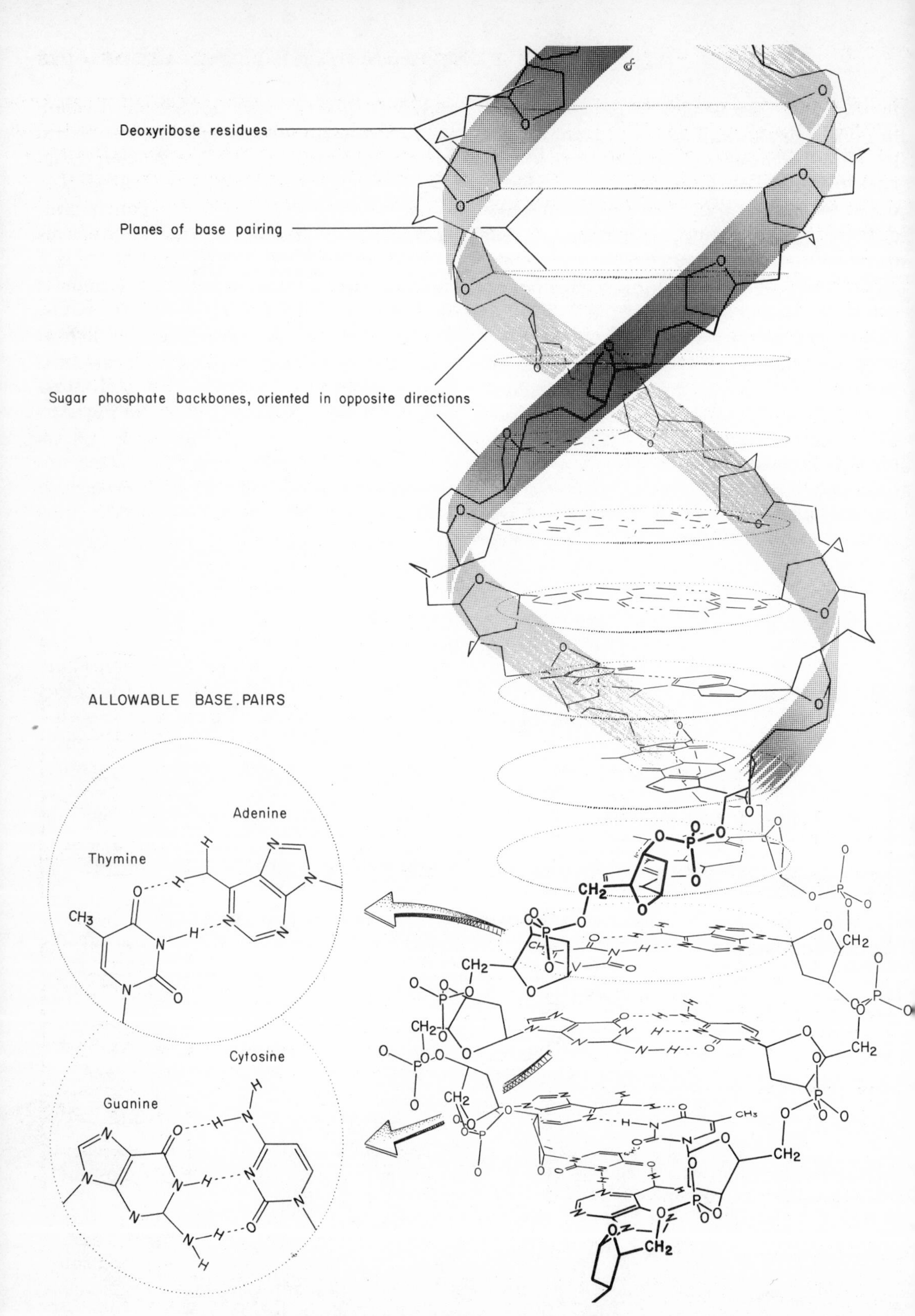


FIG. 20. *Three-dimensional representation of DNA structure.*

helixes like the original. The biological implications of this model (Fig. 21) are profound. Genetic information could be carried in the sequence, or units of sequence, of the four-letter alphabet (that is, the four different bases). Such a means of self-duplication allows the passage of identical information to both daughter cells during cell division. The occurrence of mutations can be explained in terms of an incorrect base pairing, or other aberration, during cell division.

The presence of DNA in the nuclei of plants and animal cells, coupled with many observations of its activity [7], has fully implicated DNA as a component of fundamental genetic material. Moreover, results from experiments designed to follow isotopically labeled DNA during cell division [8, 9] are in agreement with the theoretical ideas concerning DNA replication (Fig. 21). Recent studies tend to implicate RNA, rather than DNA, as a component immediately involved in protein biosynthesis [10]. These studies suggest that there is a transfer of information from the genetic material, DNA, to the messenger, RNA, giving rise to the eventual cytological expression in the form of synthesized proteins. The nature of this transfer of information is only partially known. Perhaps studies of nucleic acid biosynthesis [11, 12, 13], model building [14], or other approaches [15] will provide the answers to this major problem in the field of molecular biology.

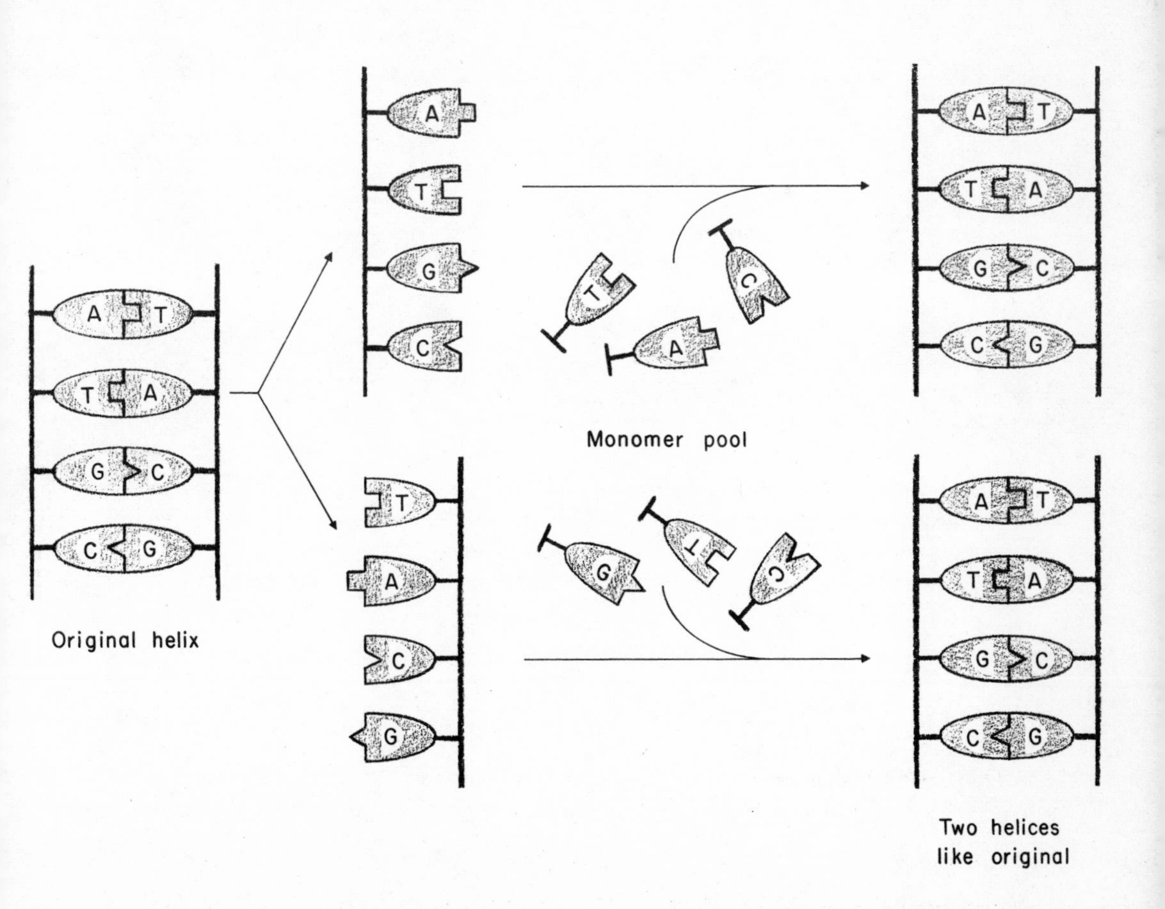


FIG. 21. *Schematic representation of postulated replication of DNA.*

## General References


Chargaff, E., and J. N. Davidson (Editors), *The Nucleic Acids* (2 Vols.), Academic, New York, 1955.

Davidson, J. N., *The Biochemistry of Nucleic Acids* (3rd ed.), Wiley, New York, 1958.


## Specific References


1. Miescher, F., *Die histochemischen und physiologischen Arbeiten*, Leipzig, 1897.
2. Watson, J. D., and F. H. C. Crick, *Nature*, **171**, 737 (1953).
3. Crick, F. H. C., and J. D. Watson, *Proc. Roy. Soc. London*, **223A**, 80 (1954).
4. Chargaff, E. et al., *J. Biol. Chem.*, **177**, 405 (1949).
5. Hoagland, M. B. et al., *J. Biol. Chem.*, **231**, 241 (1958).
6. Littauer, U. Z., and H. Eisenberg, *Biochim. et Biophys. Acta.*, **32**, 320 (1959).
7. Sinsheimer, R. L., *Science*, **125**, 1123 (1957).
8. Meselson, M., and F. W. Stahl, *Proc. Natl. Acad. Sci. U.S.*, **44**, 671 (1958).
9. Taylor, J. H. et al., *Proc. Natl. Acad. Sci. U.S.*, **43**, 122 (1957).
10. Hoagland, M. B., *Scientific American*, **201**, 55 (1959).
11. Kornberg, A. et al., *J. Biol. Chem.*, **233**, 171 (1958).
12. Grunberg-Manago, M., and S. Ochoa, *Biochim. et Biophys. Acta.*, **20**, 269 (1956).
13. Furth, J. J. et al., *Biochem. Biophys. Res. Comm.*, **4**, 431 (1961).
14. Stent, G. S., *Advances in Virus Research*, **5**, 95 (1958).
15. Wahba, A. J. et al., *Proc. Nat. Acad. Sci. U.S.*, **49**, 880 (1963).

EXPERIMENT

# 25. Isolation of RNA

## 'heory of the Experiment

Before an isolation of RNA is performed, ne should decide on the type of RNA deired. If the gross RNA is desired, wholeell homogenates can serve as starting naterial. If a specific type of RNA (for xample, nuclear, microsomal, or soluble) s desired, the subcellular fractions must e fractionated by differential centrifugaion before RNA isolation.

Two methods of isolation, varying in efficiency and in the quality of the product btained, are most frequently used. The impler method, extraction with NaCl, involves the precipitation of charged macronolecules with cold trichloroacetic acid TCA), the removal of lipids and excess 'CA from the precipitate with acetone and ther washes, the extraction of the nucleic cid from the precipitate with boiling 10% NaCl, and, finally, the precipitation of the ucleic acid from the extract by the addiion of alcohol. This NaCl extraction nethod is adequate for gross studies of ucleic acid, but is not satisfactory for precise work, since DNA occurs in the product as a contaminant if it is present in he starting material. The nucleic acids btained by the procedure are usually omewhat "degraded" into fragments of lower molecular weight than the native RNA.

A new, mild, specific RNA isolation procedure, developed in recent years (1956—see Kirby in References) involves extraction with concentrated phenol solutions. When phenol is mixed with a cellular or subcellular preparation, the strong hydrogen bonding forces of the phenol rapidly denature the protein. When the turbid phenol-water emulsion is broken by centrifugation, two phases appear. The lower, or phenol, phase contains DNA, and the aqueous upper phase contains carbohydrate and RNA. Both phases contain suspended denatured protein. Removal of the aqueous phase is followed by removal of denatured protein by centrifugation; subsequent precipitation of the macromolecules (RNA) with alcohol completes the isolation. This method yields "native" RNA—free of DNA, but usually contaminated with some polysaccharide precipitated from the aqueous phase by the alcohol used in the last step. This polysaccharide can be removed by amylase treatment, by an extraction procedure involving 2-methoxyethanol, or by the use of ion exchange columns.

## Experimental Procedure

### MATERIALS FOR THE NaCl EXTRACTION METHOD

Large rat
Blendor
Acetone
0% NaCl
Aluminum foil
Dissecting tools
Cold 30% TCA
Ether
Absolute ethanol
Clinical centrifuge

### MATERIALS FOR PHENOL EXTRACTION METHOD

Frozen or iced fresh liver
Cheesecloth
Table-model high-speed centrifuge
20% Potassium acetate (pH 5.0)
90% Phenol (w/v)
Blendor
Absolute ethanol

### *Isolation of Nucleic Acid by NaCl Method*

Stun a large rat with a blow on the head, and decapitate it. Quickly remove and cool the liver in an ice bath before weighing it to the nearest gram (50–100 g of cold, fresh, beef liver may also serve as a RNA source). Thoroughly homogenize the liver with five volumes of ice-cold distilled $H_2O$ in an ice-cold blendor (approximately 20 sec). Quickly pour the homogenate into a cooled container, and add an equal volume of cold 30% TCA. Stir gently, and allow the denatured homogenate to stand for 10 min in ice before collecting the precipitate by centrifugation at 1000 × g (clinical centrifuge) for 2 min in the cold (0–4°C). Wash the precipitate twice at low temperature by suspension in, and centrifugation from, 3 ml of cold acetone. Next, wash once with 3 ml of acetone:ether (50:50) at room temperature, once in 3 ml of ether at room temperature, and then air-dry.

Suspend the dry powder in a volume of 10% NaCl solution equal to twice the volume of the powder, cover the test tube with aluminum foil, and place it in a boiling water bath for 40 min. Add distilled water during the heating period to replace water that has evaporated.

Remove the remaining precipitate by centrifuging the cooled extract. Slowly add 2 volumes of absolute ethanol to the supernatant solution, and after cooling in an ice bath for 5 min, collect the fine precipitate by centrifugation. Wash the precipitate first with ethanol and then with ether, and air-dry. Weigh the precipitate, and store it for use in Experiment 26.

*Preparation of RNA by Phenol Method*

Homogenize 25 g of frozen liver (beef, pork, or 12-hr fasted rat) with 100 ml of cool (15°C) $H_2O$ in a blendor for 1 min. Quickly pour the homogenate into a Büchner funnel covered with a layer of cheesecloth, and vacuum-filter. Stir the filtrate into 100 ml of 90% phenol at room temperature. (Work quickly up to this stage. **Caution:** Phenol can cause painful skin burns; immediately wash off all burned areas with 95% ethanol.) Stir the solution while raising it to room temperature in a water bath, and then stir at room temperature for 30 min. Next cool the suspension in an ice bath for 5 min, and then break the emulsion by centrifugation at 2500–3000 × g for 15 min at 0–5°C. Decant the cloudy, supernatant, aqueous layer and most of any intermediate layer containing denatured protein from the brown phenol phase. Remove the remaining denatured protein from the aqueous fraction by additional centrifugation at 10,000 × g for 5 min at 0–5°C. Record the volume of the cloudy aqueous layer obtained, and add a 1/10 volume of 20% potassium acetate (*p*H 5); then precipitate the RNA by adding 2 volumes of cold absolute ethanol. Cool the cloudy solution for 5 min in an ice bath before collecting the precipitate by centrifugation at 1000 × g (clinical centrifuge) at 0–5°C. At room temperature, wash the precipitate once with ethanol:water (3:1), once with absolute ethanol, and once with ether by repeated suspension and centrifugation at 1000 × g. Air-dry and weigh the RNA for use in Experiment 26.

## Report of Results

Calculate the percentage of RNA obtained, and compare the yield with that of the glycogen obtained in Experiment 6.

## Discussion

Owing to their pyrimidine and purine bases, nucleic acids have characteristic absorption spectra in the ultraviolet region. These bases show $E_{MAX}$ values in the region 255–270 m$\mu$; the absorption maxima values vary somewhat in wavelength and in intensity as a function of *p*H. At neutral *p*H the combined bases of nucleic acid exhibit an $E_{MAX}$ of 260 m$\mu$, with a minimum absorption at 240 m$\mu$ and additional absorption at shorter wavelengths (<240 m$\mu$). This physical property of ultraviolet absorption is extremely useful in the location, identification, and isolation of nucleic acids.

**Exercises**

1. In the NaCl isolation procedure, what is the purpose of the TCA treatment? Why is it necessary to work at low temperatures during this step?
2. Prepare a flow diagram showing the route of proteins, nucleic acids, polysaccharides, and smaller molecules in both isolation procedures of this experiment.
3. In the phenol procedure, the product obtained is contaminated with glycogen. Propose a series of steps, using ion exchange, by which the product can be freed of glycogen and reisolated.
4. In calf thymus tissue, approximately 70% of the volume of the cells is occupied by nuclei. Which of the two methods above would you use to isolate RNA from calf thymus?

**References**

Chargaff, E. et al., *J. Biol. Chem.*, **186,** 51 (1950).
Kirby, K. S., *J. Biochem.*, **64,** 405 (1956).
Magasanik, B., *In* Chargaff, E., and J. N. Davidson (Editors), *The Nucleic Acids: Chemistry and Biology*, Vol. 1, Academic, New York, p. 373, 1955.

EXPERIMENT

# 26. Partial Characterization of RNA

## Theory of the Experiment

Dilute alkali (0.3–1.0$N$) readily hydrolyzes the ester bonds (but not the N-glycosyl bonds) of RNA to yield nucleotides. This ease of hydrolysis of the ester bonds is apparently due to the presence of the free 2′-hydroxyl of ribose in RNA, allowing the formation of intermediate 2′,3′ phosphodiester nucleotides, which subsequently break down to a mixture of 2′- and 3′-nucleotides.

Because the deoxyribose in DNA lacks a 2′-hydroxyl group, DNA is incapable of forming such an intermediate. Since DNA is relatively stable in the presence of dilute alkali, acidification of the preparation after alkaline treatment precipitates intact DNA. Such a procedure, as performed in this experiment, allows separation of RNA (as nucleotides) from DNA and allows study of the chemical properties of the nucleotides.

Paper electrophoresis is useful for the identification of the nucleotides released by alkaline hydrolysis of RNA. At *p*H 3.5 the overall net charges on the four RNA nucleotides are distinctly different (see Table VI below).

Since the extent of movement of small sized molecules in an electric field is largely a function of their net charge, paper electrophoresis at *p*H 3.5 serves as a convenient means of separation. Final identification of the nucleotides in question requires comparison with known nucleotides.

The quantity of RNA in a given sample can be determined by measurement of the amount of ribose and subsequent calculations. Mejbaum's quantitative test for pentoses (see p. 129), often called the orcinol determination, can be applied to the nucleotides released by alkaline hydrolysis.

**TABLE VI.** *p*Ka Values of Ionizable Groups of Nucleotides

| Nucleotide | $NH_3^+$ | First OH | Primary Phosphate | Secondary Phosphate |
|---|---|---|---|---|
| Adenylic acid | 3.7 | — | 0.9 | 6.0 |
| Guanylic acid | 2.3 | 9.7 | 0.7 | 5.9 |
| Cytidylic acid | 4.3 | 13.2 | 0.8 | 6.0 |
| Uridylic acid | — | 9.4 | 1.0 | 5.9 |

## Experimental Procedure

### MATERIALS

RNA (from Experiment 25)
20% $HClO_4$
Paper-electrophoresis apparatus
0.01*N* $I_2$:0.01*N* KI
Nucleotide standard ($1 \times 10^{-4}M$ nucleotides)
6% Alcoholic orcinol
0.5*N* KOH
2*N* HCl
Ultraviolet lamp
0.05*M* Ammonium formate
Clinical centrifuge
1.0*N* KOH
10% TCA
1% Starch
Orcinol acid reagent
0.02*M* Nucleotides

#### *Hydrolysis*

Dissolve the material obtained from Experiment 25 in sufficient 0.5*N* KOH to give a concentration of 20 mg/ml. Allow hydrolysis to proceed at room temperature for 24–48 hr. Cool the solution in ice, and titrate to *p*H 1–2 with 20% $HClO_4$ (a few drops of $HClO_4$ per milliliter of original 0.5*N* KOH solution). Remove the precipitated $KClO_4$ and DNA and/or protein by centrifugation at 1000 × g. Adjust the supernatant solution to about *p*H 3.5 with 1.0*N* KOH, and remove any additional precipitate ($KClO_4$) by centrifugation. Use the supernatant solution (crude ribonucleotides) for electrophoretic separation and the orcinol test described below.

#### *Electrophoresis*

Precise instructions for operation of the paper-electrophoresis apparatus depend on the particular design of the unit available.

However, the following steps are common to all units.

1. Sufficient RNA hydrolysate and separate nucleotides are placed on the paper(s) to allow ready location of the material with an ultraviolet lamp. Preliminary spotting of electrophoresis paper with samples assures correct aliquot sizes. (*Note:* The hydrolysate is separated into four components, therefore appropriate allowances should be made for the subdivision of the original U.V. spot.)
2. At *p*H 3.5 nucleotides are anions, therefore samples will be spotted on the electrophoresis paper near the negative pole.
3. When possible, it is advisable to run a control containing a spot of 1% starch solution. Subsequent location of the starch by spraying the control with 0.01*N* $I_2$:0.01*N* KI determines the extent of solvent migration during the separation. Appropriate corrections can then be made.
4. The papers are saturated with 0.05*M* ammonium formate buffer (*p*H 3.5). If this is done after placing the nucleotides on the paper, it is done in such a manner as to minimize solvent migration and accompanying movement of the nucleotides on the papers. To minimize water loss, the papers are then placed either in a closed box saturated with water vapor or in a bath of $CCl_4$.
5. After separation of the materials by the application of a predetermined voltage for an appropriate time interval (for example, 1000 volts for 1 hr), the papers are dried and the compounds located (ultraviolet lamp), and the areas circled with a pencil. The material that moves the greatest distance in any given direction is assigned a relative mobility of 1.0. The relative migration distances for all other components are determined with respect to the farthest moving material and expressed as relative mobilities ($R_f$ values) or decimal fractions of the distance moved by the farthest moving component.

### *Quantitative Measurement: Orcinol Test for RNA*

Dilute the acidified hydrolysate with water to obtain a final concentration of approximately 500 μg of original RNA per milliliter. Remove 0.5-, 1.5-, and 3.0-ml aliquots of the diluted RNA hydrolysate and of the nucleotide standard ($1 \times 10^{-4}M$ in total nucleotides), and make up all samples to 3.0 ml with $H_2O$. In addition, prepare a blank containing 3.0 ml of $H_2O$. Add 6 ml of orcinol acid reagent and 0.4 ml of 6% alcoholic orcinol to each tube, and heat all tubes in a boiling water bath for 20 min. Cool the tubes, and determine the optical density of each at 660 mμ against the blank.

## Report of Results

Determine the relative mobilities of the four nucleotides after correcting for solvent migration, as measured with the starch control. Compare the relative mobilities with those predicted for the four common ribonucleotides of RNA on the basis of their respective charges at *p*H 3.5. To confirm the identity of the RNA isolated, compare the relative mobilities of the U.V.-absorbing compounds released by alkaline hydrolysis with the corresponding values for the known nucleotides.

Using the orcinol reaction data and a residue weight for a nucleotide (within RNA) of 330, calculate the percentage of RNA in the original sample. Discuss what possible errors exist in these calculations.

## Discussion

The instability of RNA and the stability of DNA and phosphoproteins in alkali demonstrated in this experiment form the basis of the Schmidt-Thannhauser procedure for the quantitative determination of RNA, DNA, and phosphoproteins in tissues. In this procedure macromolecules are separated from compounds of low molecular weight by acid precipitation and washing with TCA. The precipitate is then freed of phospholipids and residual TCA by extraction with alcohol:ether and chloroform:methanol. The remaining precipitate, containing RNA, DNA, and phosphoproteins, is treated with 0.3–1.0*N* alkali at room temperature for 15–20 hr. Reacidification of this mixture with HCl results in the precipitation of the alkali-stable DNA, whereas the nucleotides, resulting from RNA hydrolysis, and the inorganic phosphorus, resulting from phosphoprotein hydrolysis, remain in solution. Inorganic phosphorus and total phosphorus determinations on this supernatant solution, and total phosphorus determination on the precipitate, give data which can be used to calculate the amount of RNA, DNA, and phosphoproteins in the tissue.

In part, the success of this procedure is due to the relative stability of nucleic acids in the presence of dilute acid. Only in the presence of concentrated acid solution (or heated dilute acid solutions) are nucleic acids broken down to their sugar pho phates and purine and pyrimidine base

In addition to the methods used in th experiment, there are several other pr cedures for the isolation and characteriz tion of mononucleotides. Of these, t method most widely used for quantitati separation and identification of nucleotid consists of an ion exchange separation the nucleotides and estimation of t amount of each by ultraviolet spectr scopy. At *p*H values near neutrality, all t commonly occurring nucleotides are r tained by strong anion exchange resi such as Dowex-1-formate. Elution of t resin with gradually increasing concentr tions of formic acid produces a sequenti elution and separation of the 2′,3′-nucle tides in the order CMP, AMP, UM GMP. The eluted fractions are examin for the presence of nucleotides by measur ment of their ultraviolet absorbance 260 mμ and 280 mμ. The peaks which a observed are identified by comparison the observed ratios, $A_{280}/A_{260}$, with t known extinction ratios of the nucleotid in acid solution.

After identification, the quantity of n cleotide present in each peak is estimat from the molar extinction coefficient. T composition of the nucleotide mixture ma then be calculated as the mole percent each nucleotide, or in any other units.

Ultraviolet Absorption of 2′,3′-Ribonucleotides at *p*H 2

| Compound | $E_{260} \times 10^{-3}$ | $A_{280}/A_{260}$ |
|---|---|---|
| Adenylic acid | 14.2 | 0.22 |
| Guanylic acid | 11.8 | 0.68 |
| Cytidylic acid | 6.8 | 1.90 |
| Uridylic acid | 9.8 | 0.30 |

**Exercises**

1. Why is the time-consuming $HClO_4$-KOH neutralization necessary in the electrophoresis procedure but unnecessary in the other assays? Could HCl be used to neutralize samples of alkaline-hydrolyze RNA prepared for the electrophores procedure?
2. Using the known *p*Ka values of the n

cleotides of RNA, (a) choose a *p*H at which adenylic and uridylic acids would theoretically move to opposite electrodes in an electric field; (b) discuss the electrophoretic separation which might be expected at *p*H 2 and 8; and (c) calculate the net change of each nucleotide at *p*H 3.5.

3. A tissue preparation is subjected to the Schmidt-Thannhauser procedure for RNA, DNA, and phosphoprotein determination. No precipitate appears upon reacidification (HCl) of the alkaline hydrolysate; the reacidified solution contains the same amount of inorganic phosphate both before and after $H_2SO_4$ hydrolysis (total phosphorus assay). What do you conclude about the extract? Why?

**References**

Schmidt, G., and S. J. Thannhauser, *J. Biol. Chem.*, **161,** 83 (1945).

Mejbaum, W., *Z. physiol. Chem.*, **258,** 117 (1953).

Dische, Z., *J. Biol. Chem.*, **204,** 983 (1953), *In* Chargaff, E., and J. N. Davidson (Editors), *The Nucleic Acids: Chemistry and Biology*, Vol. 1, Academic, New York, pp. 267, 285, 1955.

EXPERIMENT

# 27. Isolation of DNA

## Theory of the Experiment

Necessary steps in the isolation of DNA are:

1. Release of the DNA in soluble form by destruction of cell membranes and membranes of subcellular particles, such as nuclei.
2. Dissociation of DNA-protein complexes by denaturation of protein.
3. Separation of DNA from other macromolecules.

In the present experiment, whole cells of the bacterium *E. coli* are disrupted by the action of the detergent Duponol C (sodium lauryl sulfate), which also frees the DNA from nucleoproteins. Then, because DNA is insoluble in alcohol and soluble in salt solutions, a series of alcohol precipitations and salt extractions may be performed to obtain a purified DNA.

During these manipulations, certain conditions must be observed in order to avoid profound alteration in the DNA structure during isolation. Structural changes in isolated DNA may be produced by a variety of agents. The following general categories of breakdown are relevant to the conditions of DNA isolation:

1. Splitting of phosphodiester bonds
   (a) by DNAase,
   (b) by acid (*p*H $<2$),
   (c) by heat (temperatures above 90°C);
2. Splitting of N-glycosyl linkages between deoxyribose and purines by acid (*p*H $<2$);
3. Destruction of the hydrogen bonds holding together the two strands of DNA
   (a) by alkali (*p*H $>10$),
   (b) by acid (*p*H $<3$),
   (c) by heat (temperatures above 80°C),
   (d) by reduced ionic strength. The denaturation temperature of DNA depends upon the ionic strength of the solution. In distilled water, the denaturation temperature of DNA is less than 25°C.
4. Double-chain scissions caused by mechanical shear.

Extremes of *p*H, high temperature, and low ionic strength are avoided in this experiment. Although *E. coli* contains an active DNAase, as do most animal and plant tissues, DNAase action during the preparation is completely inhibited by (1) Duponol, which denatures the enzyme, and (2) the presence of a chelating agent, citrate ion. Because DNAase requires a divalent cation for activity, the citrate effectively inhibits the enzyme by chelating $Ca^{+2}$ and $Mg^{+2}$.

Degradation of DNA by shearing undoubtedly does occur during this preparation, particularly during homogenization but this type of degradation does not alter the type of primary and secondary structure present in DNA; it merely shortens the molecules.

## Experimental Procedure

### MATERIALS

*E. coli* cells
15% Duponol C in 0.14*M* NaCl:0.01*M* sodium citrate
1.4*M* NaCl
5% Recrystallized Duponol C in 45% aqueous ethanol
75% Ethanol
Potter-Elvehjem homogenizer
0.01*M* Sodium citrate
Absolute ethanol
0.14*M* NaCl:0.015*M* sodium citrate
Solid NaCl
Stirring motor
High-speed centrifuge

### *Isolation*

Slowly stir 8 ml of 0.01*M* sodium citrate into 8 g of freshly thawed *E. coli* cell paste so as to obtain an even suspension. Add 80 ml of 15% Duponol C in 0.14*M* NaCl: 0.01*M* sodium citrate. Homogenize the thick suspension once in a Potter-Elvehjem homogenizer before stirring for 30 min by hand or with a stirring motor. (The initial solution is quite viscous, but this viscosity decreases during the stirring period.)

Slowly add 2 volumes of absolute ethanol, stirring constantly. Collect the resultant precipitate by centrifugation at 2000 × g for 5 min (room temperature). Decant the turbid supernate, and save it. (Do not discard any preparation prior to obtaining a good product). Using a glass homogenizer, blend the precipitate (consider this as one volume) in 1.5 volumes of 1.4*M* NaCl. Centrifuge the resultant milky suspension at room temperature for 25 min at 8000–10,000 × g (high-speed centrifuge). Decant and save the supernate; resuspend the precipitate, by stirring in an additional volume of 1.4*M* NaCl; and centrifuge at 8000–10,000 × g as before. Then combine the first and second clear salt extracts of the precipitate, and slowly add an equal volume of absolute ethanol to the combined extracts, stirring with a glass rod. The crude DNA will begin to precipitate, usually in the form of long fibers which collect on the stirring rod. Continue stirring until precipitation no longer occurs. Store this crude DNA under 75% ethanol, or dry it under vacuum, and store in a freezer. Usually, the sticky and fibrous nature of the crude DNA will indicate that the level of purity is satisfactory for Experiment 28. Only when the crude DNA does not adhere to the stirring rod, or when extreme purity is desired, is the subsequent purification performed.

The crude DNA may be further purified by dissolving and reprecipitating it in the following manner. Remove the alcohol from the crude DNA by decanting or draining and dissolve the residue in a minimum of 0.14*M* NaCl:0.015*M* sodium citrate buffer (*p*H 7.1). Then slowly add $\frac{1}{9}$ volume of 5% Duponol C in 45% aqueous ethanol, and stir the mixture for 1 hr with a stirring motor. Add solid NaCl until the solution contains 5% NaCl. Stir the solution for 30 min, and then store it at 0–4°C in a closed vessel for at least 12 hr. Duponol will precipitate on prolonged standing (more than 12 hr) at the same temperature. Remove any precipitate by centrifugation in the cold (0–4°C) at 20,000–25,000 × g for 30 min. Decant the supernate, and maintain it at 0–4°C while slowly adding 2 volumes of ice-cold absolute ethanol, stirring constantly with a glass rod. The precipitate of fibrous DNA will collect on the rod. Wash the precipitate once in 75% ethanol, then dry it under vacuum before storing in a freezer.

## Report of Results

Calculate the percent yield of DNA from the original wet weight of *E. coli* cells.

## Discussion

In the DNA isolation procedure of this experiment, use is made of the fibrous character of the DNA product. The various salt extractions and alcohol quantities added are somewhat selective towards DNA rather than RNA, but it is the fibrous quality of the DNA during the alcohol precipitation, and the resultant DNA deposition upon the stirring rod, that assures a relatively RNA-free product. RNA forms a flocculent precipitate in the presence of alcohol. The fibrous character of DNA has also allowed application of X-ray crystallographic techniques to the structural analysis of the DNA molecule.

### Exercises

1. What is meant by the terms double-stranded helix and base pairing with respect to the proposed molecular structure of DNA?
2. What is the cytological form in which DNA exists in *E. coli* cells?
3. Do all dividing cells contain DNA? Compare the DNA content of the red blood cells of humans, fish, reptiles, and birds.

### References

Zamenhoff, S. et al., *J. Biol. Chem.*, **219,** 165 (1956).

Lehman, I. R., *J. Biol. Chem.*, **235,** 1479 (1960).

EXPERIMENT

# 28. Partial Characterization of DNA

## Theory of the Experiment

The characterization of DNA can be achieved either directly, by use of the colorimetric diphenylamine assay specific for deoxypentoses, or indirectly, by use of enzymes specific for the hydrolysis of DNA. The commercially available DNAase (deoxyribonuclease) from beef pancreas catalyzes the specific hydrolysis of the 3,5-phosphodiester linkages of DNA, yielding smaller oligonucleotides containing 5′-terminal phosphate groups (see p. 134).

The progress of such a DNA-specific hydrolysis can be observed by study of the release of acid-soluble or dialyzable diphenylamine-positive material (see Allfrey and Mirsky in References). In addition to these methods, changes in viscosity and ultraviolet absorption can be used to measure changes in the structure of DNA upon treatment with DNAase.

The rigid molecular configuration of DNA, which causes DNA solutions to be highly viscous, is a consequence of three types of interactions which act as constraints upon the nucleotides in the DNA chains. These interactions are the van der Waals forces between the "stacked" bases, the phosphodiester bonds that join the nucleotides in each chain, and the hydrogen bonds that join the bases in the two chains.

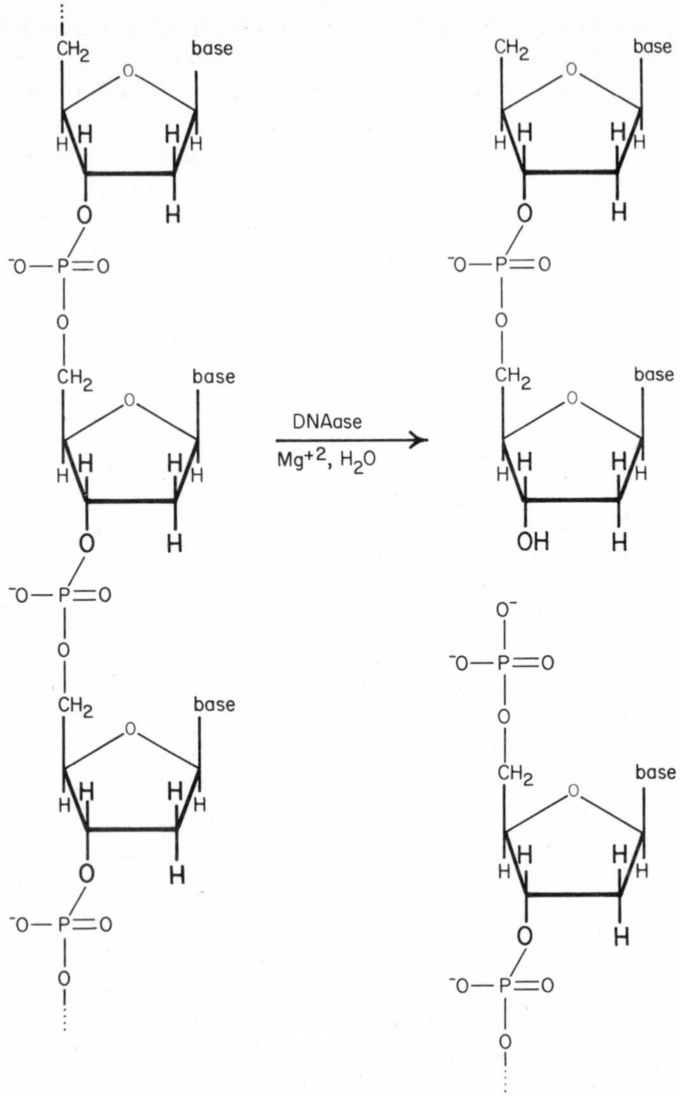

DNAase

$Mg^{+2}$, $H_2O$

The splitting of a small proportion of the phosphodiester bonds by DNAase creates points of flexibility in the DNA molecule, allowing it to assume a more folded configuration. As a result, the viscosity of DNA solutions decreases quite rapidly when DNAase is allowed to act. The decrease in viscosity upon addition of DNAase to a solution thus constitutes a most sensitive qualitative test for DNA.

An alternative method of measuring DNAase action and of studying the secondary structure of DNA is based on the fact that DNA hydrolysis results in up to a 40% increase in the absorption in the ultraviolet between 225 and 300 mμ (see Kunitz in References). This "hyperchromic effect," due either to the disruption of the hydrogen bonding between DNA base pairs or to the disruption of the stacked bases in the core of the DNA molecule, may be brought about by enzymatic hydrolysis of DNA to oligonucleotides or by uncoiling of the double-stranded molecules by heat or by treatment with a strong acid.

In this experiment, three methods—the diphenylamine assay plus the methods based on changes in viscosity and changes in ultraviolet absorption following DNAase treatment—are used to characterize the DNA isolated in Experiment 27.

## Experimental Procedure

MATERIALS

DNA from Expt. 27
10% TCA
$5 \times 10^{-4}M$ Methylene blue in $0.1M$ $MgSO_4$
DNAase (solid or freshly prepared soln.)
$0.1M$ Sodium acetate (*p*H 5.5)
DNA standard (1 mg/ml, pretreated with TCA)
Diphenylamine reagent
$0.01M$ $MgSO_4$:$0.5M$ sodium acetate (*p*H 5.5)

### *Diphenylamine Assay*

Weigh out 5 mg ±0.1 mg of DNA isolated in Experiment 27, and suspend this in 5 ml of 10% TCA. Heat this suspension in a water bath at 90–95°C for 15 min. Next, make up to 3.0 ml with $H_2O$ the following: a water blank, 0.4 and 0.8 ml of authentic DNA standard, and 0.2-, 0.6-, and 1.0-ml aliquots of the TCA-treated, isolated DNA. Add 6 ml of the diphenylamine reagent to each tube, and after mixing, heat the tubes in a boiling water bath for 10 min. Cool the tubes, and then determine the optical densities at 600 mμ after setting the colorimeter at zero optical density with the reagent blank.

### *Viscosity Measurement*

Add sufficient dry or moist DNA to 2–4 ml of $0.1M$ acetate buffer (*p*H 5.5) to make a 2% solution. Gently stir the suspension to dampen all the DNA, adjust to *p*H 5–6 if necessary (*p*H paper), and then homogenize the suspension in a small hand homogenizer until a thick suspension is obtained. Pour the suspension into a test tube, and mix in 1 or 2 drops of methylene blue solution. Draw the solution into a 0.1-ml pipette or small-bore glass tube about 1 mm in diameter. Record the time, in seconds, required for the solution to fall some fixed distance (for example, 10–15 cm) in the upper part of the tube. Hold the end of the pipette or tube in the solution during the viscosity determination. Compare this time with the time required for $0.1M$ acetate buffer to fall the same distance, and increase the DNA concentration if the DNA solution is not appreciably more viscous than the buffer blank.

When you have obtained reproducible results, add 1 μg of freshly prepared DNAase in $0.1M$ sodium acetate buffer (*p*H 5.5—that is, 0.01 ml of a solution containing 1 mg DNAase/10 ml) to the DNA solution, and measure the viscosity (as above) at 1- to 5-min intervals over a 20-min period. If the enzyme concentration is too high or too low, repeat the process with another enzyme concentration so as to obtain satisfactory results.

### *Spectrophotometric Assay*

Place 2.25 ml of an approximately 0.01% solution of the isolated, highly polymerized DNA in $0.01M$ $MgSO_4$:$0.5M$ sodium acetate (*p*H 5.5). This should yield $OD_{260}$ values (quartz cuvettes) in the range 0.2–0.5 if you use a water blank. (Make appropriate dilutions or additions if necessary). Add 0.75 ml of water to the first cuvette, and after mixing, record the optical density. Add 0.75 ml of a solution containing 5–10 μg of DNAase to the second cuvette, and record the optical density at 260 mμ at 1-min intervals for 10 min. If the enzyme concentration is too high or too low, make additional assays with different enzyme concentrations until a readily measurable rate is obtained.

## Report of Results

Using the diphenylamine data, calculate the purity of your isolated DNA with reference to the DNA standard. Present the viscosity data by plotting the time in seconds required for the DNA solution to fall 1 cm versus the duration of DNAase treatment. From these data, estimate the order of the hydrolysis of DNA under the influence of DNAase. To illustrate the hyperchromic effect, plot the $OD_{260}$ against the duration of DNAase treatment, and calculate the percent increase of the $OD_{260}$.

## Discussion

The hyperchromic effect and the decrease in viscosity observed in this experiment resulted from the enzymatic hydrolysis of DNA to oligonucleotides. During the course of this hydrolysis, both the secondary structure (interchain hydrogen bonds) and primary structure (intrachain phosphodiester bonds) were largely destroyed. Both of these changes contribute to the hyperchromic effect, whereas the viscosity decrease is mainly a reflection of the loss of secondary structure.

The relation of the hyperchromic effect to DNA secondary structure may be studied directly by using heat, rather than DNAase, to destroy this structure. If measurements of the U.V. absorption of a DNA solution are made as a function of temperature, three distinct regions are observed in the curve of $OD_{260}$ versus temperature:

1. a region in which the U.V. absorption is constant (at low temperature);
2. a sudden 40% rise of $OD_{260}$ occurring within a 5° temperature range at about 85°C (dependent upon ionic strength);
3. a "plateau," in which no further $OD_{260}$ increase occurs.

If measurements of the viscosity of the DNA solution are made *after* heating to each temperature for one hour, it is ob served that heating DNA has no effec upon the viscosity until a temperature o 85°C is reached. Thereupon, the viscosit decreases greatly as the temperature in creases, falling to less than 10% of th original value by 100°C.

The changes in DNA secondary struc ture produced by heat have been cor related with a loss of biological activity DNA extracted from streptomycin-resis tant *Diplococcus pneumoniae* has the abilit to transmit drug resistance to strepto mycin-sensitive *pneumococci*. This "trans forming" DNA loses all of its ability t confer streptomycin resistance when it i heated at temperatures of 90°C or abov and then cooled quickly.

Recently it has been possible to revers the thermal inactivation of transformin DNA. By a process of slow cooling, Mar mur and Lane were able to restore 25% of the transforming activity to a DN preparation which had been inactivated b heating at 100°C.

The physical studies of Doty et al would suggest that the U.V. absorptio and the viscosity of DNA change wit heating because the two strands of th DNA molecule are separated by heating A recombination of the two strands t form a native DNA helix is suggested a an explanation for restoration of trans forming activity by slow cooling.

### Exercises

1. What are the products of acid hydrolysis of DNA?
2. What components of the DNA molecule are responsible for the observed ultraviolet absorption at 260 mμ?
3. As implied in the discussion, allowing heated (and therefore single-stranded) streptomycin-resistant DNA to cool slowly in the presence of heated, or single-stranded, streptomycin-susceptible DNA increases the amount of active DNA available for bacterial transformation. To what do you attribute this be havior?

### References

Allfrey, V., and A. E. Mirsky, *J. Gen Physiol.*, **36,** 227 (1952).
Doty, P., *Rev. of Mod. Phys.*, **31,** 107 (1959)
———, et al., *Proc. Natl. Acad. Sci. U.S* **46,** 461 (1960).
Kunitz, M., *J. Gen. Physiol.*, **33,** 349 (1950)
Marmur, J., and D. Lane, *Proc. Natl. Acad Sci. U.S.*, **46,** 453 (1960).

# PART FIVE **METABOLISM**

## The Cell and Cellular Requirements

Metabolism deals with the ways in which cells derive and use chemical compounds to sustain life. The general pattern of this process involves the uptake of extracellular compounds, followed by intracellular utilization of the compounds to satisfy *energetic* and *synthetic* requirements of the cells, and the eventual excretion from the cell of the end products of intracellular reactions.

The number of compounds needed by cells to support intracellular reactions is often small; from these few materials, all of the complex cellular components are synthesized. For example, plants and certain microorganisms can live in environments containing a very limited number of compounds—$CO_2$, salts, and water in plants; salts and a single organic compound as an energy source in certain microorganisms.

Higher forms of animal life, and certain microorganisms, also satisfy their energy requirements and many of their synthetic requirements by use of a limited number of organic compounds, but these forms also require small quantities of other materials (nutrients) to support growth and maintain cells. In general, these more diversified needs reflect the lack of capacity to synthesize particular compounds needed for the intracellular reactions of the cells. For example, all animal cells need small quantities of many amino acids (see Experiment 18) and smaller quantities of certain vitamins.

## Biochemical Pathways

Many organic compounds can serve as the source of energy and of carbon for biosynthetic processes. This is particularly true in microorganisms, where adaptation to a particular energy source or substrate is common. The reactions involved in glucose utilization illustrate one of the pathways commonly used by cells to satisfy the energy needs and to supply appropriate compounds for cellular synthesis.

In cells, glucose is usually converted to compounds of a lower energy state (for example, lactate or ethanol plus $CO_2$) by a sequence of stepwise reactions. These reactions are called anaerobic glycolysis (see Experiment 30).

The term $\Delta F'$ refers to the free energy change with all reactants and products at unit activity except $H^+$, which is at $1 \times 10^{-7} M$ (pH 7).

There is only a small overall free energy change ($-\Delta F'$) in these reactions when compared to the energy made available by the complete oxidation of glucose (reaction *3*).

Accordingly, only a small fraction of the

$$(1)\quad C_6H_{12}O_6 \xrightarrow{\text{animal tissue}} 2CH_3\text{—}\underset{\substack{|\\ H}}{\overset{\substack{OH\\ |}}{C}}\text{—}COOH \qquad \Delta F' = -47.4'\ \text{kcal/mole}$$

L (+) Lactic acid

$$(2)\quad C_6H_{12}O_6 \xrightarrow{\text{anaerobic yeast}} 2C_2H_5OH + 2CO_2 \qquad \Delta F' = -56.1'\ \text{kcal/mole}$$

$$(3)\quad C_6H_{12}O_6 + 6O_2 \longrightarrow 6CO_2 + 6H_2O \qquad \Delta F' = -686\ \text{kcal/mole}$$

energy released is made available by glycolysis. Such glycolytic reactions are used by anaerobic organisms as a source of energy. Often the products of such anaerobic glycolysis are more varied than lactate or ethanol. For example, formic acid, acetic acid, $H_2$, and acetylmethylcarbinol are metabolic end products in certain microorganisms.

Most aerobic organisms contain, in addition to a glycolytic system, a collection of enzymes which catalyze the complete aerobic oxidation of glucose (reaction *3*). Such organisms use the products of glycolysis as substrates for a series of oxidative steps. These steps involve the conversion of the organic glycolytic products into $CO_2$ and $H_2O$ by means of a series of decarboxylations, hydrations, and dehydrogenations (oxidations). These steps proceed in a cyclic manner, leading to a continuing utilization of glycolytic products. This cycle is called the tricarboxylic acid cycle, the citric acid cycle, or the Krebs cycle.

During the course of the steps mentioned above, a series of dehydrogenations or oxidations takes place in which electrons and hydrogen ions are removed from the substrates and transferred to a series of electron carriers (reactions *4*, *5*, and *6*). This passage of electrons from substrate through biological carriers to oxygen is called electron transport (see Experiment 31). Electron transport to oxygen results

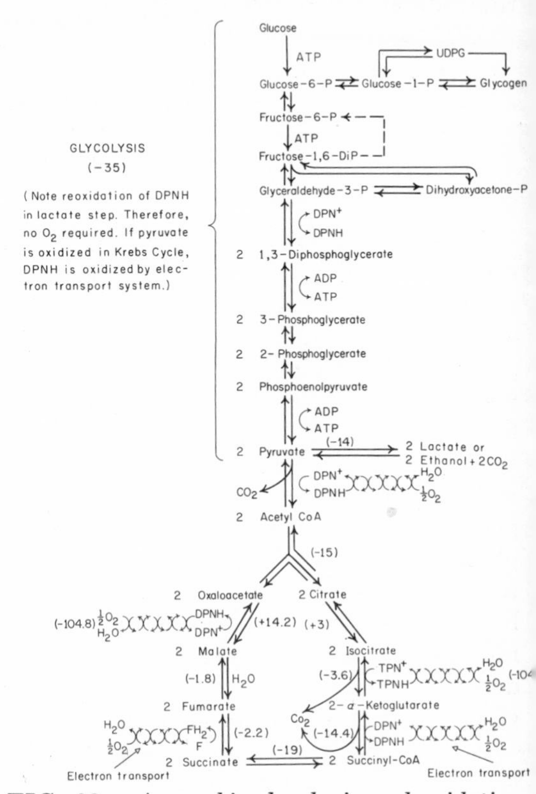


FIG. 22. *Anaerobic glycolysis and oxidation.*

$$(4)\qquad AH_2 \rightleftarrows A + 2H^+ + 2e^-$$

$$(5)\qquad 2H^+ + 2e^- + DPN^+ \rightleftarrows DPNH + H^+$$

$$(6)\qquad \text{(Sum)}\ \underset{\text{(reduced substrate)}}{AH_2} + \underset{\text{(oxidized carrier)}}{DPN^+} \rightleftarrows \underset{\text{(oxidized substrate)}}{A} + \underset{\text{(reduced carrier)}}{DPNH} + H^+$$

Electrons are then passed on from the initial electron acceptor to other acceptors and eventually to oxygen to yield water in the reaction:

$$(7)\qquad 2e^- + 2H^+ + \tfrac{1}{2}O_2 \rightleftarrows H_2O$$

These reactions are summarized schematically in reaction *8*.

(8)

| reduced substrate → oxidized substrate | oxidized first carrier → reduced first carrier | reduced second carrier → oxidized second carrier | oxidized third carrier → reduced third carrier | $\tfrac{1}{2}O_2$ → $H_2O$ |
|---|---|---|---|---|

in the requirement for oxygen by aerobic organisms and leads to the formation of the water in the aerobic respiration of reaction *3*.

The interrelationships of these pathways—glycolysis, the Krebs cycle, and electron transport—are illustrated in Fig. 22.

### *Energetics of Pathways*

As can be seen, there is energy release in each of the pathways. For a mole of glucose this breaks down as follows:

| | $\Delta F'$ |
|---|---|
| Sum of energy released during glycolysis (excluding lactate formation) | −35.0 |
| Sum of energy released during Krebs cycle | −57.6 |
| Sum of energy released during electron transport | −594.4 |
| | −687.0 kcal/mole[a] |

It is this energy released during the exergonic processes of metabolism that is used by cells and organisms to drive endergonic reactions. During the course of glucose utilization by means of these pathways, some of this energy is released as heat. The remainder is transformed into a special kind of biologically utilizable chemical energy which is stored in the phosphate anhydride bonds of adenosine triphosphate (ATP). These bonds have been termed "energy-rich," or "high-energy," bonds because their hydrolysis releases much larger amounts of free energy than does the hydrolysis of amide, ester, or glycosidic bonds. For example:

*(9)* $ATP + H_2O \longrightarrow ADP + Pi$ $\Delta F' = -8.9$ kcal/mole at $p$H 7.5

*(10)* $\text{Glycylglycine} + H_2O \longrightarrow 2\ \text{glycine}$ $\Delta F' = -3.6$ kcal/mole

Reaction *9* describes the hydrolysis of an energy-rich bond (an anhydride), whereas reaction *10* is an example of the hydrolysis of a low-energy bond (an amide or ester as a rule).

Specific enzymes within cells catalyze the endergonic reactions essential for life (see Experiments 32 and 33). During the course of enzyme-controlled "coupled reactions," an anhydride or high energy bond of ATP is ruptured while the endergonic reaction (for example, the synthesis of a new bond) proceeds. The energy difference between the two reactions is released as heat. Reactions *11*, *12*, and *13* illustrate the energetics (but not the mechanism) of the two half-reactions and the entire reaction catalyzed by such a single enzyme, aceto-CoA kinase of yeast and liver.

*(11)* $ATP^{-4} + H_2O \longrightarrow AMP^{-2} + HP_2O_7^{-3} + H^+$ $\Delta F' = -9.5$ kcal/mole

*(12)* $\text{Acetate}^- + H^+ + CoA \longrightarrow \text{acetyl CoA} + H_2O$ $\Delta F' = +8.2$ kcal/mole

*(13)* $ATP^{-4} + \text{acetate}^- + CoA \longrightarrow \text{acetyl CoA} + AMP^{-2} + HP_2O_7^{-3}$ $\Delta F' = -1.3$ kcal/mole

### *Other Reactions*

The reactions outlined above, which briefly depict the means of glucose degradation, ATP synthesis, and the driving of endergonic reactions, are an important part of metabolism; but by no means do they reflect the entire process. Many other compounds can lead to or be synthesized from the compounds mentioned in this scheme. In fact, metabolism includes all the chemical reactions by which amino acids and proteins, fatty acids and lipids, nucleotides and nucleic acids, sugars and polysaccharides, and so on, are synthesized and degraded.

### *Coenzymes Involved in Metabolism*

Several enzymatic steps involved in metabolic pathways require as chemical reactants specific small organic molecules—for example, $DPN^+$, $TPN^+$, Coenzyme A, Coenzyme Q, ADP, ATP, FMN, and FAD in addition to the substrate derived from the previous step in the pathway (see p. 140).

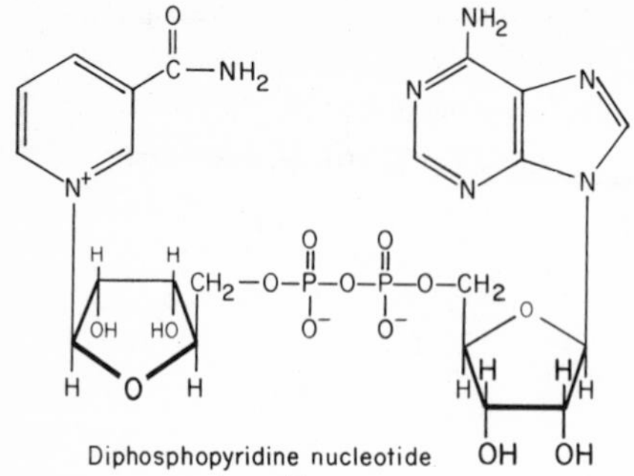

Diphosphopyridine nucleotide ($DPN^+$)

DPNH

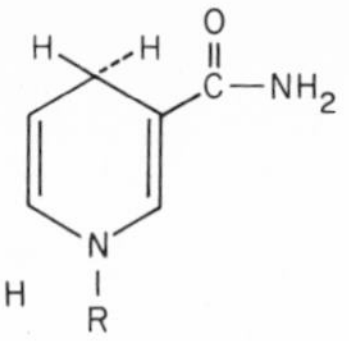

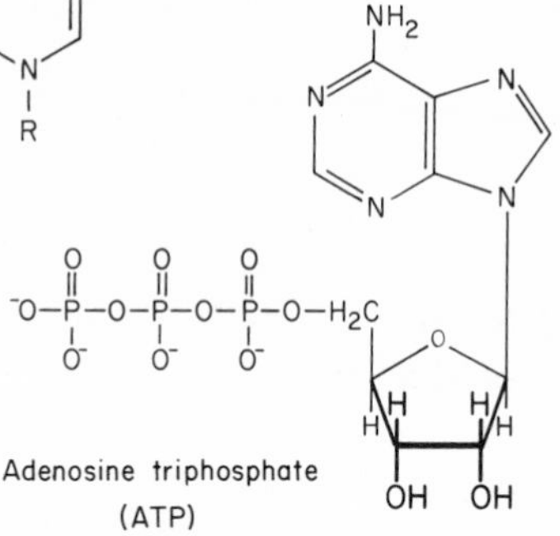

Adenosine triphosphate (ATP)

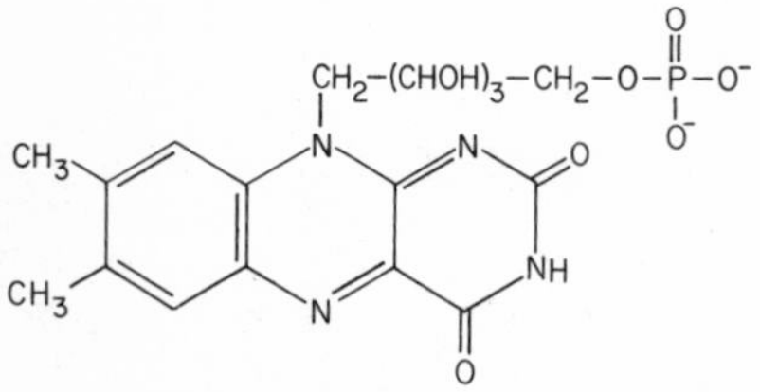

Flavin mononucleotide (FMN)

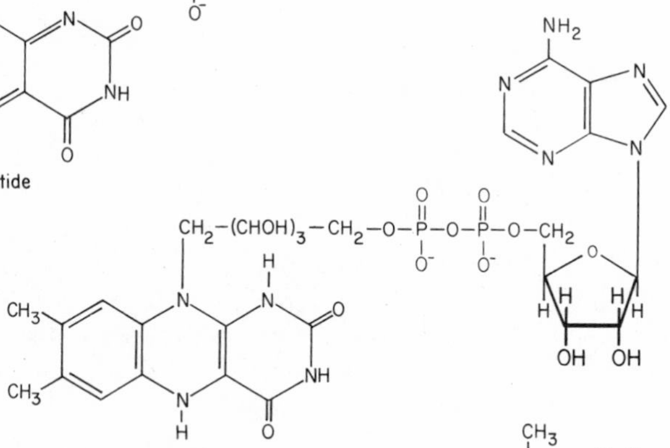

Reduced flavin Adenine dinucleotide ($FADH_2$)

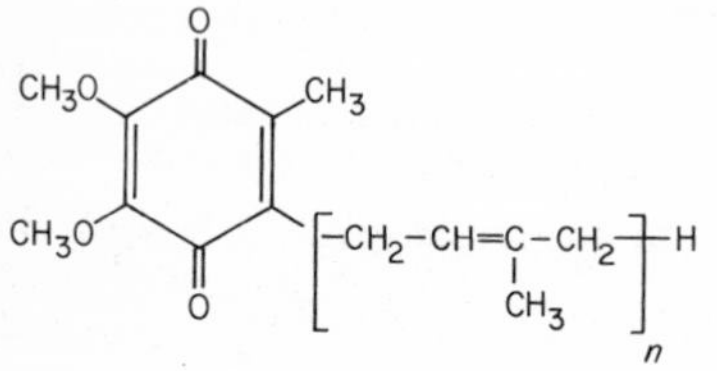

Coenzyme Q (ubiquinone)

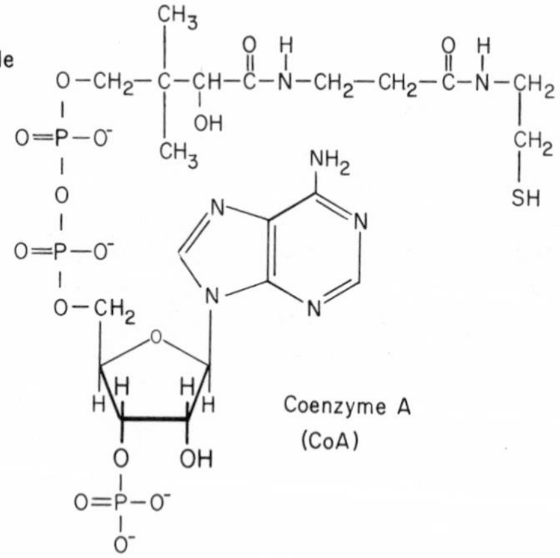

Coenzyme A (CoA)

Some of these materials (particularly FAD and FMN) are tightly bound to the enzymes involved (for example, they are not removed by dialysis) and have therefore been considered a part of the enzyme and given the designation *prosthetic groups* to distinguish them from the protein or apoenzyme portion of the enzyme molecule. Other of these accessory molecules, however, are not much more tightly bound than the normal substrate.

These small molecules (coenzymes) act as specific reactants in the particular reactions involved—for example, ADP in the pyruvate kinase reaction of glycolysis (see p. 141).

$$\underset{\text{Phosphenolpyruvate (PEP)}}{CH_2{=}C(COO^-){-}O{-}PO_3^{2-}} + ADP^{-2} \underset{}{\overset{Mg^{+2}}{\rightleftharpoons}} \underset{\text{Pyruvate}}{CH_3{-}C({=}O){-}COO^-} + ATP^{-4}$$

or $DPN^+$ in the malic dehydrogenase reaction of the Krebs cycle,

$$\underset{\text{Malic acid}}{{}^-OOC{-}CH_2{-}CH(OH){-}COO^-} + DPN^+ \rightleftharpoons \underset{\text{Oxaloacetic acid}}{{}^-OOC{-}CH_2{-}C({=}O){-}COO^-} + DPNH + H^+$$

and are therefore as necessary as the substrate in a reaction.

In general, the supply of these accessory molecules in any given cell need not be high, for each is being continually regenerated by other metabolic steps in the cell. For example, in electron transport the supply of each member of the chain is constantly cycling between oxidized and reduced forms. The supply of ATP is depleted by the various ATP-utilizing steps in cells (see Experiment 33) and repleted by the steps of glycolysis and oxidative phosphorylation (see Experiments 30 and 32).

## References

Krebs, H. A., and H. L. Kornberg, *Energy Transformations in Living Matter*, Springer-Verlag, Berlin, 1957.

Fruton, J. S., and S. Simmonds, *General Biochemistry* (2nd ed.), Chapters 15, 19–21, Wiley, New York, 1958.

White, A., P. Handler, E. L. Smith, and D. J. Stetten, *Principles of Biochemistry* (2nd ed.), Chapters 16 and 17, McGraw-Hill, New York, 1958.

Conn, E. E., and P. K. Stumpf, *Outlines of Biochemistry*, Wiley, New York, 1963.

Karlson, P., *Introduction to Modern Biochemistry*, Academic, New York, 1963.

EXPERIMENT

# 29. Manometry: Calibration of Warburg Flasks and Manometers

## Theory of the Experiment

The Warburg constant-volume respirometer is used to measure gas changes. This device, shown in Fig. 23, consists of a Warburg flask attached to the inner arm of a manometer, which is fitted with a three-way stopcock. The Warburg flask consists of a main vessel with a center well. Attached to the main vessel is a side arm, which is equipped with a gas vent. The flask may be opened to the air by twisting the gas vent so that the hole is lined up with the slot in the ground-glass joint of the side arm. In operation the three-way stopcock and the gas vent are turned so that the system is sealed; before reading, the fluid reservoir thumbscrew is adjusted so that the fluid level in the inner arm of the manometer (the closed end) comes to a fixed position, usually 150 mm on the scale. Thus the volume of the system remains constant during all observations. Gas evolution or uptake is measured by *changes* in the height of the column of fluid in the outer arm of the manometer. The volume of gas exchanged is obtained by multiplying the change in the height of the fluid column of the outer arm by a factor known as the flask constant, or $k$. This factor, relating gas exchange and change in the outer arm of a constant-volume respirometer, is a function of the density of the manometer fluid, the temperature, the particular gas exchanged, and the volume of the gas and liquid phases.

### *Derivation of Flask Constant Expression*

From Boyle's law:

$$PV = RT$$

where $P$ is the gas pressure in millimeters of manometer fluid, $V$ is the volume of gas, $R$ is the gas constant, and $T$ is the absolute temperature (273°K + t°C), we have

$$\frac{PV}{T} = R$$

$$\frac{PV}{T} = \frac{P_0V_0}{T_0}$$

where the subscript zero refers to *standard conditions* (one atmosphere, (760 mm of Hg) and 273°K). We can then write

$$V_{0g} = V_g\left(\frac{T_0}{T}\right)\left(\frac{P}{P_0}\right)$$

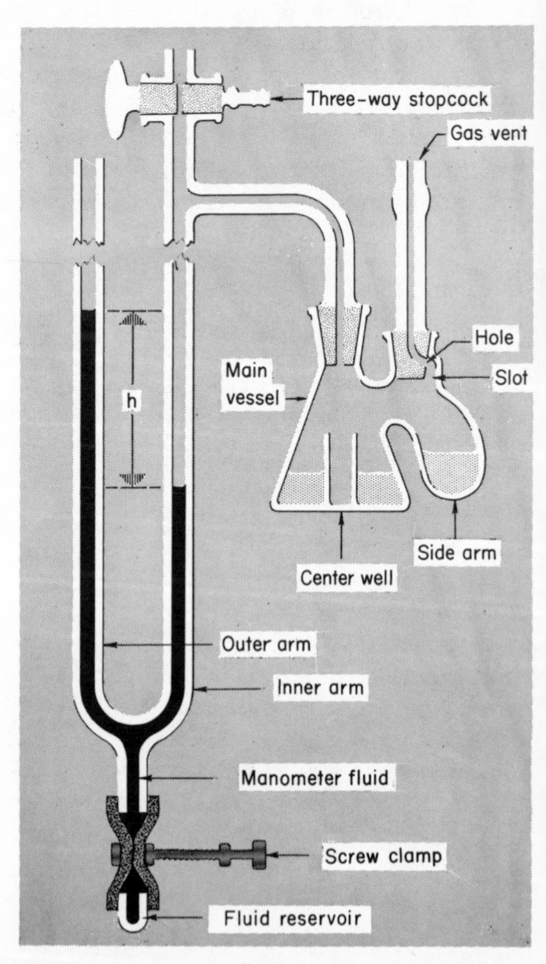


FIG. 23. *Warburg constant-volume respirometer.*

where $V_{0g}$ is the volume of the gas phase at standard conditions, and $V_g$ is the volume of the gas phase under experimental conditions.

Because of the vapor pressure, $r$, of the fluid in the Warburg flask, the actual value of $V_{0g}$ is

$$V_{0g} = V_g\left(\frac{T_0}{T}\right)\left(\frac{P-r}{P_0}\right)$$

Gases will dissolve in the fluid in the flask. The concentration of a dissolved gas is directly proportional to the partial pressure of that gas above the fluid (Henry's law).

Thus

$$V_{0f} = (V_f)\ (\alpha)\ \left(\frac{P-r}{P_0}\right)$$

where $V_{0f}$ is the volume of the gas dissolved in a fluid having a volume $V_f$ at standard conditions. Here $\alpha$ is the Bunsen solubility coefficient, an empirically determined constant which is a function of temperature and of the concentration of dissolved salts. This coefficient has the units of milliliters of gas at standard conditions dissolved in one milliliter of fluid at a given temperature. It should be noted that the solubility of a gas, although affected by the presence of salts, is practically independent of the presence of other dissolved gases.

The total volume of gas in the Warburg flask and manometer is

$$V_0 = V_{0g} + V_{0f} = V_g\left(\frac{T_0}{T}\right)\left(\frac{P-r}{P_0}\right) + V_f\ (\alpha)\ \left(\frac{P-r}{P_0}\right)$$

After a gas exchange occurs and the fluid in the inner arm of the manometer is readjusted to the original reference point (for example, 150 mm), the height of fluid in the outer arm of the manometer will change by $h$ mm, and the pressure in the closed system will be $(P - r) - h$ (written as a gas *uptake*). The total volume of gas at the conclusion of the gas exchange will be

$$V_0' = V_g\left(\frac{T_0}{T}\right)\left(\frac{(P-r)-h}{P_0}\right) + V_f\ (\alpha)\left(\frac{(P-r)-h}{P_0}\right)$$

Accordingly, the volume of gas exchanged, $X$, is equal to the difference between $V_0$ and $V_0'$.

$$X = V_0 - V_0' = \left[V_g\left(\frac{T_0}{T}\right)\left(\frac{P-r}{P_0}\right) + V_f(\alpha)\left(\frac{P-r}{P_0}\right)\right]$$
$$- \left[V_g\left(\frac{T_0}{T}\right)\left(\frac{(P-r)-h}{P_0}\right) + V_f(\alpha)\left(\frac{(P-r)-h}{P_0}\right)\right]$$

$$X = \left(\frac{V_g}{P_0}\right)\left(\frac{T_0}{T}\right)(h) + \left(\frac{V_f}{P_0}\right)(\alpha)\,(h)$$

$$X = h\left[\frac{V_g\left(\frac{T_0}{T}\right) + V_f(\alpha)}{P_0}\right] = hk$$

The quantity $k$ is the flask constant. The term $h$ is expressed as millimeters of manometer fluid. The temperatures $T_0$ and $T$ are 273°K and 273° + t°C, respectively. The volumes $V_g$ and $V_f$ are expressed as microliters, and $P_0$ is expressed as millimeters of manometer fluid. Since the manometer fluid (Krebs formulation) has a density of 1.033 g/ml, and the density of Hg is 13.6 g/ml,

$$P_0 = 760 \times \frac{13.6}{1.033} = 10{,}000 \text{ mm}$$

The constant $\alpha$ is expressed in terms of microliters of gas per millimeter of fluid. The value of $X$ is thus the volume change of gas *at standard conditions* and has the dimension of microliters ($\mu$l). This means that $X$ divided by 22.4 is equivalent to the number of micromoles of gas exchanged.

## *Evaluation of Flask Constant*

The flask constant $k$ can be determined by an evaluation of $V_g$, the volume of the gas phase, since the other terms are known (see Table VII, p. 145, for $\alpha$). The volume of the gas phase is the difference between the total volume of the system (including the volume of the portion above 150 mm in the inner arm of the manometer) and the volume of the fluid in the flask, $V_f$.

The most accurate method of determining the total volume is to weigh the amount of mercury required to fill the system, but this is tedious and time consuming. A simpler method is to find $h$ (the change in millimeters of fluid in the outer arm) when a known amount of gas ($X$) is exchanged. Since $h$ and $X$ are known, as are $T$, $\alpha$, $V_f$, then $V_g$, the total volume of the flask, can be calculated.

In this experiment, the oxidation of hydrazine by ferricyanide is used to determine $V_g$, and hence $k$.

$$N_2H_4 + 4Fe(CN)_6^{-3} \longrightarrow N_2 + 4Fe(CN)_6^{-4} + 4H^+$$

Since this is a quantitative conversion, the quantity of nitrogen evolved ($X$) in the presence of excess hydrazine is calculated from the amount of ferricyanide added to the reaction mixture.

## Experimental Procedure

### MATERIALS

Warburg constant-temperature bath and shaker
Potassium ferricyanide reagent (8.23 g/l)
Krebs manometer fluid
Stopcock grease
Pipe cleaners
Manometer, manometer supports, and Warburg flasks
Hydrazine reagent
Anhydrous lanolin
Detergent powder
$CHCl_3$

Place 2.0 ml (quantitative) of the potassium ferricyanide reagent (8.23 g/l) and 0.6 ml of water in the main vessel (not the center well) of a Warburg flask. Place 0.5 ml of the hydrazine reagent in the side arm. After removing old grease with $CHCl_3$, apply sufficient anhydrous lanolin to the sides of the glass joint of the manometer to insure a tight seal. Seal the joint of the flask to the manometer with a rotary motion. Be sure that the anhydrous lanolin is evenly distributed on the joint with no lines or streaks. Then secure the flask to the manometer with a rubber band or steel spring. Grease the gas vent joint with lanolin, insert, and seal by rotating to the appropriate position to close the system from the atmosphere. Apply stopcock grease, not lanolin, to the three-way stopcock, taking care not to plug the bore of the stopcock with grease. Remove any grease plugs with a pipe cleaner dampened with $CHCl_3$. Turn the stopcock so that the system is open to the air, and place the assembled respirometer in the constant-temperature bath in such a way as to avoid mixing the two reagents in the Warburg flask. Prepare a control flask, termed a thermobarometer, by placing 3.1 ml of $H_2O$ in a Warburg flask of similar size and attaching the flask to a manometer. Place the thermobarometer beside the first respirometer. In practice, one thermobarometer is sufficient for several "experimental" respirometers in the same bath. Allow the respirometer and thermobarometer to equilibrate in the constant-temperature bath while shaking them for 5 min with the stopcocks open.

Seal the respirometer, but not the thermobarometer, by turning the stopcock. Since gas is to be evolved in this experiment, it is advisable here to utilize the entire scale of the manometer; therefore, seal the system, and adjust the level of the manometer fluid so that the fluid level in the outer arm is near the lowest mark on the scale while the fluid level of the inner arm is at 150 mm. To accomplish this, run the fluid (in both arms) up as high as possible with the stopcock open by compressing the rubber fluid reservoir, or by attaching a small rubber suction bulb to the stem of the stopcock. Close the stopcock, thereby sealing the system, and then adjust the fluid level of the inner arm to 150 mm by loosening the thumbscrew clamp on the fluid reservoir.

Make an initial reading of the respirometer. Since all readings are constant volume readings, all readings of the fluid level of the outer arm must be preceded by an adjustment of the inner arm to some fixed point, in this case 150 mm. Continue to note and record the manometer reading at 2-min intervals until the fluid level of the outer arm remains essentially constant. Then adjust the inner arm of the thermobarometer to 150 mm, close the stopcock, and record the fluid level of the outer arm of the thermobarometer. Both the respirometer and the thermobarometer are now temperature equilibrated. Any changes in gas volume due to temperature and atmospheric pressure variation will be reflected by the thermobarometer readings, which can be used to correct the corresponding respirometer reading.

After the respirometer and thermobarometer are equilibrated, tip the contents of the side arm of the calibration flask into the main vessel in the following manner. First place the index finger firmly over the tip of the outer arm of the manometer. Then quickly withdraw the respirometer from the bath in an upright direction, and, still covering the tip of the outer arm, tilt the flask so that the contents of the side arm run into the main vessel. Then tilt the flask in the reverse direction, allowing the contents of the main vessel to flow into the side arm. Quickly repeat emptying the side arm into the main vessel, allowing the entire contents to drain into the main vessel, and return the respirometer to the constant-temperature bath. This entire operation should be performed quickly; practice with an empty respirometer if you are not sure of the procedure.

After shaking the flask for 2 min in the bath, take readings of both the thermobarometer and respirometer after adjusting the fluid level to 150 mm as before. Repeat readings at 2-min intervals until the reaction is complete, as indicated by the cessation of gas evolution.

At the conclusion of the experiment *open the stopcock*, and note that the fluid levels equalize. (It is essential to open the respirometers before removing them from the bath. Why?) Remove the respirometers from the bath, and clean the flask by removing the gas vents and rinsing the flasks with water. Then immerse the flasks and gas vents for 5 min in hot detergent solution. Avoid prolonged immersion: alkaline detergent solutions etch glass. Rinse the flasks with distilled water, and allow them to drain until dry.

## Report of Results

Prepare a graph showing the volume of gas exchange in millimeters versus time in minutes.

Using the known volume ($X$) for the gas evolved [calculated from the known quantity of $K_3Fe(CN)_6$ in the flask], $V_f$, $\alpha$ from Table VII, the temperature, and the change in the volume of fluid in the outer arm in millimeters ($h$), calculate the volume of gas ($V_g$) in the system (the flask and the inner arm back to the 150-mm mark). Then using this value of $V_g$, the $\alpha$ values given in Table VII, and a fluid volume of 3.1 ml, calculate values for the flask constant at the temperature of the bath for $O_2$ and $CO_2$. Record these values for use in future experiments.

**TABLE VII.** Values of $\alpha$ (Bunsen Solubility Coefficients) for $O_2$, $CO_2$, and $N_2$. Solubility of Gases in Water.

| Temperature °C | $O_2$ | $CO_2$ | $N_2$ |
|---|---|---|---|
| 20 | 0.0310 | 0.878 | 0.0152 |
| 25 | 0.0283 | 0.759 | 0.0143 |
| 30 | 0.0261 | 0.665 | 0.0134 |
| 35 | 0.0244 | 0.592 | 0.0126 |
| 37 | 0.0239 | 0.567 | 0.0123 |
| 40 | 0.0231 | 0.530 | 0.0118 |

## Discussion

Warburg measurements are of great value in metabolic studies of many reactions which involve gas exchanges. For example, glycolysis can be assayed by measuring $CO_2$ evolution (see Experiment 30), whereas aerobic respiration of mitochondria is assayed by measuring $O_2$ uptake (see Experiments 31 and 32). Decarboxylations are assayed by measurement of $CO_2$ evolution, and organic acid production can be measured as $CO_2$ released from a bicarbonate buffer.

When studying these reactions, it is best to use the maximum possible scale on the outer manometer arm in order to allow greater latitude in enzyme and substrate concentration. In this experiment—a gas evolution—the outer arm was initially adjusted to a low position to make use of the full scale. In gas consumption work, such as Experiments 31 and 32, the outer arm should be started at a high position (with the inner arm still at 150 mm) for full-scale use.

### Exercises

1. Why is it necessary to open the respirometer stopcock before removing the respirometer from the constant-temperature bath?
2. A student has lost the value of his respirometer flask constant for $O_2$ consumption at 37°C in a system containing 2.0 ml of liquid, but he remembers that the value for the same flask at 37°C and 3.1 ml of fluid volume for a $CO_2$ evolving system is 1.37. Can he calculate the desired $O_2$ flask constant, or must he recalibrate the respirometer?
3. Any given flask constant is a function of several variables. List four of these that are unique to each individual flask constant (that is, if changed, would make a new assay of the flask constant necessary).

### Reference

Umbreit, W. W., R. Burris, and J. F. Stauffer, *Manometric Techniques*, Burgess, Minneapolis, 1957.

EXPERIMENT

# 30. Glycolysis in a Cell-free Yeast Extract

## Theory of the Experiment

Many biological systems utilize glucose or other monosaccharides as an energy source, and concomitantly produce simpler molecules, such as lactic acid, pyruvic acid, ethanol, and $CO_2$, which are passed into the extracellular medium or consumed by other enzyme systems. The transformation of glucose and other monosaccharides and the production of simpler molecules by cleavage of C—C bonds is called glycolysis. This series of experiments is designed to demonstrate features of this multienzyme system.

The products of glycolysis are many and varied, depending upon the enzyme systems and the prevailing conditions. Glycolysis occurs in anaerobic organisms without any change in the overall oxidation state of the substrates. Thus in intact yeast cells under anaerobic conditions, the overall glycolytic reaction (or fermentation) may be written

(1) Glucose $\longrightarrow$ 2 ethanol + $2CO_2$

$$\Delta F' = -56 \text{ kcal/mole}$$

whereas in intact anaerobic mammalian muscle and liver the following overall stoichiometry is observed:

(2) Glucose $\longrightarrow$ 2 lactic acid

$$\Delta F' = -47 \text{ kcal/mole}$$

Aerobic organisms usually oxidize the products of glycolysis (for example, pyruvate), by means of additional enzyme systems, to $CO_2$ and $H_2O$.

In intact cells, the ATP produced by glycolysis and other pathways is rehydrolyzed to ADP and inorganic phosphate during other metabolic reactions (see p. 139), therefore ATP does not appear in the above equations. The actual glycolytic processes in anaerobic yeast (reaction *3*) and anaerobic muscle (reaction *4*) are:

(3) Glucose + 2ADP + 2Pi $\rightleftarrows$ 2 ethanol + $2CO_2$ + 2ATP $\quad \Delta F' = -39$ kcal/mole

(4) Glucose + 2ADP + 2Pi $\rightleftarrows$ 2 lactic acid + 2ATP $\quad \Delta F' = -31$ kcal/mole

In the multienzyme system of glycolysis, the specific steps that lead to reactions *3* and *4* are listed below. It can be seen that glycolysis in yeast, muscle, or liver is identical except for the final steps in the sequence. Examination of this sequence of enzymatic steps confirms equations *3* and *4* and points out that a complete anaerobic glycolytic system in the absence of any ATPase or ATP-utilizing systems requires:

1. A supply of inorganic phosphate for triose phosphate dehydrogenase (enzyme 6).
2. A supply of ADP to accept phosphate in the phosphoglycerokinase (enzyme 7) and pyruvic kinase (enzyme 10) reactions.
3. An initial supply of ATP to start the hexokinase (enzyme 1) and phosphofructokinase (enzyme 3) reactions.
4. A continuous supply of $DPN^+$. In the absence of $O_2$, the DPNH produced by triose phosphate dehydrogenase (enzyme 6) is reoxidized to $DPN^+$, and the pyruvate is reduced to either lactate or ethanol and $CO_2$ (enzymes 12 and 13). Some microorganisms attain this necessary reoxidation of DPNH by reduction of dihydroxyacetone phosphate with glycerol phosphate dehydrogenase (enzyme 14).

A failure to meet any one of these requirements will either halt or alter the overall glycolytic process.

The predominant features of this pathway were first discovered by Harden and Young. These investigators, working with dialyzed cell-free yeast extracts lacking ATPase and other ATP consuming activities, first observed the requirement of the glycolytic system for inorganic phosphate and dialyzable heat-stable cofactors (for example, $DPN^+$). Consequently, the dependence of glycolysis upon inorganic phosphate has been called the Harden-Young effect. The actual experiments of Harden and Young revealed an interesting alteration of the glycolytic process. These workers incubated their extract with glucose and inorganic phosphate, but in the absence of added ADP. They found that, with their crude extract, part of the glucose was converted to ethanol and $CO_2$ according to enzymes 1–12 in the diagram of glycolysis, while an equivalent amount was transformed to the hexose diphosphate stage according to enzymes 1–4.

The summation of these activities (reaction 7), known as the Harden-Young equation, describes the stoichiometric inorganic phosphate uptake during glycolysis when the system is lacking added ADP and competing enzymes active in converting ATP to ADP. Stated in another way, the Harden-Young equation depicts glycolysis in an ADP-limiting system in which glucose serves as a means of regenerating the small amount of ADP present. Thus in

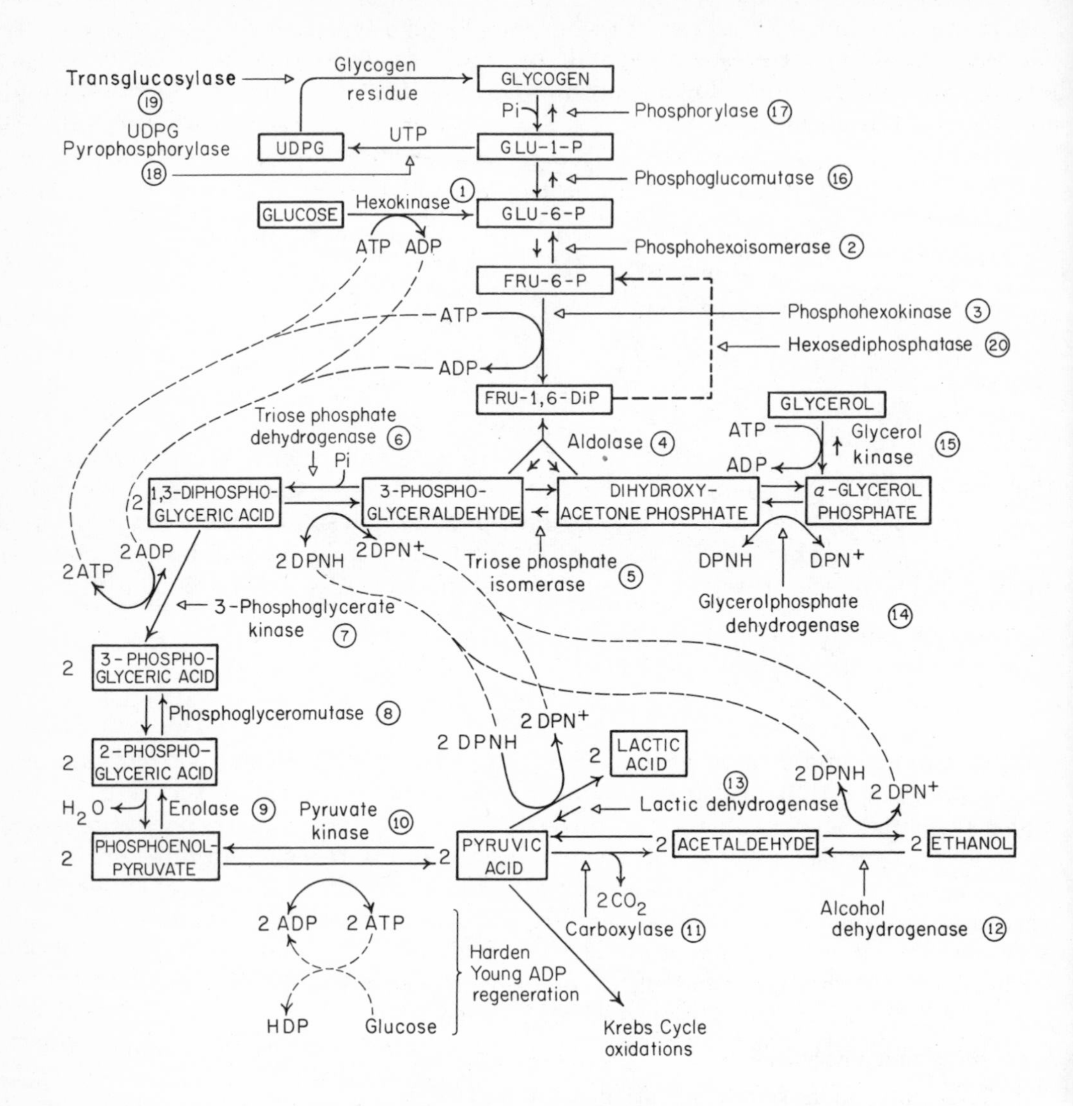


normal anaerobic yeast glycolysis (reaction *3*), the high-energy phosphates of 1,3-diphosphoglycerate and phosphoenolpyruvate are transferred to an adequate supply of ADP. In Harden-Young glycolysis (reaction *7*) the ADP is limiting; therefore the supply of ADP is cycled, first accepting the high-energy phosphate and then being regenerated by hexokinase and phosphofructokinase, with the resultant utilization of a second unit of glucose to yield fructose-1,6-diphosphate (HDP).

In this experiment on glycolysis, a cell-free extract of yeast with little or no ATPase is obtained by autolysis of yeast and removal of cellular debris. This clear

*(5)* $$\text{Glucose} + 2\text{ADP} + 2\text{Pi} \rightleftharpoons 2 \text{ ethanol} + 2CO_2 + 2\text{ATP}$$

*(6)* $$\text{Glucose} + 2\text{ATP} \rightleftharpoons \text{fructose-1,6-diphosphate} + 2\text{ADP}$$

*(7)* $$2\text{Glucose} + 2\text{Pi} \rightleftharpoons 2 \text{ ethanol} + 2CO_2 + \text{fructose-1,6-diphosphate}$$

extract is then preincubated with certain substrates to "spark" the reaction—that is, to eliminate any lag in reaction period and to deplete the extract of endogenous substrates. Glycolysis as predicted by the Harden-Young equation (reaction 7) and normal glycolysis—that is, with added ADP (equation *3*)—are then measured in this preincubated preparation by manometric evolution rate of measurement of the $CO_2$. Correlation of the quantity of $CO_2$ evolved with the quantity of glucose and inorganic phosphate consumed (obtained from the Nelson's and modified Fiske-Subbarow procedures) allows an analysis of the stoichiometry of the glycolytic reaction and the Harden-Young equation.

Further, the effects and sites of action of various inhibitors of the glycolytic reaction are studied, particularly, the effects of: (1) fluoride ion, known to inhibit enolase (enzyme 9) by formation of $MgFPO_4$; (2) iodoacetate (IAA), known to inhibit triose phosphate dehydrogrenase (enzyme 6) by reaction with the active site of the enzyme; and (3) arsenate, known to replace phsophate in the triose phosphate dehydrogenase (enzyme 6) reaction, leading to a loss in phosphorylation of ADP because of the spontaneous nonenzymatic hydrolysis of 1-arseno-3-phosphoglyceric acid to 3-phosphoglyceric acid plus arsenate.

Scheduling of this experiment, normally a one-period experiment, depends upon the availability of equipment, the number of students, and so on. Two considerations are important in establishing a schedule. First, the yeast autolysate is active for only 24 hr. Therefore, a new autolysate will be necessary if the manometry is handled on separate days. Second, the assays may be stopped and safely stored at 1–4°C once protein denaturation is achieved.

## Experimental Procedure

### MATERIALS

Fleischmann's 20-40 dried yeast
0.1$M$ $KHCO_3$
Clinical centrifuge
0.2$M$ Glucose
0.08$M$ Potassium phosphate buffer ($p$H 6.0)
0.2$M$ Fructose-1,6-diphosphate (HDP), Mg salt (or 0.2$M$ HDP, K salt, 0.2$M$ $MgCl_2$)
0.2$M$ Potassium phosphoglycerate ($p$H 6–7)
0.05$M$ Iodoacetate
5% $ZnSO_4 \cdot 2H_2O$
Nelson's reagent A
Arsenomolybdate reagent
10% TCA
Reducing reagent
Blendor (Waring type)
Table model high-speed centrifuge
1.0$M$ Glucose
0.04$M$ Potassium phosphate buffer ($p$H 6.0)
0.1$M$ ADP, K salt ($p$H 6.0)
0.1$M$ $KH_2A_3O_4$
0.2$M$ KF
0.3$N$ $Ba(OH)_2$
Nelson's reagent B
Glucose standard (100 $\mu$g/ml)
Acid molybdate reagent
Phosphate standard (1 $\mu$mole/ml)

### *Preparation of Cell-free Yeast Extract*

Blend 80 g of Fleischmann's 20-40 dried yeast with 334 ml of 0.1$M$ $KHCO_3$ in a Waring blendor for 60 sec, allow the slurry to stand for 5 min, and then blend it for an additional 30 sec. Allow the yeast slurry to autolyze at 37°C for 4 hr, centrifuge it at 25,000 × g for 15 min in the cold, and decant off the clear supernatant extract. Use immediately if possible. This extract retains activity for up to 24 hr if stored at 1–3°C. *Do not freeze.*

### *Preincubation Procedure*

Pipette 5 ml of the cell-free yeast extract into a test tube, and add 0.5 ml of 0.2$M$ fructose-1,6-diphosphate, Mg salt (or 0.2$M$ HDP, K salt, 0.2$M$ $MgCl_2$ if Mg salt not available), and 0.4 ml of 1.0$M$ D-glucose. Incubate the mixture for 45 min at 37°C in a water bath. After preincubation, active preparations bubble ($CO_2$) vigorously when in a 1-ml pipette. Plan your schedule to avoid delay between the end of the preincubation and the start of the manometric assay. Keep the preincubated extract in an ice bath until used.

### *Incubation Procedures*

In order to establish the stoichiometry of the Harden-Young equation, you will need to ascertain both the initial (endogenous and added) and final concentrations of glucose and inorganic phosphate. You will determine the initial concentrations in all experiments by analyzing two control tubes, A and B. Set up these tubes along with the Warburg flasks and incubate them at 37° for 5 min; that is a time period equal to the temperature equilibration time of the reactions in the Warburg flasks prior to the addition of the contents of the sidearms.

Prepare a series of reaction mixtures in calibrated Warburg flasks (see Experiment 29) and two zero-time control test tubes, A and B, as described in the following table, adding all the reagents except the cell-free yeast extract. Tube A yields the value for endogenous inorganic phosphate; tube B serves as a check on the assay of added inorganic phosphate. The sum of the inorganic phosphate in tubes A and B is equivalent to the level of inorganic phosphate present initially in certain of the assays (*Note:* The level of initial phos-

PART I. Stoichiometry of Glycolysis (volumes in milliliters)

| Substance | Vessel Number | | | | | Tube Number | |
|---|---|---|---|---|---|---|---|
| | 1 | 2 | 3 | 4 | 5 | A | B |
| Main Vessel | | | | | | | |
| 0.04*M* Potassium phosphate (pH 6.0) | — | 1.0 | — | 1.0 | — | — | 1.0 |
| 0.08*M* Potassium phosphate (pH 6.0) | — | — | — | — | 1.0 | — | — |
| 0.1*M* ADP | — | — | — | — | 0.8 | — | — |
| $H_2O$ | 2.4 | 1.4 | 2.4 | 1.4 | 0.6 | 2.6 | 1.9 |
| Preincubated cell-free yeast extract | 0.5 | 0.5 | 0.5 | 0.5 | 0.5 | 0.5 | — |
| Side Arm | | | | | | | |
| 0.2*M* Glucose | — | — | 0.2 | 0.2 | 0.2 | — | 0.2 |
| $H_2O$ | 0.2 | 0.2 | — | — | — | — | — |

PART II. Roles of Glycolytic Inhibitors (volumes in milliliters)

| Substance | Vessel Number | | | | | | | Tube Number | |
|---|---|---|---|---|---|---|---|---|---|
| | 1 | 2 | 3 | 4 | 5 | 6 | 7 | A | B |
| Main Vessel | | | | | | | | | |
| 0.04*M* (pH 6.0) | 1.0 | 1.0 | 1.0 | 1.0 | — | — | — | — | 1.0 |
| 0.2*M* KF | — | 0.5 | — | — | — | — | — | — | — |
| 0.05*M* Iodoacetate | — | — | 1.0 | 1.0 | — | — | — | — | — |
| $H_2O$ | 1.4 | 0.9 | 0.4 | 0.2 | 2.2 | 2.3 | 2.5 | 2.6 | 1.9 |
| Preincubated cell-free yeast extract | 0.5 | 0.5 | 0.5 | 0.5 | 0.5 | 0.5 | 0.5 | 0.5 | — |
| Side Arm | | | | | | | | | |
| 0.2*M* D-Glucose | 0.2 | 0.2 | 0.2 | 0.2 | 0.2 | — | 0.1 | — | 0.2 |
| 0.2*M* 3-Phosphoglycerate | — | — | — | 0.2 | — | — | — | — | — |
| 0.1*M* Arsenate | — | — | — | — | 0.2 | 0.2 | — | — | — |
| 0.2*M* Fructose-1,6-diphosphate | — | — | — | — | — | 0.1 | — | — | — |

phate in flask 5, Part I, equals A + 2B). When all of the components except the extract have been added, lubricate the manometer joints with lanolin, and check to see that the manometer stopcocks are not plugged with grease. Add the cell-free yeast extract to the proper Warburg vessels and control tubes, and attach the flasks to their respective manometers.

Place the flasks (stopcocks open) in the Warburg bath maintained at 37°C, and shake them to allow temperature equilibration. At the same time, place the zero-time control tubes A and B in a 37°C bath. After a 5-min temperature equilibration period raise the manometer fluid in both arms, close the stopcocks, and adjust the inner arm of the manometer to 150 mm so that the outer arm reads in the range 25–75 mm. Gas is evolved in this system, therefore maximum use of the manometer requires an initial low reading on the outer manometer arm (see Fig. 23). Record the outer arm setting (take all readings after setting the inner arm at 150 mm), and then initiate the enzymatic reaction by tipping the contents of the side arm into the main vessel (see Experiment 29, p. 142). When you initiate the reaction (that is, after 5 min at 37°C), remove the zero-time control tubes A and B from their 37°C bath, and place them in an ice bath.

Make and record manometer readings on all flasks at 5-min intervals by plotting the millimeters of gas ($CO_2$) produced (ordinate) versus time in min (abcissa). In Part I (Stoichiometry) continue manometer readings until flask 4 or 5 has evolved 20 μmoles of $CO_2$ (usually between 45 and 60 min). In Part II, continue readings at 5-min intervals for 60 min. After the final reading, *open* the stopcocks, and remove the manometers and flasks from the Warburg apparatus. Disconnect the flasks from the manometers in order to obtain aliquots for glucose and inorganic phosphate analysis.

### *Inorganic Phosphate Analysis*

Add a 0.5-ml aliquot of each incubation mixture and of the zero-time controls A and B to separate, marked, clinical centrifuge tubes containing 2 ml of 10% TCA. Add an additional 2.5 ml of $H_2O$ to each tube, and mix the contents by shaking. The substrates and denatured protein present at this stage are stable on storage at 1–4°C for future assay if necessary. Remove the protein precipitates by centrifugation (clinical centrifuge), and withdraw duplicate 1-ml aliquots from each supernatant solution. Add 3 ml of $H_2O$ to each aliquot, and then determine the concentration of inorganic phosphate present in each sample by the modified Fiske-Subbarow procedure—(that is, add 1 ml of acid molybdate reagent; 1 ml of reducing reagent; $H_2O$ up to 10 ml, and mix; wait 20 min; and read the $OD_{660}$ values, and compare them with that of a standard containing 0.5 μmoles of inorganic phosphate treated in a similar manner).

### *Glucose Analysis*

Add a 0.5-ml aliquot of each incubation mixture and of the zero-time controls A and B to separate, marked, clinical centrifuge tubes containing 1.0 ml of 0.3*N* $Ba(OH)_2$. Then add, in this order, 1.0 ml of 5% $ZnSO_4$ and 2.5 ml of $H_2O$. Mix the contents of each tube thoroughly. The substrates and denatured protein present at this stage are stable if stored at 1–3°C either before or after centrifuging. Separate the precipitate by centrifugation (clinical centrifuge), and remove the following aliquots of the supernatant solutions for glucose determination.

PART I

| Vessel or Tube Number | 1 | 2 | 3 | 4 | 5 | A | B |
|---|---|---|---|---|---|---|---|
| Aliquot size in ml | 1.0 | 1.0 | 0.5 | 1.0 | 1.0 | 1.0 | 0.5 |

PART II

| Vessel or Tube Number | 1 | 2 | 3 | 4 | 5 | 6 | 7 | A | B |
|---|---|---|---|---|---|---|---|---|---|
| Aliquot size in ml | 1.0 | 0.5 | 0.5 | 0.5 | 1.0 | 1.0 | 1.0 | 1.0 | 0.5 |

Determine the glucose concentration in the 3.1 ml of each flask and control tube by the Nelson procedure. That is, add 1 ml of Nelson's combined reagent (A:B = 25:1), and heat it at 100°C for 20 min; then cool, and add 1 ml of arsenomolybdate reagent. Shake this occasionally for 5 min; add water to 10 ml; and read the $OD_{540}$, and compare it with the $OD_{540}$ value of standards of 50 μg and 100 μg of glucose treated in the same manner (see Experiment 1, p. 12).

## Report of Results

First, using your own data and flask constant(s) and those of other students (if necessary), prepare two graphs, one of Part I and one of Part II, plotting μmoles of $CO_2$ produced in each flask versus time in minutes.

Second, using values for inorganic phosphate and glucose present in each flask at zero time and at the end of the incubation, calculate the stoichiometry observed in each reaction flask in Parts I and II. This process is frequently simplified by completion of tables similar to those below. It should be noted that this crude extract contains some phosphatase activity. Thus, for all enzymatic assays, some inorganic phosphate will be released from the HDP added during preincubation. The extent of this release is evident in flask 1 in Part I. Therefore, when determining the change in phosphate (ΔPi) levels in the flasks, make an appropriate correction for this phosphate release (that is, add the positive ΔPi of flask 1 to the negative ΔPi values of other assays).

Third, using the kinetic graphs of $CO_2$ production and the calculated stoichiometries, interpret the data from all assays in the light of our knowledge of glycolysis. In particular, answer these questions:

1. Which assays depict a Harden-Young reaction, and does the stoichiometry verify the equation?
2. Which assays depict the Harden-Young effect?
3. Do the rates and stoichiometry expressed by the preparations coincide with our concepts of the mechanism of action of these inhibitors?
4. Why is the rate of $CO_2$ production from phosphoglyceric acid not optimal in the presence of iodoacetate?
5. Does the stoichiometry of the system $AsO_4^{-3}$ = HDP differ from Harden-Young glycolysis?

## Discussion

As mentioned earlier, aerobic organisms usually oxidize the products of glycolysis to $CO_2$ and $H_2O$. This is accomplished by the Krebs cycle, electron transport, and oxidative phosphorylation pathways (see Experiments 31 and 32). Thus a steady metabolic flow of compounds through these pathways requires that each pathway operate at a rate commensurate with but not in excess of that needed to satisfy the needs of the other pathways. Numerous controls, or "checks and balances," relating the rate of one pathway to the rates of others exist in cells. These phenomena have been named after the workers who originally observed these controls in action; thus the name Pasteur effect, for the inhibition of glycolysis by aerobic oxidation, and the name Crabtree effect, for the inhibition of aerobic oxidation by glycolysis. The nature of these controls in whole cells is not fully understood, but very plausible theories exist. For example, the Pasteur effect may be related to the localized depletion of the inorganic phosphate supply (necessary for glycolysis) which

RESULTS: PART I. Stoichiometry of Glycolysis

| Vessel Number | Initial Reactants: μmoles Glu | μmoles Pi | ml Enz | Other | Determined Phosphate/3.1 ml: Initial | Final | Δ | Determined Glucose/3.1 ml: Initial | Final | Δ | Total μmoles $CO_2$ formed |
|---|---|---|---|---|---|---|---|---|---|---|---|
| A | — | — | 0.5 | — | | | | | | | |
| B | 40 | 40 | — | — | | | | | | | |
| 1 | — | — | 0.5 | — | | | | | | | |
| 2 | — | 40 | 0.5 | — | | | | | | | |
| 3 | 40 | — | 0.5 | — | | | | | | | |
| 4 | 40 | 40 | 0.5 | — | | | | | | | |
| 5 | 40 | 80 | 0.5 | 80 μmoles ADP | | | | | | | |

RESULTS: PART II. Roles of Glycolytic Inhibitors

| Vessel Number | Initial Reactants: μmoles Glu | μmoles Pi | ml Enz | Other | Determined Phosphate 3.1 ml: Initial | Final | Δ | Determined Glucose 3.1 ml: Initial | Final | Δ | Total μmoles $CO_2$ formed |
|---|---|---|---|---|---|---|---|---|---|---|---|
| A | — | — | 0.5 | — | | | | | | | |
| B | 40 | 40 | — | — | | | | | | | |
| 1 | 40 | 40 | 0.5 | — | | | | | | | |
| 2 | 40 | 40 | 0.5 | KF | | | | | | | |
| 3 | 40 | 40 | 0.5 | IAc | | | | | | | |
| 4 | 40 | 40 | 0.5 | IAc + PGA | | | | | | | |
| 5 | 40 | — | 0.5 | $AsO_4$ | | | | | | | |
| 6 | — | — | 0.5 | $AsO_4$ + HDP | | | | | | | |
| 7 | 20 | — | 0.5 | — | | | | | | | |

accompanies oxidative phosphorylation. The full understanding of the control of pathways and individual enzymes in cells is one of the challenging fields in biochemistry and molecular biology today.

### Exercises

1. In an ATPase-free glycolyzing system with glycogen as a substrate and with excess ADP and Pi, what would be the net yield of ATP per glucose residue reacted?
2. Given a cell-free yeast extract such as the one used in this experiment (that is, lacking ATPase), predict the nature and amount of each end product (for example, EtOH, $CO_2$, pyruvate) or remaining reactant that would be found upon completion of the glycolysis reaction in each of the following mixtures. (Values expressed in $\mu$moles added to incubation mixture.)
3. In your research it becomes necessary to prepare fructose-1,6-diphosphate and 3-phosphoglyceric acid labeled with $P^{32}$. Using concepts developed in this and other experiments, describe a series of steps that would lead to the preparation and isolation of these labeled compounds.

### References


Harden, A., *Alcoholic Fermentation* (4th ed.) Longmans Green and Co., London, 1932.

Fruton, J. S., and S. Simmonds, *General Biochemistry* (2nd ed.), Chapter 19, Wiley, New York, 1958.

Nord, F. F., and S. Weiss, *In* Sumner and Myrbäck (Editors), *The Enzymes* (Vol. 2), Part I, Academic, New York, p. 684, 1950.


| | Glucose | Pi | PGA | ATP | ADP | Other |
|---|---|---|---|---|---|---|
| a | 60 | 150 | — | 1 | 150 | — |
| b | 60 | 50 | — | 1 | — | — |
| c | 60 | 120 | — | 1 | — | — |
| d | 60 | — | — | 1 | — | arsenate |
| e | 60 | 30 | 10 | 10 | — | iodoacetate |

EXPERIMENT

# 31. Electron Transport in a Heart-particle Preparation

## Theory of the Experiment

Pyruvate, produced by glycolysis or other processes, is subsequently oxidized to $CO_2$ and $H_2O$ by means of the Krebs cycle and an electron transport system (see p. 000). The Krebs cycle is a sequence of enzymatic steps, including decarboxylations, hydrations, oxidations, and transfers. The oxidation steps involve the removal of pairs of electrons and, concomitantly, pairs of protons from the substrates, as shown by the following half reaction.

$$AH_2 \rightleftharpoons A' + 2H^+ + 2e^-$$

Here $A$ is the reduced substrate; $A'$, the oxidized dehydrogenated substrate. These electrons and protons are accepted by the first of a series of electron carriers or reversible, oxidizable substances. That is, the oxidation of the substrate is coupled with the reduction of the electron carrier, $B$.

$$AH_2 + B \rightleftharpoons A' + B'H_2$$

where $B$ is the oxidized carrier and $B'$ the reduced carrier.

The reduction of the electron carrier

does not always include acceptance of the protons, but may involve only electron transfer, whereas the protons become a part of the aqueous medium (for example, during the reduction of ferric iron, which may be associated with a particular protein).

$$BH_2 + 2Fe^{+3} \rightleftarrows B' + 2Fe^{+2} + 2H^+$$

Eventually the electrons are passed to molecular oxygen, as shown by the half reaction

$$\tfrac{1}{2}O_2 + 2H^+ + 2e^- \rightleftarrows H_2O$$

This final reaction accounts for most of the known oxygen consumption by aerobic organisms. Thus the oxidation of substrates by many aerobes includes (1) the removal of electrons (and, concomitantly, hydrogens) from a substrate such as a Krebs-cycle intermediate; (2) the transfer of electrons through a series of electron carrier agents (the electron transport system)—a transfer which may or may not include acceptance of the hydrogens by the carriers; and (3) the final transfer of the electrons to molecular oxygen. This process is illustrated by the following diagram.

$2H^+$                      $2H^+$

$AH_2$ / $A'$ — $B'$ / $BH_2$ — $Fe^{+2}$ / $Fe^{+3}$ — $Fe^{+3}$ / $Fe^{+2}$ — $Fe^{+2}$ / $Fe^{+3}$ — $\tfrac{1}{2}O_2$ / $H_2O$

Carriers transferring hydrogens and electrons — Carriers transferring only electrons

When substrates are being oxidized by such a system, the extent of substrate oxidation is directly dependent upon, and can be measured by, the uptake of molecular oxygen.

Study of the nature of the electron carriers began in the 1880's when MacMunn reported the existence of one or more pigments whose spectra changed during respiration. He claimed that the pigments resembled, but were distinct from, hemoglobin. The significance of MacMunn's observations was not discovered until the 1920's, when Keilin and Warbu[rg inde]pendently provided clear experime[ntal evi]dence for the function and chemic[al nature] of these cellular pigments.

In 1926, Warburg demonstrated the presence of the respiratory component (electron carrier) involved in the reaction with $O_2$. Noting that CO inhibited $O_2$ utilization in yeast, he theorized that CO must compete with $O_2$ for the same respiratory component. To test his hypothesis, Warburg measured the relative effectiveness of various wavelengths of light in counteracting the inhibition of yeast respiration by CO. Since the photochemical efficiency of a given wavelength of light in counteracting CO inhibition is proportional to the extent of absorption of this light, the photochemical action spectrum should be equivalent to the absorption spectrum of the substance(s) undergoing the photochemical reaction—in this case, the CO compound or the respiratory component that reacts with CO. The "action spectrum" found by Warburg exhibited maxima at 590, 550, and 432 mμ. In 1955, Warburg's observations were confirmed by Castor and Chance (Fig. 24). Such a spectrum suggested a heme protein similar to hemoglobin.

At the same time, and independently of Warburg, Keilin was studying reversible oxidation and reduction of pigments in a variety of organisms. For reduction, Keilin excluded air, or added substrates or chemical reducing agents. For his spectrophotometric studies, he used a hand spectroscope, a device which detects only zones of intense light absorption. He noted a strong absorption band at 590 mμ, a weak band at 550 mμ, and a broad, intense band at 430–435 mμ (see Fig. 24). Keilin found that, in the reduced state, yeast exhibited a multibanded absorption spectrum, and he clearly distinguished at least three components; cytochromes *a*, *b*, and *c*, according to the position of their absorption bands when reduced (oxidized pigments failed to show many of the bands).

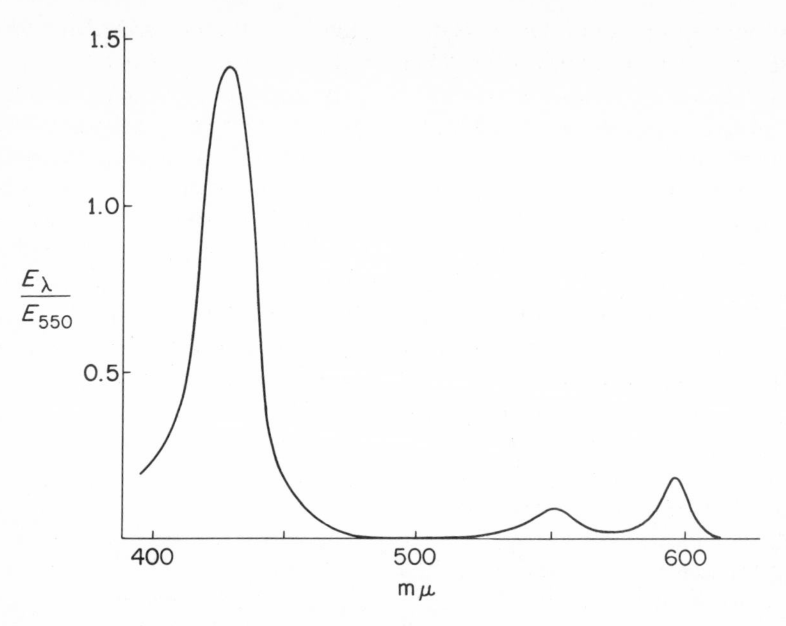

FIG. 24. *Relative photochemical absorption spectrum of the CO compound of yeast.* [*Adapted from Castor and Chance,* J. Biol. Chem., **217,** *453* (*1955*).]

Subsequent workers have identified a wide assortment of cytochromes from various plants, animals, and microorganisms. All of these pigments have absorption spectra similar to that shown in Fig. 24 when in the reduced state, but differ as to the exact location of their maxima. By convention, the three peaks, or bands, are called $\alpha$, $\beta$, and $\gamma$ (Soret) bands, in order of decreasing wavelength of absorption maxima. Several of the more important cytochromes and the location of their absorption maxima are listed in Table VIII.

It must be pointed out that the dual

**TABLE VIII.** Absorption Bands ($\lambda$max) of the Cytochromes (position of bands in millimicrons)

| Cytochrome | Name of Band | Absorption Maxima of Pigment in Reduced Form | Maxima of CO Complex |
|---|---|---|---|
| a | $\alpha$ | 605 | —— |
| | $\gamma$ | 452 | —— |
| $a_3$ | $\alpha$ | 600 | 590 |
| | $\gamma$ | 445 | 432 |
| b | $\alpha$ | 564 | —— |
| | $\beta$ | 530 | —— |
| | $\gamma$ | 431 | —— |
| c | $\alpha$ | 550 | —— |
| | $\beta$ | 520 | —— |
| | $\gamma$ | 415 | —— |

nature of cytochromes $a$ and $a_3$ is not a settled matter at present. It may be that a complex involving a copper-protein is involved, in addition to a heme-protein, in "cytochrome oxidase."

The effects of oxygen tension, substrate concentration, and respiratory inhibitors upon the oxidation-reduction changes of cytochromes have demonstrated conclusively that these pigments function as components of an electron transport system mediating the oxidation of substrates by molecular oxygen in aerobic cells. Cytochromes have also been shown to be involved in electron transport systems of photosynthetic cells and of a variety of anaerobic microorganisms.

In general, the cytochromes and other pigments involved in electron transport are associated with subcellular particles such as mitochondria or sarcosomes (muscle mitochondria). Thus studies of this type of biological oxidation usually involve the use of isolated, subcellular particles or fragments of particles. However, not all isolated mitochondrial or particle preparations retain their entire activity toward all substrates. The heart-muscle preparation (Keilin-Hartree particle fragments) studied in this experiment contains the complete cytochrome system of heart muscle, coenzyme Q, nonheme iron, various flavoproteins (including the succinate and DPNH flavoproteins), and various dehydrogenases associated with DPNH and succinate. Most of the dehydrogenases and other enzymes of the Krebs cycle are absent or inactive. The particle fragments also lack pyridine nucleotides and substrates. These systems are studied in this experiment by means of conventional manometric techniques, with $O_2$ consumption as the expression of activity.

The activity of particle preparations can be measured by methods other than manometry. Although oxygen is the terminal electron acceptor in manometric experiments, several different reducible dyes can also be electron acceptors. Some of these dyes are "autoxidizable"—that is, they may be reoxidized by molecular oxygen—thus providing a "shunt," or second means of electron transfer to oxygen. Use of such dyes for quantitative studies requires anaerobic conditions. Other dyes, such as 2,6-dichlorophenol-indophenol (DPIP), are nonautoxidizable. Thus the reduction of DPIP and the concomitant loss of color can be used as a measure of biological oxidation.

Oxidized DPIP (blue)

$2H^+ + 2e^-$

Reduced DPIP (colorless)

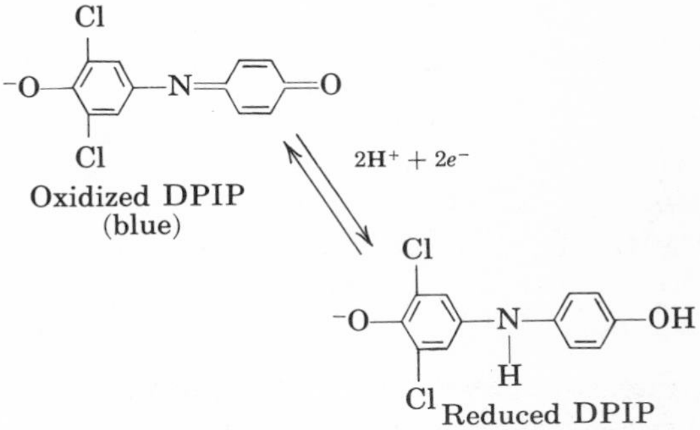

In this experiment the hand spectroscope, used for examination of the particle preparation for reduced cytochrome absorption bands, affords an excellent means for the qualitative determination of absorption spectra in the visible range. The heart of this simple device is the Amici prism, which consists of two crown prisms and one flint prism mounted so that the dispersion of light is limited to the area covered by the eyepiece (Fig. 25). For the sake of comparision, two objects can be viewed simultaneously, one of which may be a standard substance whose absorption spectrum is known. A numbered wavelength scale is superimposed upon the spectra by means of a small telescope containing the scale. A total reflecting prism is mounted so that the image of the scale is reflected onto the face of the Amici prism and thereby into the eyepiece lens.

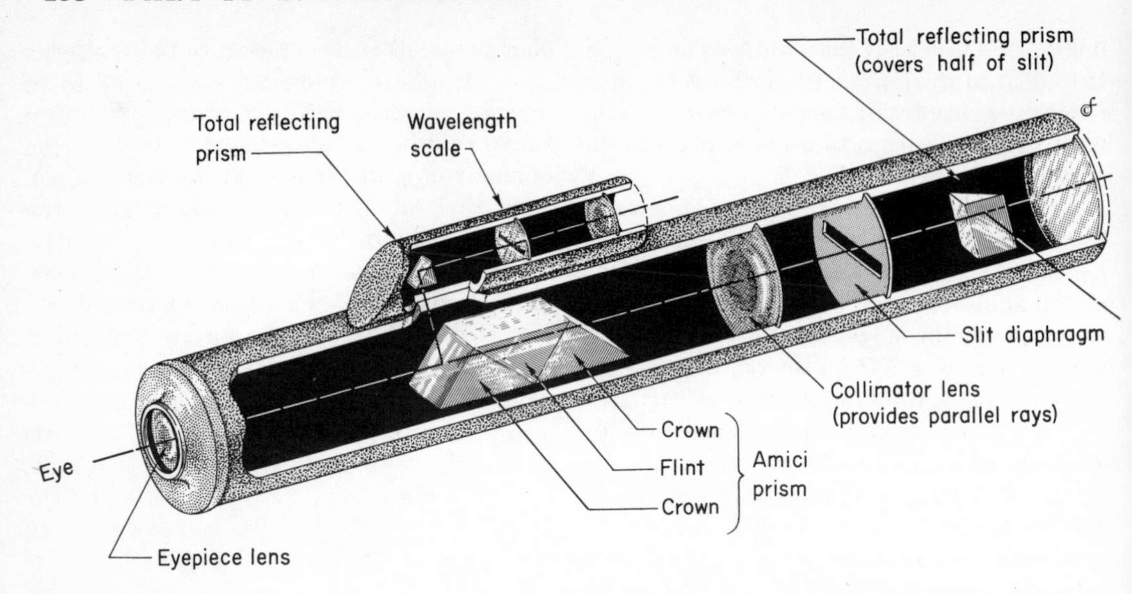


FIG. 25. *The hand spectroscope.*

## Experimental Procedure

MATERIALS

Warburg respirometer and constant-temperature bath
Hand spectroscope
Pig heart
Blendor
Table model high-speed centrifuge
$0.25M$ Potassium phosphate:$0.001M$ EDTA buffer ($p$H 7.4)
$0.001M$ $DPN^+$
$0.25M$ Malate (Na)
$0.5M$ Succinate (Na)
$6 \times 10^{-4}M$ Cytochrome *c*
$6M$ KOH
$0.1M$ KCN
Antimycin A (180 $\mu$g/ml)
DPIP solution
Calibrated manometers and Warburg flasks
Meat grinder
$0.001M$ EDTA ($p$H 7.4)
$1M$ Acetic acid
$0.02M$ Potassium phosphate:$0.001M$ EDTA ($p$H 7.4)
$Na_2S_2O_4$
$0.3M$ Potassium phosphate buffer ($p$H 7.4)
$0.05M$ Succinate (Na)
$0.25M$ $\beta$-Hydroxybutyrate (Na)
Sodium ascorbate (Na)
$0.5M$ Malonate (Na)
$0.01M$ Methylene blue
95% Ethanol

### *Preparation of Heart-muscle Particles*

All of the following operations are performed in the cold room (0–4°C).

Free a fresh pig heart of visible fat and fibrous tissue. Pass the muscle twice through a precooled (2°C) meat grinder fitted with a fine-grind plate. The minced muscle may be stored at −20°C for months without risking loss of activity. Wash the minced heart by stirring in 10 volumes of cool (not more than 20°C) $0.001M$ EDTA ($p$H 7.4) for about 4 hr, replacing the buffer six to eight times during this interval. Squeeze excess fluid from the minced muscle with four layers of cheesecloth after each washing.

Homogenize the minced muscle in a chilled blendor for 7 min at 0–4°C with $0.02M$ potassium phosphate:$0.001M$ EDTA ($p$H 7.4), using 500 ml per 200 g of original, trimmed heart muscle. Centrifuge the homogenate at 800 × g for 15–20 min. Decant the cloudy supernatant solution, and save it. Rehomogenize the precipitate in a chilled blendor for 2 min with 300 ml of cold $0.02M$ potassium phosphate:

0.001*M* EDTA (*p*H 7.4), and centrifuge as before. Add this supernatant liquid to that from the first centrifugation. (Usually 500–600 ml of suspension is obtained from 200 g of heart.)

Reduce the *p*H of the cold (<5°C) combined supernatant suspension to 5.6 by cautiously adding (with stirring) cold 1*M* acetic acid. Then centrifuge at 4000–5000 × g for 15 min. The clear, reddish supernatant contains cytochrome *c* and hemoglobin. Save it for examination in the hand spectroscope. Wash the precipitate once by suspending it in an equal volume of ice water and then centrifuge for 5 min at 1500 × g.

Finally, suspend the precipitate in about an equal volume of 0.25*M* potassium phosphate:0.001*M* EDTA (*p*H 7.4). The final volume should be about 70 ml from 200 g of heart. Disperse any lumps of precipitate by shaking gently in a flask or by gentle homogenization in a hand-operated glass homogenizer. Store this preparation at 1–4°C until used (stable for at least a week).

### *Examination of Absorption Bands with the Hand Spectroscope*

Place 4 ml of the supernatant solution obtained from the 4000–5000 × g centrifugation (above) in a test tube close to the end of the hand spectroscope—between the spectroscope and a 60-watt bulb serving as a light source. Compare the spectrum of the light passing through the solution with that of a water blank. Absorption bands appear as dark zones in the spectra. Add 5–10 mg of sodium hydrosulfite to the crude heme-protein solution; mix gently, avoiding aeration, and repeat the observation.

Perform the same operations with a dilute (pink) solution of cytochrome *c*.

Finally, prepare 2 tubes containing 1:5 dilutions of heart particles in 0.25*M* potassium phosphate:0.001*M* EDTA buffer. Add 5–10 mg sodium hydrosulfite to the first; mix gently, avoiding aeration; and examine for absorption bands. To the second tube add 0.1 ml of 0.5*M* succinate, gently swirl the tube, and examine it in the spectroscope. Allow the tube to stand with occasional swirling for 5 min, and reexamine. Next add 0.1 ml of 0.1*M* KCN and an additional 0.1 ml of 0.5*M* succinate, and repeat the observations. (*Note:* Absorption bonds are frequently difficult to detect in particle preparations due to excessive turbidity).

### *Manometric Measurement of Substrate Specificity and Cofactor Requirement*

Prepare a series of calibrated Warburg flasks in the following manner.

Series I (volumes in milliliters)

| Substance | Vessel Number | | | | | |
|---|---|---|---|---|---|---|
| | 1 | 2 | 3 | 4 | 5 | 6 |
| Main Vessel | | | | | | |
| 0.001M $DPN^+$ | — | — | — | 0.3 | 0.3 | 0.3 |
| 0.3M Potassium phosphate buffer (pH 7.4) | 1.5 | 1.5 | 1.5 | 1.5 | 1.5 | 1.5 |
| Heart-particle suspension | 0.5 | 0.5 | 0.5 | 0.5 | 0.5 | 0.5 |
| $H_2O$ | 0.85 | 0.95 | 0.90 | 0.65 | 0.5 | 0.5 |
| Side Arm | | | | | | |
| 0.25M Malate | — | — | — | — | 0.2 | — |
| 0.5M Succinate | — | 0.05 | 0.10 | 0.05 | — | — |
| 0.05M Succinate | 0.15 | — | — | — | — | — |
| 0.25M $\beta$-Hydroxybutyrate | — | — | — | — | — | 0.2 |

Series II (volumes in milliliters)

| Substance | Vessel Number | | | | |
|---|---|---|---|---|---|
| | 1 | 2 | 3 | 4 | 5 |
| Main Vessel | | | | | |
| $6 \times 10^{-4}M$ Cytochrome c | — | 0.3 | 0.3 | — | 0.3 |
| 0.3*M* Potassium phosphate buffer (pH 7.4) | 1.5 | 1.5 | 1.5 | 1.5 | 1.5 |
| Heart particles | 0.5 | 0.5 | — | 0.5 | 0.5 |
| Boiled heart particles* | — | — | 0.5 | — | — |
| $H_2O$ | 0.95 | 0.65 | 0.4 | 0.7 | 0.4 |
| Side Arm | | | | | |
| 0.5*M* Succinate | 0.05 | 0.05 | — | — | — |
| Fresh 0.1*M* sodium ascorbate† | — | — | 0.3 | 0.3 | 0.3 |

*Immerse in a boiling water bath for 10 min.

†198 mg of sodium ascorbate/10 ml or 176 mg of ascorbic acid and 1 ml of 1*N* NaOH made up to 10 ml with $H_2O$. Prepare just before use.

For each incubation mixture, add the heart particles and enough water to make the total volume 3.0 ml. After all additions have been made, put 0.1 ml of 6*M* KOH in the center well (total fluid volume 3.1 ml). Then grease the top of the center well with a minimum of lanolin so as to inhibit spillage of the KOH.

Next, place a piece of pleated filter paper in each center well so that the top of the paper just projects above the top of the center well. After all additions have been made, seal the flasks to their respective manometers, and place them in the constant-temperature bath (36–37°C). After 5 min in the bath, close the stopcocks, adjusting the fluid levels so that the inner arm is at 150 mm and the outer arm is high (gas to be consumed). Take readings at 2-min intervals to check temperature equilibration (some autoxidation of ascorbate takes place during equilibration). When the flasks are equilibrated, tip them carefully to transfer the contents of the side arm, but prevent contact with the KOH in the center well. Allow vessel 1 of Series I to go to completion, but take about six readings at 5-min intervals on all the other manometers of each series.

### *Inhibitors of Respiration*

This procedure is similar to that used in Series I and II, with the following excep-

Series III (volumes in milliliters)

| Substance | Vessel Number | | | | | | |
|---|---|---|---|---|---|---|---|
| | 1 | 2 | 3 | 4 | 5 | 6 | 7 |
| Main Vessel | | | | | | | |
| 0.3*M* Potassium phosphate buffer (pH 7.4) | 1.5 | 1.5 | 1.5 | 1.5 | 1.5 | 1.5 | 1.5 |
| 0.5*M* Malonate | — | — | — | — | 0.2 | 0.2 | 0.2 |
| 0.1*M* KCN (added last) | — | 0.3 | 0.3 | — | — | — | — |
| 0.01*M* Methylene blue | — | — | 0.3 | 0.3 | — | — | 0.3 |
| Heart particles | 0.5 | 0.5 | 0.5 | 0.5 | 0.5 | 0.5 | 0.5 |
| $H_2O$ | 0.95 | 0.65 | 0.35 | 0.65 | 0.75 | 0.4 | 0.45 |
| Side Arm | | | | | | | |
| 0.5*M* Succinate | 0.05 | 0.05 | 0.05 | 0.05 | 0.05 | 0.4 | 0.05 |

Series IV (volumes in milliliters)

| Substance | Vessel Number | | | | | |
|---|---|---|---|---|---|---|
| | 1 | 2 | 3 | 4 | 5 | 6 |
| Main Vessel | | | | | | |
| 0.3*M* Potassium phosphate buffer (pH 7.4) | 1.5 | 1.5 | 1.5 | 1.5 | 1.5 | 1.5 |
| $6 \times 10^{-4}$*M* Cytochrome c | — | — | — | 0.3 | 0.3 | 0.3 |
| Antimycin A in 95% EtOH (180 $\mu$g/ml) | — | 0.03 | 0.03 | — | 0.03 | — |
| 95% EtOH | 0.03 | — | — | 0.03 | — | 0.03 |
| 0.01*M* methylene blue | — | — | 0.3 | — | — | — |
| 0.1*M* KCN (added last) | — | — | — | — | — | 0.3 |
| Heart particles | 0.5 | 0.5 | 0.5 | 0.5 | 0.5 | 0.5 |
| $H_2O$ | 0.92 | 0.92 | 0.62 | 0.37 | 0.37 | 0.07 |
| Side Arm | | | | | | |
| 0.5*M* Succinate | 1.5 | 1.5 | 1.5 | — | — | — |
| Fresh 0.1*M* Sodium ascorbate* | — | — | — | 0.3 | 0.3 | 0.3 |

*198 mg of sodium ascorbate/10 ml or 176 mg of ascorbic acid and 1 ml of 1N NaOH made up to 10 ml with $H_2O$. Prepare just before use.

tions: (1) add the KCN immediately before placing the flask on the manometer, because KCN is converted to volatile HCN at *p*H 7 and distills into the alkali of the center well; (2) read all flasks at 5-min intervals for 30 minutes.

### *Measurement of Respiration by Dye Reduction*

Prepare a series of calibrated colorimeter tubes in the manner outlined in the table below, adding all of the components to the tubes except the heart particles and the succinate.

Add the heart-particle suspension to tube 1 (the blank) and to *one* of the other tubes. Mix the contents of each of these two tubes, and read the optical density of the experimental tube against the blank (tube 1) at 600 m$\mu$ at 30-sec intervals for 2 min to check for endogenous substrate. Then add the succinate to the assay tube. Mix the contents, and read the optical density at 600 m$\mu$ against the blank at 30-sec intervals for 5 min. Repeat the entire procedure with each of the remaining four tubes, one at a time, reading each against the single blank.

Series V (volumes in milliliters)

| Substance | Tube Number | | | | | |
|---|---|---|---|---|---|---|
| | 1 (blank) | 2 | 3 | 4 | 5 | 6 |
| DPIP Solution | — | 0.4 | 0.4 | 0.4 | 0.4 | 0.4 |
| 0.3*M* Potassium phosphate buffer (pH 7.4) | 4.5 | 4.5 | 4.5 | 4.5 | 4.5 | 4.5 |
| 0.5*M* Malonate | — | — | 0.2 | 0.2 | — | — |
| 0.1*M* KCN | — | — | — | — | 0.9 | — |
| Antimycin A (180 $\mu$g/ml) | — | — | — | — | — | 0.01 |
| $H_2O$ | 4.4 | 3.9 | 3.7 | 3.3 | 3.0 | 3.9 |
| Heart particles (added later) | 0.1 | 0.1 | 0.1 | 0.1 | 0.1 | 0.1 |
| 0.5*M* Succinate (added later) | — | 0.1 | 0.1 | 0.5 | 0.1 | 0.1 |

## Report of Results

Using the flask constants calculated in Experiment 29 for an $O_2$ system containing 3.1 ml of fluid, convert all of your data into microliters of $O_2$ consumed. Then plot the microliters of $O_2$ consumed versus time in minutes. Prepare a similar graph of the dye experiment, plotting decrease in optical density versus time.

## Discussion

The heart-particle preparation of this experiment represents only one of many mitochondrial preparations which have been used to study electron transport. So far, attempts at further simplification or solubilization of these particle systems have been only partially successful. Various active mitochondrial subunits have been obtained as a result of detergent or digitonin treatment (see Crane and Glenn; Cooper and Lehninger; and Hatefi, Haavik, and Griffiths in References). Partial separations of individual soluble components, followed by reconstitution into active preparations, have been achieved.

As a result of this work, it has been possible to create schematic models such as that shown in Fig. 26. These models summarize the various known mitochondrial constituents known to be involved in electron transport, and also point out the sites of inhibition of several electron transport inhibitors.

O
$CH_3O$ — — $CH_3$
$CH_3O$ — — $-[CH_2—CH=C—CH_2—H]-_n$
O — $CH_3$

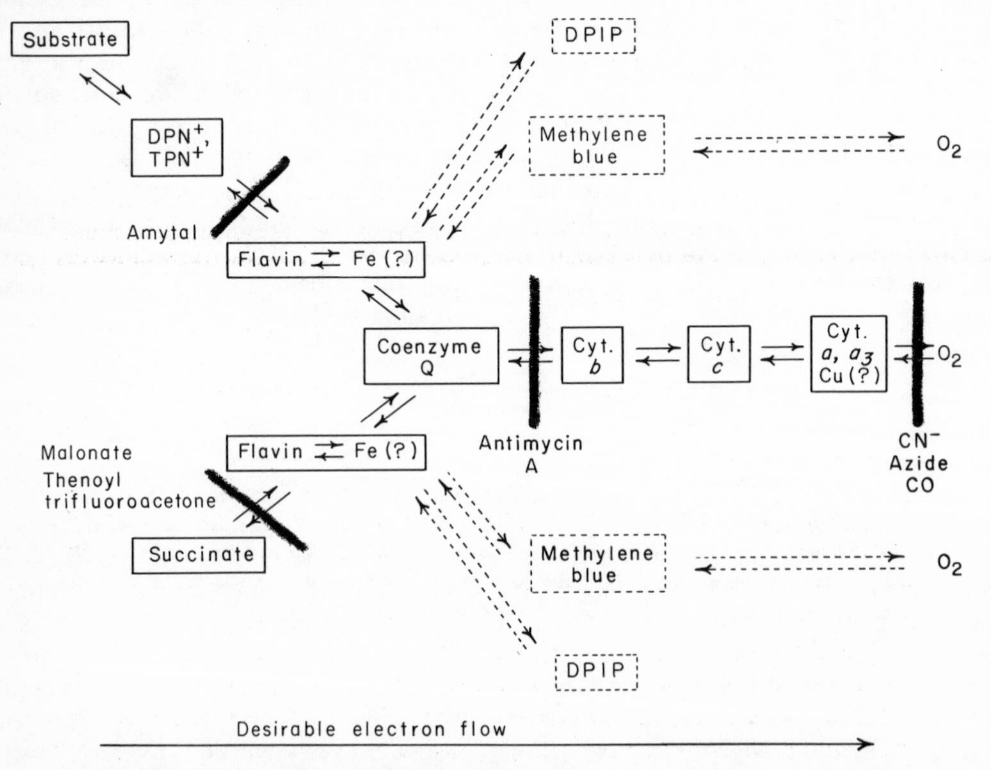


FIG. 26. *Schematic representation of electron transport.*

Note the suggested participation of coenzyme Q (ubiquinone). For beef-heart mitochondrial CoQ, the value of $n = 10$. This substance presumably participates by oxidation-reduction of the quinone system (possibly by semiquinone involvement). In the working scheme above, the actions of site-specific inhibitors are localized.

### Exercises

1. Define (a) dehydrogenase, (b) oxidase, (c) cytochrome.
2. What is the purpose of the KOH in the center well of the flasks used in this experiment?
3. Ascorbic acid reduces cytochrome *c* but does not affect cytochrome *b*. Explain this observation in terms of the redox potentials of these substances (see p. 000).
4. Predict which of the following systems, in combination with the heart particles (succinoxidase) of this experiment, would be expected to consume $O_2$: (a) succinate, DPIP, and cyanide; (b) succinate, antimycin A, cyanide, and methylene blue; (c) ascorbate, antimycin A, DPIP, and malonate; (d) succinate, methylene blue, cytochrome *c*, and malonate (concentration equal to that of succinate).

### References

Beinert, H., and W. Lee, *Biochem. Biophys. Res. Comm.*, **5**, 40 (1961).
Keilin, D., and E. F. Hartree, *Biochem. J.*, **41**, 500 (1947).
Cleland, K. W., and E. C. Slater, *Biochem. J.*, **53**, 547 (1953).
Crane, F. L., and J. L. Glenn, *Biochim. et Biophys. Acta.*, **24**, 100 (1957).
Cooper, C., and A. L. Lehringer, *J. Biol. Chem.*, **219**, 489 (1956).
Hatefi, Y., A. G. Haavik, and D. E. Griffiths, *Biochem. Biophys. Res. Comm.*, **4**, 441, 447 (1961).

EXPERIMENT

# 32. Oxidative Phosphorylation

## Theory of the Experiment

The process of oxidative degradation of foodstuffs to carbon dioxide and water is exergonic (for example, glucose oxidation, reaction *1*).

$$(1)\quad \text{Glucose} + 6O_2 \longrightarrow 6CO_2 + 6H_2O \qquad \Delta F' = -686 \text{ kcal/mole}$$

(Here $\Delta F'$ is identical with $\Delta F°$, except that the standard condition of $H^+$ ion is *p*H 7.) In cells, this energy is released during the passage of substrates through the various biochemical pathways. The extent of energy release in various pathways varies. In the complete oxidation of glucose by the glycolytic system, Krebs cycle, and electron transport system, by far the greatest energy release occurs in electron transport (approximately −590 kcal/mole glucose). This becomes obvious when we consider the amount of electron transport per glucose molecule and the energy release during the oxidation of the initial electron carriers (for example, DPNH oxidation, reaction *2*).

$$(2)\quad H^+ + \text{DPNH} + \tfrac{1}{2}O_2 \longrightarrow \text{DPN}^+ + H_2O \qquad \Delta F' = -52.4 \text{ kcal/mole}$$

Part of this energy released from foodstuffs during their degradation is retained by cells as a special kind of chemical energy (chemical reactivity) which is stored in the phosphate anhydride bonds of ATP. The phosphorylations of ADP which result in ATP formation or "energy storage" are classified into two types: substrate-level phosphorylations and electron-transport phosphorylations.

In substrate-level phosphorylation, ATP is formed as a result of an interconversion of a substrate other than an electron transport material. The glycolytic steps catalyzed by phosphoglycerate kinase and pyruvic kinase (see Experiment 30) are two examples of substrate-level phosphorylation. Another example is the formation of ATP during the conversion of succinyl CoA to succinate in the Krebs cycle.

(3)

$$\mathrm{O{=}C(-O-PO_3^{2-})-CH(OH)-CH_2-O-PO_3^{2-}} + \mathrm{ADP^{-3}} \xrightleftharpoons[\text{kinase}]{\text{phosphoglycerate}} \mathrm{O{=}C(-O^-)-CH(OH)-CH_2-O-PO_3^{2-}} + \mathrm{ATP^{-4}}$$

(4)

$$\mathrm{H^+} + \mathrm{O{=}C(-O^-)-C(=CH_2)-O-PO_3^{2-}} + \mathrm{ADP^{-3}} \xrightleftharpoons[\text{kinase}]{\text{pyruvate}} \mathrm{O{=}C(-O^-)-C(=O)-CH_3} + \mathrm{ATP^{-4}}$$

The second kind of phosphorylation of ADP was demonstrated by Ochoa and others in the early 1940's. These workers firmly established the concept that substrate oxidation is accompanied by reduction of an initial electron acceptor such as $DPN^+$, $TPN^+$, or a flavin (FAD or FMN) and that, during the subsequent transfer of these electrons to $O_2$, phosphorylations of ADP can occur. A scheme depicting this electron flow and the sites of these phosphorylations is shown below.

Many experiments involving a variety of approaches have been performed, notably by Lehninger and Chance and their coworkers, to develop this scheme. One approach has been the determination of the relationship between the uptake of inorganic phosphate and the oxygen consumed when various substrates are oxidized. The resultant P/O ratios ($\mu$moles of inorganic phosphate taken up per $\mu$atom of oxygen consumed) indicate the locations of the phosphorylative steps. For example, succinate yields a P/O ratio of nearly 2, whereas efficient preparations oxidizing malate yield a P/O ratio of nearly 3. The exact location of the phosphorylations has developed from the study of "partial reactions," in which only parts of the sequence are operative. Thus Lehninger has obtained P/O ratios of nearly 1 for the oxidation of ferrocytochrome *c* by liver mitochondria in the presence of oxygen. Similar studies

$$\text{Substrate} \rightleftharpoons \begin{matrix}DPN^+\\TPN^+\end{matrix} \xleftrightarrow{\text{ADP, Pi} \to \text{ATP}} \text{flavoprotein} \rightleftharpoons \text{CoQ} \rightleftharpoons \text{Cyt } b \xleftrightarrow{\text{ADP, Pi} \to \text{ATP}} \text{Cyt } c \rightleftharpoons \text{Cyt } (a + a_3) \xleftrightarrow{\text{ADP, Pi} \to \text{ATP}} O_2$$

$$\text{CoQ} \rightleftharpoons \text{flavoprotein} \rightleftharpoons \text{succinate}$$

with other parts of the scheme, along with the spectrophotometric studies by Chance and his co-workers, have established this scheme.

It is useful to consider a working scheme similar to that proposed by Lehninger in connection with what actually happens during the phosphorylation of ADP to form ATP that may accompany a given step in the electron transport system (let substrate $= AH_2$).

$$AH_2 + DPN^+ \rightleftarrows A + DPNH + H^+$$
$$DPNH + X \rightleftarrows DPNH{-}X$$
$$DPNH{-}X + H^+ + FAD \rightleftarrows DPN^+{\sim}X + FADH_2$$
$$DPN^+ \sim X + Pi \rightleftarrows DPN^+ + X \sim P$$
$$X \sim P + ADP \rightleftarrows X + ATP$$

In this scheme the hypothetical substance $X$ is introduced. It forms a complex with DPNH. When this complex is dehydrogenated by reaction with FAD, a reactive complex is formed which can "activate" the inorganic phosphate through the formation of $X{\sim}P$. This reactive substance then phosphorylates ADP to form ATP.

Even if this scheme is only partially correct, it is useful, since it compels you to think about the kinds of chemical mechanisms which may be involved. It also provides a working basis for the explanation of "uncoupling" phosphorylation from oxidation. Substances which would compete with ADP for phosphate in the last of the reactions above might well serve as "uncouplers."

Several important facts with regard to oxidative phosphorylation are these: First, the system is particulate. Efforts to completely "solubilize" it have been unsuccessful to date. As isolated, these particles, termed mitochondria, or sarcosomes when from muscle, contain all the oxidative enzymes (cytochromes), certain dehydrogenases, and the other enzymes of the Krebs tricarboxylic acid cycle. Only in brain mitochondria have the enzymes of glycolysis also been found.

Second, these particulate systems are enclosed in a semipermeable membrane, as shown by the sensitivity of the particles to the osmotic pressure of the medium. If the concentration of the solutes in a suspension of particles is decreased appreciably below isotonicity — below $0.25M$ sucrose or 0.9% sodium chloride—the particles swell and eventually burst. Therefore mitochondria, or sarcosomes, are generally prepared in at least $0.25M$ sucrose.

Third, the oxidation of substrates by carefully prepared suspensions of mitochondria may depend markedly on whether the ingredients for ATP synthesis (ADP and inorganic phosphate) are present. This phenomenon is known as "coupling."

Fourth, the phosphorylative mechanism is generally more sensitive to inhibitory agents than is the oxidative mechanism. Low concentrations of compounds such as substituted phenols (notably, 2,4-dinitrophenol, pentachlorophenol, dibromophenol), azides, and Dicumarol block (or "uncouple") phosphorylation from electron transport; therefore ATP synthesis is not required for electron transport. Uncoupling can also be induced by a decrease in the osmotic pressure of the medium. Often the effect of these uncoupling agents on "tightly coupled" mitochondria is an increase in the rate of substrate oxidation, possibly indicating that the phosphorylation steps are slower than the oxidation reactions, and therefore are rate limiting. The basic nature of the mechanism of oxidative phosphorylation remains obscure, constituting a major problem in biochemical research.

In this experiment, two preparations exhibiting oxidative phosphorylation are examined: (1) rat-liver mitochondria containing activity towards a rather wide range of substrates and with a limited stability, and (2) beef-heart particles (sarcosomes or sarcosome fragments) containing a limited range of substrate activities, but unusual in that they can be stored in the frozen state for months without loss of activity.

The P/O ratios of each of the preparations with various substrates are determined from the manometric data and the uptake of inorganic phosphate or ATP synthesis. The ATP formed is trapped by the action of hexokinase, which regenerates ADP. It is possible to use catalytic amounts of ADP and to avoid degradation of the ATP by ATPases in these preparations (high P/O values require addition of excess hexokinase to offset the ATPase action). Tightly coupled mitochondria often shown very little ATPase activity.

## Experimental Procedures

### MATERIALS COMMON TO BOTH PARTS

Warburg equipment and calibrated flasks
6*N* KOH
5% TCA
Acid molybdate reagent
0.5*M* Pyruvate
Clinical centrifuge
1% Hexokinase solution in 0.1% glucose
Reducing reagent
Phosphate standard (1 μmole/ml)
Table-model high-speed centrifuge

### ADDITIONAL MATERIALS FOR RAT-LIVER MITOCHONDRIA ASSAY

Rat
0.25*M* Sucrose in 0.001*M* EDTA (*p*H 7.4)
Drive motor
0.2*M* Sodium hydrogen glutamate
Mixture of ATP, $MgCl_2$, glucose, $DPN^+$, and EDTA at 3, 15, 150, 0.3 and 3 μmoles/ml respectively
0.1*M* Potassium phosphate (*p*H 7.4)
Dissecting tools
Potter-Elvehjem homogenizer with Teflon pestle
Ice
0.005*M* 2,4-Dinitrophenol
2.5*M* NaOH

### ADDITIONAL MATERIALS FOR BEEF-HEART PARTICLES

0.25*M* Sucrose in 0.185% $K_2HPO_4$ solution
Biuret reagent
1*M* Potassium phosphate (*p*H 7.2)
0.1*M* $MgSO_4$
0.5*M* Sodium malate
0.1*M* KCN
0.25*M* Sodium β-hydroxybutyrate
50% TCA
0.25*M* Sucrose
1*M* Sucrose
0.05*M* ATP (*p*H 7)
1*M* Glucose
$1 \times 10^{-4}M$ 2,4-Dinitrophenol
0.5*M* α-Ketoglutarate
0.5*M* Sodium succinate
0.5*M* Sodium malate

## *Part I. Rat-liver Mitochondria*

### PREPARATION OF MITOCHONDRIA (*maintain at 1–3°C where possible*)

Stun a rat with a blow on the head, and quickly decapitate it. Allow the animal to bleed for 30 sec, and then rapidly remove the liver, weigh it to the nearest 0.1 g, and place it in ice-cold 0.25*M* sucrose in 0.001*M* EDTA. When cold, place the liver in a 100-ml beaker in an ice bath, add 4 ml of ice-cold 0.25*M* sucrose:0.001*M* EDTA for each gram of liver, and cut the liver into 15–30 fine pieces. Then pour the cold suspension into an iced Potter-Elvehjem homogenizer with a loose Teflon pestle, and homogenize the suspension in the iced homogenizer for 2–3 min. Centrifuge the homogenate at 600 × g for 15 min in the cold at 0–5° to remove the cellular debris and cell nuclei. Then centrifuge the 600 × g supernate at 9000 × g for 10 min. Decant off the supernate, measure its volume, and discard it. Using a glass rod, gently resuspend the precipitate, composed predominantly of mitochondria, in a volume of ice-cold 0.25*M* sucrose in 0.001*M* EDTA buffer equal to the volume of the discarded 9000 × g supernate, and again centrifuge the mitochondria at 9000 × g for 10 min. Discard the washings. Finally, resuspend the washed mitochondria in a volume of ice-cold 0.25*M* sucrose in 0.001*M* EDTA equal to the original 9000 × g supernate. Store the preparation in an ice bath not longer than 1–2 hr before using.

INCUBATION

Prepare a series of calibrated Warburg flasks in the manner shown below, quickly adding the 6.0*M* KOH, the mitochondria preparation, and the hexokinase solution last. Place a small fluted wick in the center well to hasten absorption of $CO_2$.

Immediately after adding the KOH, hexokinase, and mitochondria, gently swirl the contents of the main vessel, and transfer 0.3 ml from each flask to separate marked clinical centrifuge tubes containing 4.7 ml of ice cold 5% TCA. Mix the contents of each tube to assure proper enzyme denaturation. Then quickly seal the flasks to their respective prelubricated manometers, and close the side arm vents. Then, noting the time, place all the flasks (stopcocks open) in a Warburg bath maintained at 37°C. Allow the flasks to equilibrate for 5 min before adjusting the inner arms to 150 mm and the outer arms to 250–300 mm (gas is consumed in these operations). Make the first reading, and continue readings at 5-min intervals over a 30-min period. At the end of the incubation period, open the stopcocks, remove the manometers, and transfer a 0.3-ml aliquot from each flask to separate marked clinical centrifuge tubes containing 4.7 ml of 5% TCA. Mix the contents of each tube. Then centrifuge the precipitates in both the zero-time and 35-min TCA samples at full speed on the clinical centrifuge.

Analyze duplicate 0.5-ml aliquots from each of the deproteinized preparations for the remaining inorganic phosphate by the modified Fiske-Subbarow procedure (1.0 ml of acid molybdate reagent; 1.0 ml of reducing reagent; $H_2O$ up to 10 ml; mixing; and, after 20 min, OD determination at 660 mμ) against a blank and standards (0.1–1.0 μmoles inorganic phosphate).

## *Part II. Beef-heart Mitochondria*

PREPARATION

Obtain a beef heart from a freshly slaughtered animal, and pack the heart in ice to cool it. Perform all further operations at 1–3°C. When cold, promptly trim off all fat and connective tissue, and cut the muscle into 2-cm cubes. Pass the cold, diced muscle through a precooled meat grinder fitted with a fine-grind plate (see Crane et al in References).

Weigh out 200 g of minced muscle, and place it in a precooled blendor; add 600 ml of ice-cold 0.25*M* sucrose containing 1.85 g of $K_2HPO_4$ per liter, and then add 1 ml of 6*N* KOH. Immediately homogenize the preparation for 30 sec; then stir in an

Rat liver Mitochondria Assay (volumes in milliliters)

| Substance | Vessel Number | | | | |
|---|---|---|---|---|---|
| | 1 | 2 | 3 | 4 | 5 |
| Main Vessel | | | | | |
| Mixture of ATP, $MgCl_2$, glucose, $DPN^+$, and EDTA | 1.0 | 1.0 | 1.0 | 1.0 | 1.0 |
| 0.1*M* Potassium phosphate (pH 7.4) | 0.5 | 0.5 | 0.5 | 0.5 | 0.5 |
| 0.5*M* Sodium pyruvate | 0.1 | — | 0.1 | — | 0.1 |
| 0.2*M* Sodium hydrogen glutamate | 0.1 | 0.1 | — | — | 0.1 |
| 0.0005*M* 2,4-Dinitrophenol | — | — | — | — | 0.1 |
| 0.25*M* Sucrose in 0.001*M* EDTA | 0.25 | 0.35 | 0.35 | 0.45 | 0.15 |
| 1% Hexokinase solution in 0.1% glucose | 0.05 | 0.05 | 0.05 | 0.05 | 0.05 |
| Mitochondria | 0.5 | 0.5 | 0.5 | 0.5 | 0.5 |
| Center Well | | | | | |
| 6.0*M* KOH | 0.1 | 0.1 | 0.1 | 0.1 | 0.1 |

additional 1 ml of 6*N* KOH, remove a small aliquot, and determine the *p*H. Quickly adjust the *p*H to 7.2–7.4 with additional 6*N* KOH (0.1–0.5 ml) if necessary. When the *p*H of the preparation is adjusted, immediately centrifuge the preparation at 1600 × g for 13 min. Decant off only the upper red-colored supernatant solution (discarding the sediment and a flocculant buff-colored layer above the sediment), and pour the supernatant solution through a single layer of cheesecloth.

Add an amount of ice-cold 0.25*M* sucrose equal to one-fourth the volume of the red supernatant solution, and then centrifuge the diluted supernatant solution at 30,000 × g for 30 min. Resuspend the residue in a volume equal to one-fourth the volume of the diluted red supernatant solution by stirring with a glass rod until a uniform suspension is obtained, and then reisolate the particles by centrifuging the suspension at 30,000 × g for 30 min.

Uniformly resuspend the sedimented mitochondria in about 30 ml of ice-cold 0.25*M* sucrose by stirring with a glass rod. Then remove 0.1-, 0.4-, and 0.8-ml aliquots of the suspension, and determine the protein concentration in this preparation by the Biuret method of Gornall et al. (aliquots plus $H_2O$ up to 2.0 ml; add 8 ml of Biuret reagent; mix; wait 30 min; and determine OD at 660 mμ against a water blank and standards; and 15 mg of protein treated in an identical manner—see p. 000). If necessary, dilute the mitochondrial suspension with cold 0.25*M* sucrose to obtain a final protein concentration of about 25 mg protein/ml.

Pour the suspension into a number of small test tubes, and freeze the samples in a dry ice-acetone bath. Store the frozen preparations at −20°C until needed (stable for months). Thaw the preparation in cold tap water immediately before use. Do not refreeze or reuse a thawed preparation.

### *Assay*

Prepare a series of calibrated Warburg flasks in the following manner, quickly adding the 6.0*N* KOH, the mitochondrial preparation, and the hexokinase solutions last. Place a small fluted wick in the center well to enhance absorption of $CO_2$.

Group I (volumes in milliliters)

| | Vessel Number | | | | |
|---|---|---|---|---|---|
| | 1 | 2 | 3 | 4 | 5 |
| Main Vessel | | | | | |
| 1*M* Sucrose | 0.65 | 0.65 | 0.65 | 0.65 | 0.65 |
| 1*M* Potassium phosphate (pH 7.2) | 0.05 | 0.05 | 0.05 | 0.05 | 0.05 |
| 0.05*M* ATP (pH 7) | 0.10 | 0.10 | — | 0.10 | 0.10 |
| 0.1*M* $MgSO_4$ | 0.05 | 0.05 | 0.05 | 0.05 | 0.05 |
| 1*M* Glucose | 0.10 | 0.10 | 0.10 | 0.10 | 0.10 |
| 0.5*M* Malate | 0.10 | 0.10 | 0.10 | — | 0.10 |
| 0.5*M* Pyruvate | 0.10 | 0.10 | 0.10 | — | 0.10 |
| $1 \times 10^{-4}$*M* 2,4-Dinitrophenol | — | 0.40 | — | — | — |
| 0.1*M* KCN | — | — | — | — | 0.04 |
| $H_2O$ | 1.65 | 1.25 | 1.80 | 1.85 | 1.61 |
| 1% Hexokinase solution | 0.05 | 0.05 | — | 0.05 | 0.05 |
| Heart mitochondria | 0.45 | 0.45 | 0.45 | 0.45 | 0.45 |
| Side Arm | | | | | |
| 50% TCA | 0.20 | 0.20 | 0.20 | 0.20 | 0.20 |
| Center Well | | | | | |
| 6*N* KOH | 0.10 | 0.10 | 0.10 | 0.10 | 0.10 |

Group II (volumes in milliliters)

| | Vessel Number | | | | | | |
|---|---|---|---|---|---|---|---|
| | 1 | 2 | 3 | 4 | 5 | 6 | 7 |
| Main Vessel | | | | | | | |
| 1M Sucrose | 0.65 | 0.65 | 0.65 | 0.65 | 0.65 | 0.65 | 0.65 |
| 1M Potassium phosphate (pH 7.2) | 0.05 | 0.05 | 0.05 | 0.05 | 0.05 | 0.05 | 0.05 |
| 0.05M ATP (pH 7) | 0.10 | 0.10 | 0.10 | 0.10 | 0.10 | 0.10 | 0.10 |
| 0.1M $MgSO_4$ | 0.05 | 0.05 | 0.05 | 0.05 | 0.05 | 0.05 | 0.05 |
| 1M Glucose | 0.10 | 0.10 | 0.10 | 0.10 | 0.10 | 0.10 | 0.10 |
| 0.5M Malate | 0.10 | 0.10 | — | — | — | — | — |
| 0.5M Pyruvate | 0.10 | — | — | — | — | — | — |
| 0.5M $\alpha$-Ketoglutarate | — | 0.10 | 0.10 | 0.10 | — | — | — |
| 0.25M $\beta$-Hydroxybutyrate | — | — | — | — | 0.20 | — | — |
| 0.5M Succinate | — | — | — | — | — | 0.10 | 0.10 |
| $1 \times 10^{-4}$M 2,4-Dinitrophenol | — | — | — | 0.40 | — | — | 0.40 |
| $H_2O$ | 1.65 | 1.65 | 1.75 | 1.35 | 1.65 | 1.75 | 1.35 |
| 1% Hexokinase solution | 0.05 | 0.05 | 0.05 | 0.05 | 0.05 | 0.05 | 0.05 |
| Heart mitochondria | 0.45 | 0.45 | 0.45 | 0.45 | 0.45 | 0.45 | 0.45 |
| Side Arm | | | | | | | |
| 50% TCA | 0.20 | 0.20 | 0.20 | 0.20 | 0.20 | 0.20 | 0.20 |
| Center Well | | | | | | | |
| 6.0M KOH | 0.10 | 0.10 | 0.10 | 0.10 | 0.10 | 0.10 | 0.10 |

Immediately after adding the KOH, hexokinase, and mitochondria, gently swirl the contents of the main vessel (do not admix the contents of center well and side arm), and transfer 0.5 ml from each flask to separate, marked clinical centrifuge tubes containing 9.5 ml of ice-cold 5% TCA. Mix the contents of each centrifuge tube to assure proper enzyme denaturation. Next quickly seal the flasks to their respective prelubricated manometers, and close the side arm vents. Then, noting the time, place all the flasks (stopcocks open) in a Warburg bath maintained at 37°C. Allow the flasks to equilibrate for 5 min before adjusting the inner arms to 150 mm and the outer arms to 250–300 mm (gas is consumed in these operations). Make the first reading, and continue readings at 5-min intervals over a 15-min period. Immediately after the 15-min reading, open the stopcocks, remove the manometers from the bath, and tip the contents of the side arm into the main vessel. Then remove the flasks from the manometers, and, using a wide-tip pipette, transfer 0.5-ml aliquots of the contents of the main vessels into marked centrifuge tubes containing 9.5 ml of ice-cold 5% TCA. Centrifuge the precipitates in both the zero-time and 20-min TCA samples at full speed in the clinical centrifuge.

Analyze duplicate 0.1-ml aliquots of the zero-time TCA supernatant solution and 0.2-ml aliquots of the 20-min TCA supernatant solution for the remaining inorganic phosphate by the modified Fiske-Subbarow procedure (1.0 ml of acid molybdate reagent; 1.0 ml of reducing reagent; $H_2O$ up to 10 ml; mixing; and, after 20 minutes, OD determination at 660 m$\mu$) against a blank and standards (0.1–1.0 $\mu$mole inorganic phosphate).

## Report of Results

Using the data of Experiment 29, calculate the flask constant appropriate for the fluid volume used in the assay. Then using the new flask constants, plot for all assays the microliters of $O_2$ consumed versus time in minutes (note that the first reading was taken 5–6 min after the reaction started). Extrapolate each curve to zero time on the graph.

Calculate the P/O ratio exhibited by each preparation by using the extrapolated values for $O_2$ consumption and the data from the inorganic phosphate determinations. (*Note:* P/O determinations will require consideration of the aliquot sizes assayed and conversion of $\mu$liters of oxygen to $\mu$atoms of oxygen.)

Compare your P/O ratios with those expected from the substrates tested.

## Discussion

Each of the components in the electron-transport pathway has a particular oxidation-reduction (redox) potential at $p$H 7 and 30°C ($E_0'$ value). Since electrons flow toward systems with higher potentials, it is possible to predict a possible scheme for electron transport even if we know only the $E_0'$ values of the components.

It is possible to estimate, on a theoretical basis, the sites of ATP synthesis in oxidative phosphorylation from the $E_0'$ values of the components. The greater the magnitude of the difference between the $E_0'$ values of two redox components, the greater the energy release during the passage of electrons from one component to another. This relationship is expressed by the equation

$$\Delta F' = -n\mathfrak{F}\Delta E_0'$$

where $\Delta F'$ is the change in free energy with all reactants at unit activity except the $H^+$ ion, which is at $1 \times 10^{-7} M$; $n$ is the number of electrons transferred; $\mathfrak{F}$ is the Faraday constant (23,063 cal/volt equivalent), and $\Delta E_0'$ is the difference between the two standard potentials in question. Examination of Table IX, with the $E_0'$ values of electron-transport components and use of the above formula, reveals that certain steps in the scheme involve a large energy release.

**TABLE IX.** Correlation of Energy Release With $E_0'$ Values of Component in Electron Transport

| System | Approximate $E_0'$ at $p$H 7 (volts) | $\Delta E_0'$ volts | $\Delta F'$ (303°K) per electron pair (kcal/mole) |
|---|---|---|---|
| Oxygen | +0.8 | | |
| | | 0.5 | −23 |
| Cytochrome *a* | +0.3 | | |
| | | 0.05 | −2.3 |
| Cytochrome c | +0.25 | | |
| | | 0.25 | −11.6 |
| Cytochrome *b* | 0.0 | | |
| | | 0.12 | −5.5 |
| Flavoprotein | −0.12 | | |
| | | 0.18 | −8.3 |
| $DPN^+$-linked dehydrogenase | −0.3 | | |

Since the synthesis of a high-energy phosphate bond in ATP appears to require at least 6–9 kcal/mole,

$$ADP + Pi \longrightarrow ATP + H_2O \qquad \Delta F' = 6\text{–}9 \text{ kcal/mole}$$

three of the steps in Table IX show sufficient energy release during electron transport to allow a coupled ATP synthesis to occur. It therefore follows that the steps

$$DPN^+ \longrightarrow \text{flavoprotein}$$
$$\text{Cyt } b \longrightarrow \text{Cyt } c$$
$$\text{Cyts } a + a_3 \longrightarrow O_2$$

are the most likely steps for ATP synthesis by oxidative phosphorylation. (It should be pointed out that this analysis does not prove that these sites are phosphorylating sites. The relative concentrations of oxidized and reduced electron carriers determine the actual $\Delta F$ involved (see the Nernst equation from quantitative analysis), and this analysis is based on concentrations of unit activity except for the $H^+$ ion.)

**Exercises**

1. Estimate the maximum P/O ratio expected from the oxidation of (a) succinate to fumarate, (b) pyruvate to acetyl-CoA, (c) citrate to $\alpha$-ketoglutarate, (d) $\beta$-hydroxybutyrate to acetoacetate, (e) glutamate to $\alpha$-ketoglutarate, and (f) $\alpha$-ketoglutarate to succinate.
2. Would DNP (2,4-dinitrophenol) completely uncouple phosphorylation from oxidation when $\alpha$-ketoglutarate is used as a substrate? When succinate is used? Explain your answers.
3. Discuss and compare the possible effects of arsenite, arsenate, malonate, and DNP upon oxidation and phosphorylation by mitochondria.

**References**

Chance, B., and G. R. Williams, *Adv. in Enzymol.*, **17,** 65 (1956).
Crane, F. L. et al., *Biochim. et Biophys. Acta.*, **22,** 475 (1956).
Hatefi, Y., and R. L. Lester, *Biochim. et Biophys. Acta.*, **27,** 83 (1958).
Krebs, H. A., and H. L. Kornberg, *Energy Transformations in Living Matter*, Springer-Verlag, Berlin, 1957.
Lehninger, A. L., *Scientific American*, **202,** 102 (1960).

EXPERIMENT

# 33. Acyl Activation Reactions

## Theory of the Experiment

As seen in Experiments 31 and 32, energy released by respiration is in part conserved as chemical energy in the form of ATP. Among other functions, this ATP is used by cells to drive many endergonic reactions. Examples of such endergonic reactions are the synthesis of acetyl-CoA (reaction *1*) and the synthesis of a peptide bond in proteins or peptides (reaction *2*). Since the energetics of these reactions are against synthesis, it is necessary to "activate" one of the components in order to make the reaction go. Biochemically, this activation is accomplished by the synthesis of a carboxyl anhydride at the expense of one of the phosphate anhydride bonds of ATP. The free activated species is very unstable in aqueous systems and is there-

$$(1) \qquad CH_3{-}\overset{O}{\overset{\|}{C}}{-}OH + HS{-}CoA \rightleftharpoons CH_3{-}\overset{O}{\overset{\|}{C}}{-}S{-}CoA + H_2O \qquad \Delta F' \cong 8 \text{ kcal/mole}$$

$$(2) \qquad H_2N{-}\overset{R}{\overset{|}{C}H}{-}\overset{O}{\overset{\|}{C}}{-}OH + H_2N{-}\overset{R'}{\overset{|}{C}H}{-}\overset{O}{\overset{\|}{C}}{-}OH \rightleftharpoons H_2N{-}\overset{R}{\overset{|}{C}H}{-}\overset{O}{\overset{\|}{C}}{-}N{-}\overset{R'}{\overset{|}{C}H}{-}\overset{O}{\overset{\|}{C}}{-}OH \qquad \Delta F' \cong 3.5 \text{ kcal/mole.}$$

fore assumed to be stabilized by the presence of the enzyme. The actual mechanisms of activations vary with the organism or enzyme system under test, but the net result is the same (reactions *3* and *4*). The actual synthesis of a new bond is then accomplished by one or more additional steps in which the activated species is transferred by the specific enzyme (Enz) to its final acceptor or partner (reaction 5). In a multistep acceptor process, synthesis of a new bond takes place as in *6* and *7*. The resultant overall reaction of activation and transfer is endergonic because the energy released as a result of ATP hydrolysis compensates for the energy required to synthesize the new bond.

Thus we may envision certain biochemical mechanisms involved in many otherwise endergonic processes in cells as being multistep processes. Activations that consume ATP are followed by one or more transfers to final partners. Four examples of known activations and subsequent transfer reactions catalyzed by enzymes are given on p. 173. (Note that, in general, one enzyme catalyzes both the activation and transfer reactions.)

$$(3)\qquad \text{Enzyme} + \text{R}-\overset{\overset{\displaystyle O}{\|}}{\text{C}}-\text{OH} + \text{ATP} \rightleftarrows \underbrace{\text{R}-\overset{\overset{\displaystyle O}{\|}}{\text{C}}-\text{O}-\overset{\overset{\displaystyle O}{\|}}{\underset{\underset{\displaystyle O^-}{|}}{\text{P}}}-\text{O}-\text{adenosine}}_{\text{Enzyme}} + \text{PP}$$

Acyl-AMP-enzyme complex

$$(4)\qquad \text{Enzyme} + \text{R}-\overset{\overset{\displaystyle O}{\|}}{\text{C}}-\text{OH} + \text{ATP} \rightleftarrows \underbrace{\text{R}-\overset{\overset{\displaystyle O}{\|}}{\text{C}}-\text{O}-\overset{\overset{\displaystyle O}{\|}}{\underset{\underset{\displaystyle O^-}{|}}{\text{P}}}-\text{O}^-}_{\text{Enzyme}} + \text{ADP}$$

Acyl-phosphate-Enz complex

$$(5)\qquad \text{Acyl-phosphate-Enz} + \text{H}-(\text{acceptor}) \rightleftarrows \text{R}-\overset{\overset{\displaystyle O}{\|}}{\text{C}}-(\text{acceptor}) + \text{HO}-\overset{\overset{\displaystyle O}{\|}}{\underset{\underset{\displaystyle O^-}{|}}{\text{P}}}-\text{O}^- + \text{Enz}$$

$$(6)\qquad \text{Acyl-AMP-Enz} + \text{H}-(\text{acceptor 1}) \rightleftarrows \text{R}-\overset{\overset{\displaystyle O}{\|}}{\text{C}}-(\text{acceptor 1}) + \text{AMP} + \text{Enz}$$

$$(7)\qquad \text{R}-\overset{\overset{\displaystyle O}{\|}}{\text{C}}-(\text{acceptor 1}) + \text{H}-(\text{acceptor 2}) \rightleftarrows \text{R}-\overset{\overset{\displaystyle O}{\|}}{\text{C}}-(\text{acceptor 2}) + \text{H}-(\text{acceptor 1})$$

1. Acetate activation (aceto-CoA-kinase) of yeast and liver (fatty acid activation in liver follows the same mechanism) (reaction *8*).
2. Amino acid (AA) activation in a variety of tissues (postulated as the first step in protein synthesis) (see *9*).
3. Glutamine synthetase (reaction *10*).
4. Acetokinase and phosphotransacetylase of various anaerobic microorganisms (two separate enzymes) (*10* and *11*).

In the reactions above, with the exception of acetokinase (see 4), the enzyme has a rather high affinity for the activated species, hence the inclusion of the enzyme in the activated complex. Therefore, the actual quantity of activated species will depend on the enzyme concentration. For example, glutamyl phosphate has never been isolated from a glutamine-synthesizing reaction, and is implicated only as a result of isotope studies which are consistent with its existence during the reaction.

Activations are readily measured by "trapping" the activated species with nucleophilic reagents such as hydroxylamine. The presence of such agents usually does not seriously affect the enzyme, thus, as an activated species (for example, an amino acid) is formed enzymatically, it is nonenzymatically removed from the system in the form of a hydroxamate (reaction *13*).

$$\begin{array}{lcl} \text{Enz} + \text{acetate} + \text{ATP} & \rightleftarrows & \text{Enz-AMP-acetate} + \text{PP} \\ \text{Enz-AMP-acetate} + \text{CoA} & \rightleftarrows & \text{acetyl CoA} + \text{AMP} + \text{Enz} \\ \hline \text{Acetate} + \text{ATP} + \text{CoA} & \overset{\text{Enz}}{\rightleftarrows} & \text{acetyl CoA} + \text{AMP} + \text{PP} \end{array} \tag{8}$$

$$\begin{array}{lcl} \text{Enz} + \text{AA} + \text{ATP} & \rightleftarrows & \text{Enz-AMP-AA} + \text{PP} \\ \text{Enz-AMP-AA} + \text{S-RNA} & \rightleftarrows & \text{S-RNA-AA} + \text{AMP} + \text{Enz} \\ \hline \text{AA} + \text{ATP} + \text{S-RNA} & \overset{\text{Enz}}{\rightleftarrows} & \text{S-RNA-AA} + \text{AMP} + \text{PP} \end{array} \tag{9}$$

$$\begin{array}{lcl} \text{Enz} + \text{glutamic acid} + \text{ATP} & \rightleftarrows & \text{Enz-}\gamma\text{-glutamylphosphate} + \text{ADP} \\ \text{Enz-}\gamma\text{-glutamylphosphate} + \text{NH}_3 & \rightleftarrows & \text{glutamine} + \text{HO—P(=O)(O}^-\text{)—O}^- \\ \hline \text{Glutamic acid} + \text{ATP} + \text{NH}_3 & \overset{\text{Enz}}{\rightleftarrows} & \text{glutamine} + \text{ADP} + \text{HO—P(=O)(O}^-\text{)—O}^- \end{array} \tag{10}$$

$$\text{Acetokinase: acetate} + \text{ATP} \rightleftarrows \text{acetyl phosphate} + \text{ADP} \tag{11}$$

$$\text{Phosphotransacetylase: acetyl phosphate} + \text{CoA} \rightleftarrows \text{acetyl CoA} + \text{HO—P(=O)(O}^-\text{)—O}^- \tag{12}$$

$$\text{Enz} + \text{AA} + \text{ATP} \rightleftarrows \text{Enz-AMP-AA} + \text{PP} \xrightarrow{\text{NH}_2\text{OH}} \underset{\text{AA Hydroxamic acid}}{\text{H}_2\text{N—CH(R)—C(=O)—NHOH}} + \text{AMP} + \text{Enz} \tag{13}$$

Hydroxamates form characteristic colored complexes in acid solution in the presence of ferric iron. Therefore, the extent of enzymatically synthesized hydroxamate can be determined colorimetrically by measuring (in acid solution) the absorption of the colored complex at 520 mμ. This allows an easy fixed-time assay for measuring the rate of activation or synthesis of activated species.

In this experiment, a crude homogenate of rat liver is examined by means of the hydroxamate assay for acetate and amino acid activation.

## Experimental Procedure

MATERIALS

Rat
Ice
Homogenizer
0.1*M* ATP-K salt
0.02*M* Mixture of 15 L-amino acids
0.1*M* Tris-Cl (*p*H 7.0)
2*M* $FeCl_3$
0.1*N* HCl
Dissecting tools
0.05*M* KCl
3.0*M* $NH_2OH \cdot HCl$ (*p*H 7.0)
0.1*M* $MgCl_2$
1.0*M* Potassium acetate
100% TCA
0.01*M* Hydroxamate standard (Acetylhydroxamic acid)

### *Preparation of Enzyme*

Stun a medium-sized (100–180 g) rat with a blow on the head, and quickly decapitate it. Remove the liver at once, weigh it, and cool it in a beaker of ice. When cool, place the liver in an ice-cold 50-ml beaker, cut it up into fine pieces, add 2 volumes (that is, twice the weight of the liver in grams) of ice cold 0.05*M* KCl, and pour the contents into an iced homogenizer. Maintaining the homogenizer in ice, gently turn the plunger of the homogenizer (by hand or motor) until an even suspension (no original lumps) is obtained. Use this suspension immediately for assays (that is, have all assay tubes prepared before preparing the enzyme).

### *Hydroxamate Assay*

Prepare the following mixtures in clinical centrifuge tubes (values expressed in milliliters).

Add 0.4 ml of extract to enzyme assay tubes (5–10), mix the contents of each tube thoroughly, and incubate each assay tube for 60 min (30 min if pressed for time) at 37°C in a water bath. Terminate the reaction by adding 1.4 ml of cool 100% TCA to each assay tube and each standard (tubes 2–4), mix the contents of each tube thoroughly, and then add 0.6 ml of 2.0*M* $FeCl_3$ to all tubes. After further mixing, centrifuge the coagulated protein in tubes 5–10, and decant the clear supernates and the standards into colorimeter tubes. Allow the tubes to sit for 10 min, and then read

| Substance | Tube Number | | | | | | | | | |
|---|---|---|---|---|---|---|---|---|---|---|
| | 1 | 2 | 3 | 4 | 5 | 6 | 7 | 8 | 9 | 10 |
| 0.02M Mix of 15 L-amino acid* | — | — | — | — | 0.2 | — | 0.2 | 0.2 | — | — |
| 1.0M Potassium acetate | — | — | — | — | — | 0.2 | 0.2 | 0.2 | — | — |
| 0.1M ATP | — | — | — | — | 0.3 | 0.3 | 0.3 | — | 0.3 | — |
| 0.1M Tris-Cl | — | — | — | — | 0.6 | 0.6 | 0.6 | 0.6 | 0.6 | 0.6 |
| 3.0M $NH_2OH \cdot HCl$ | 1.0 | 1.0 | 1.0 | 1.0 | 1.0 | 1.0 | 1.0 | 1.0 | 1.0 | 1.0 |
| 0.1M $MgCl_2$ | — | — | — | — | 0.3 | 0.3 | 0.3 | 0.3 | 0.3 | 0.3 |
| 0.01M Hydroxamate standard | — | 0.1 | 0.2 | 0.4 | — | — | — | — | — | — |
| $H_2O$ | 2.0 | 1.9 | 1.8 | 1.6 | 0.2 | 0.2 | — | 0.3 | 0.4 | 0.7 |

*Heat in a boiling water bath to dissolve tyrosine, and then add to other components.

and record the optical densities at 520 mμ against the blank (tube 1). (*Note:* If the final volume of 5 ml is not sufficient for colorimetry, dilute all tubes with a minimum and constant amount of 0.1*N* HCl just before OD determination).

## Report of Results

Draw a hydroxamate standard curve (OD versus concentration), and determine the actual μmoles of hydroxamate formed or compound activated in each assay tube per hour. Explain the observed results in light of our knowledge of acyl activations and the expected purity of a crude homogenate enzyme.

## Discussion

Activation reactions can also be studied by use of exchange reactions with radioisotopes. Activation systems are readily reversible; therefore if all the components on one side of an equation are incubated together, the reaction dictates that the components on the other side of the equation will be formed, even if some of these components are initially absent. Further, the dynamic state of the reaction dictates that if materials on both sides of the equation are present, a continuous interchange of reactants and products occurs. Thus if one of the reactants is radioactive at the start of the reaction, the radioisotope will accumulate, in part, in one of the products. The rate of buildup of the radioisotope in the initially unlabeled product is indicative of the rate of exchange or synthesis in the reaction.

To illustrate this point, consider the following example of an exchange assay of amino acid activation. If one incubates enzyme, ATP, and amino acid (all the reactants on one side of reaction *14*) in the presence of radioactive pyrophosphate ($PP^{32}$), a small initial synthesis of activated complex occurs (this is small, since the enzyme is in a very low molar concentration) (see p. 173 and reaction *14* below). Subsequently, the dynamic state of the reaction dictates that the reverse reaction will begin to accumulate labeled ATP (reaction *15*; the small amount of PP becomes swamped by the large pool of $PP^{32}$). The extent of labeling of the initially unlabeled reactant (ATP) increases with time as the exchange proceeds. Stoppage of the reaction at various times, isolation of the ATP and eventual determination of the specific activity of the ATP (see Appendix III) allows an assay of the rate of exchange.

The quantity measured in such exchange studies is the percent exchange (see *16*). The mechanism of the exchange dictates that a further correction must be performed to obtain a true measure of the rate of the exchange. This correction elimi-

$$\text{Enz} + \text{AA} + \text{ATP} \longrightarrow \underset{\text{activated complex}}{\text{Enz-AMP-AA}} + \text{PP} + \underset{\text{added initially in large amount}}{PP^{32}} \tag{14}$$

$$\text{Enz} + \text{AA} + ATP^{32} \longleftarrow \text{Enz—AMP—AA} + PP^{32} \tag{15}$$

$$\%\ \text{Exchange} = \frac{\text{radioactivity in isolated component at time } t \times 100}{\text{radioactivity in isolated component at complete equilibration}} \tag{16}$$

nates the error in the method that develops when a molecule labeled as a result of an exchange (ATP in reaction *15*) loses its label by a second exchange with a now unlabeled reactant (cold PP released from ATP in reaction *14*). The form of the correction, calculated on a strict theoretical basis, is

Exchange reactions are particularly useful because they are more sensitive than the hydroxamate assays and therefore require much smaller amounts of enzyme and substrate. In addition, exchange reactions do not require the addition of an unnatural trapping reagent, such as $NH_2OH$.

$$\text{(17)}\quad \text{Actual exchange} = -2.303\,\frac{\left(\begin{array}{c}\mu\text{moles labeled}\\ \text{compound}\end{array}\right)\left(\begin{array}{c}\mu\text{moles compound}\\ \text{to be labeled}\end{array}\right)}{\left(\begin{array}{c}\mu\text{moles labeled}\\ \text{compound}\end{array}\right)+\left(\begin{array}{c}\mu\text{moles compound}\\ \text{to be labeled}\end{array}\right)} \times \log_{10}\left(1-\frac{\%\text{ exchange}}{100}\right)$$

### Exercises

1. In the acetate-activation experiment, what would you expect the addition of coenzyme A to do to the results obtained from the hydroxamate assay? Would the effect be the same in a pyrophosphate exchange assay of acetate activation?
2. Is it possible to measure equilibrium constants of reactions from exchange data?
3. What steps would you perform during enzyme isolation to determine the cytological location (soluble or particulate) of the activating enzymes of this preparation?

### References

Schweet, R. S., *Biochim. et Biophys. Acta.*, **18,** 566 (1955).
Hoagland, M. B. et al., *J. Biol. Chem.*, **218,** 345 (1956).
Berg, P., *J. Biol. Chem.*, **222,** 991 (1956).

EXPERIMENT

# 34. Metabolic Transformations of Acetate

## Theory of the Experiment

The existence of various pathways may be observed by *isotopic labeling*. In this technique, a substrate, "labeled" in one or more atoms as a result of isotopic enrichment (see Appendix III) is introduced into a biological system. The isotopic substrate has similar, if not identical, biochemical properties to those of its natural counterpart, and is therefore metabolized in a manner identical, or nearly so, to that of the natural substrate. Termination of the reaction and isolation of biochemical entities, followed by detection of relative differences in mass or radioactivity of the isolated materials, allows the detection of minute quantities of the atoms of the original substrate. Thus it is possible to observe the interconversion of molecules, since the occurrence of isotope in the isolated material(s) is indicative of the metabolic flow of the particular atom(s) of the original substrate to the isolated material(s).

Acetyl-CoA, produced as a result of acetate activation (see Experiment 33), occupies a prominent position in metabolism—one that can be easily subjected to examination by use of the isotopic labeling technique. For example, acetyl-CoA can be oxidized in the Krebs cycle to yield energy

(ATP), $CO_2$, and $H_2O$. Under certain circumstances, the acetyl unit of acetyl-CoA is readily converted to glycogen. Further, acetyl-CoA serves as a direct precursor for fatty acid and steroid biosynthesis. The carbons of acetate in acetyl-CoA may also find their way by less direct pathways into other compounds such as nucleic acids and proteins.

In this experiment, the conversion of $C^{14}$ acetate is studied in the intact rat. Acetate incorporation into liver glycogen is enhanced by prior depletion of liver glycogen through fasting and cold treatment, followed by introduction of glucose along with the labeled acetate. Acetate incorporation into cholesterol is sharply increased by injecting 2 ml of 10% Triton WR 1339 per 100 g of rat prior to injection of the $C^{14}$ acetate. After a brief period following injection, the respiratory $CO_2$ and the various materials from the liver are isolated by use of the following scheme.

## Experimental Procedure

### MATERIALS

Rat (100–150 g)
Dissecting tools
1.5*M* Glucose, 0.1*M* sodium acetate-$C^{14}$, containing 10 μcuries per ml (Either 1-$C^{14}$ or 2-$C^{14}$ acetate or both may be used.)
Hypodermic syringe
1*N* NaOH ($CO_2$ free)
10% TCA
Anhydrous $Na_2SO_4$
0.5% Digitonin
Ether
Absolute ethanol
3*N* $H_2SO_4$
1*M* $BaCl_2$
Respiratory apparatus
Ice
Clinical centrifuge
95% EtOH
Chloroform: methanol (1:1)
15% Ethanolic KOH
Petroleum ether (30–60°C)
Acetone: ether (1:1)
Acetone
Towel
Acetone: absolute EtOH (1:1)
Bromcresol-green solution
10% Triton WR 1339

### *Preliminary Preparations*

Before beginning this experiment, study Appendix III, and be sure you understand the theory and purpose of the entire experiment. This experiment, as designed, does not constitute a health hazard, but you should assiduously follow all directions concerning disposal of animals, washing of instruments and glassware, collection of respiratory $CO_2$, and so on. All operations with isotopic material are performed over waxed or waxed-back paper so as to limit contamination. Report *all* breakage or spillage to the instructor.

Several days before the actual experiment, select a 100–150 g rat, and place it in a separate, marked cage. During the next few days provide the rat with food and water. Occasionally remove the rat and return it to the cage so as to familiarize it with your handling technique.

Twenty-four hours before the experiment deprive the rat of food but not water. One hour before injecting the isotope place the animal in a cold room (4°C).

While the rat is in the cold room assemble and test a respiratory apparatus similar to the one shown in Fig. 27. Using a funnel, place 50 ml of $CO_2$-free 1*N* NaOH in each absorber test tube, avoiding wetting the stoppers and tops of the tubes. Then quickly plug the absorbers with their respective stoppers, and test the apparatus by pulling a slight vacuum on the system so that the bubble rate in *all* tubes is just faster than can be counted. All tubes should bubble at the same rate; if they do not, check for a leak in the system. After the *short* check, discontinue the vacuum, and leave the system closed so as to prevent the entrance of atmospheric $CO_2$.

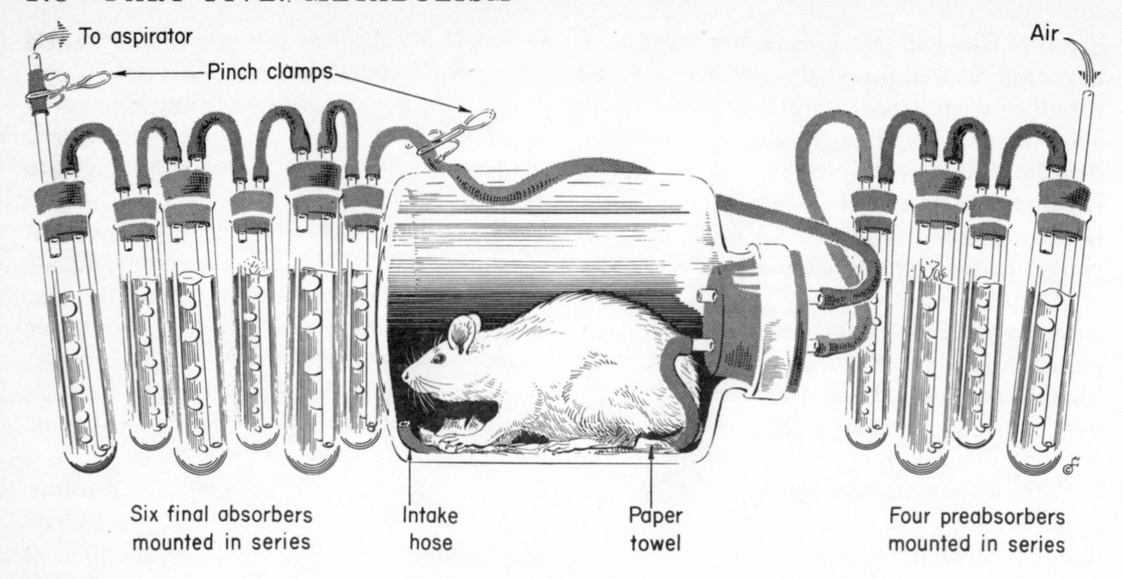


FIG. 27. *Respiratory apparatus.* [*Adapted from Cowgill and Pardee,* Biochemical Research Techniques, *New York, John Wiley & Sons, Inc., 1957.*]

## *Collection of Respiratory $CO_2$*

*With the supervision of the instructor*, inject intraperitoneally 2 ml of 10% Triton WR 1339 per 100 g of rat, and then a few minutes later inject 1 ml of a solution containing 1.5 millimoles of unlabeled glucose and 100 μmoles of either carboxyl or methyl $C^{14}$ acetate containing about 10 μcuries of $C^{14}$.

As soon as the isotope is injected place the rat in the respiratory jar, and place the closed jar on the respiratory apparatus. Often this process is simplified by holding the jar with the open end tilted down and allowing the rat to crawl into the jar. Avoid harsh treatment, for frightened rats produce little glycogen. *Immediately* after placing the rat in the closed system start the aeration, using the same bubble rate as before. Be sure the flow of air is adequate, otherwise $CO_2$ will accumulate in the jar, causing increased breathing and excitation of the animal, thus lowering the rate of glycogen synthesis. Cover the jar with a towel to reduce fright in the rat, and continue aeration for 30 min. If for any reason the flow of air is discontinued and cannot be started again, move the system to a hood, remove the rat, and notify the instructor immediately.

At the end of the 30-min period, stop the aeration, rapidly move the apparatus to a hood, and quickly and calmly remove the rat by tilting the mouth of the jar downward and pulling the rat out by its tail if necessary. Close off the open absorber hoses with pinch clamps to limit absorption of additional $CO_2$ from the atmosphere.

## *Sacrifice of the Rat and TCA Treatment of Liver*

Immediately after removing the rat from the respiratory apparatus, stun it with a blow on the head, and quickly decapitate it, collecting all the blood released in a beaker reserved for this purpose. *Rapidly* dissect out the liver and weigh it in a tared, iced petri dish to the nearest 0.1 g (triple beam balance). (Work rapidly in all steps before TCA addition to avoid glycogen breakdown.)

Then, in the hood, place the liver in a cold, dry mortar held in an ice bath. Slice the liver into small pieces with scissors and, add a volume (in milliliters) of an ice-cold 10% TCA solution numerically equal to three times the weight of the liver in grams. Still in the hood, mince and grind the liver with a pestle until you obtain an even suspension. Decant the cold suspension into precooled 50-ml centrifuge tubes, rinse the mortar and pestle with an additional volume of 10% TCA, and add the rinse to the original suspension.

Wrap the remainder of the rat, including the blood, in paper toweling and wax paper, and place it in the receiver marked for this purpose. Clean up any blood spots, and wash your hands thoroughly with soap and water.

Centrifuge the cold, combined TCA suspension at 1000 × g for 3 min ($\frac{3}{4}$ speed on clinical centrifuge), and then decant the supernate into a 100-ml graduate kept in an ice bath. Resuspend the precipitate in an additional 10 ml of cold 10% TCA, and spin down the precipitate as before. Add the supernatant TCA solution to the previous supernatant solution in the iced graduate.

At this stage, three preparations have been prepared for future isolation of important materials. The respiratory $CO_2$ is obtained from the alkaline respiratory absorbers. The glycogen is obtained from the TCA supernatant extract. The cholesterol, neutral fat, proteins, and nucleic acids are obtained from the TCA precipitate. These isolations may be performed simultaneously or independently. Only in the glycogen isolation is immediate further work necessary at this stage, although some time can be saved by starting the extraction of cholesterol and neutral fat at this stage. *Do not discard any solutions until the desired products are obtained and counted.*

### *Glycogen Isolation*

Dissolve a few small crystals of NaCl in the TCA extract, and then stir in one volume of 95% ethanol. Next, using a water bath, quickly raise the temperature of the suspension to 40°C, and then return the suspension to an ice bath until cooled to 1–3°C. Centrifuge the cloudy precipitate of glycogen at 500 × g (clinical centrifuge). Decant off the supernatant solution, measure its volume, and save it for future use. Redissolve the white flocculant precipitate (glycogen) in 10 ml of $H_2O$.

Plate out 0.5 ml of the TCA supernatant solution and 0.2 ml of the glycogen solution on separate, tared planchets, and dry the samples down with a heat lamp. (*Do not overheat or burn the planchet aliquots.*) Weigh the dry samples, and then count them as directed by the instructor (see Appendix III).

After plating out the glycogen sample, precipitate the remaining glycogen with one volume of 95% ethanol, centrifuge the precipitate (3 min, full speed on clinical centrifuge), and wash the precipitate in successive 5-ml washes of absolute ethanol, acetone, acetone:ether (1:1), and ether by suspension and centrifugation. (*Note:* Avoid using ether-soluble centrifuge tubes during these operations.) Finally, place the powdered glycogen in a tared vial, record its weight and radioactivity per milligram (obtained by counting the 0.2-ml aliquot of dissolved glycogen), and turn in your sample to the instructor.

### *Extraction of TCA-insoluble Residue—Isolation of Proteins and Nucleic Acid*

Extract the neutral fat and cholesterol from the TCA-insoluble residue by suspending the residue in 3 volumes of chloroform:methanol (1:1) and allowing it to stand in a covered container at room temperature for 12 hr or more before centrifuging the remaining residue (clinical centrifuge). Further extract the precipitate by suspending it in an additional 15 ml of chloroform:methanol (1:1) and centrifuging it as before. Combine the second chloroform:methanol extract with the original extract, and save it for isolation of cholesterol and neutral fats (see the following section).

Suspend the nucleic acid and denatured

protein residue in sufficient 95% ethanol to yield a final volume of 10 ml. Then plate out a 0.5-ml aliquot of the suspension on a tared planchet, and dry the sample down with a heat lamp. Weigh the dry sample, and then count it as directed by the instructor (see Appendix III).

### *Isolation of Cholesterol and Ether-soluble Nonsaponifiable Lipids*

Under vacuum, remove the chloroform: methanol from the combined extract obtained in the previous section by gently heating (steam bath) the extract in a round-bottom flask fitted with a still head connected to an aspirator. Then suspend the residue in 40 ml of 15% alcoholic KOH, and reflux for 5 hr. Add 5 ml of $H_2O$ to the cooled saponified mixture, transfer the solution to a small separatory funnel, and extract the aqueous solution three times with 20-ml volumes of low-boiling (30–60°C) petroleum ether. Then wash the petroleum ether fractions with 15 ml of $H_2O$, and add this aqueous extract to the aqueous phase. The aqueous phase is used for fatty acid isolation in the next section. Pour the petroleum ether phase into a 125-ml Erlenmeyer flask containing 5 g of anhydrous $Na_2SO_4$. Stopper the flask, and shake for 2 min to allow the $Na_2SO_4$ to dry the nonpolar solvent.

Decant the dried petroleum ether into a 100- or 200-ml round-bottom flask, and remove the petroleum ether by gently heating (steam bath) the extract while applying vacuum (still head and aspirator).

Extract the residue from the flask with three 2-ml portions of acetone: absolute ethanol (1:1), and place the combined extract in a 50-ml centrifuge tube. Add 24 ml of 0.5% digitonin, and heat the suspension on a steam cone for 30 sec before allowing the digitonide to precipitate in the cold room for at least 2 hr. Collect the digitonide from the cooled solution by centrifuging for 10 min at 2000 × g. Save the supernatant solution. Then suspend the digitonide in a final volume of 5 ml of 95% ethanol, plate out 0.5 ml of the suspension on a tared planchet, and dry the sample down with a heat lamp. Similarly plate out 1 ml of the digitonide supernatant solution containing other nonsaponifiable lipids, applying 0.25-ml portions at a time, if necessary. Weigh the dry samples, and then count them as directed by the instructor (see Appendix III).

### *Isolation of the Fatty-acid Fraction*

Acidify the aqueous phase of the saponified lipids (obtained in the previous section) to *p*H 4 with 3*N* $H_2SO_4$, using *p*H paper or a few drops of added bromcresol green (blue ⟶ yellow at *p*H 4) for *p*H determination. Extract the acidified solution with three 20-ml portions of 30–60°C petroleum ether. Remove the solvent of the combined petroleum ether extract with vacuum (still head and aspirator) and gentle heating (steam bath), and finally suspend the residue in 10 ml of ether. Plate out 0.5 ml of this solution on a tared planchet, and dry it down with a heat lamp. Weigh the dry sample, and then count it as directed by the instructor (see Appendix III).

### *Isolation of Respired $CO_2$*

Pour the contents of each of the final absorbers into a graduate cylinder, record the volume, and remove a 5-ml aliquot for isolation of respired $CO_2$. Place the 5-ml aliquot in a glass clinical centrifuge tube, add 5 ml of 1*M* $BaCl_2$ ($CO_2$-free), and mix with a glass stirring rod. Collect the precipitate of $BaCO_3$ by centrifugation (3 min at 1000 × g), and suspend it in 5 ml of 95% ethanol. Plate out 0.3 ml of this suspension on a tared planchet, and dry the sample with a heat lamp. Weigh the dry sample, and then count it as directed by the instructor (see Appendix III).

### *Radioactive Waste*

Place all glassware, instruments, and other equipment that has been in contact with radioactive material in a special container designated by the instructor. The instructor will explain the methods for decontamination and/or disposal.

### *Counting-efficiency Correction*

Using a nonvolatile $C^{14}$ standard containing a known activity (μcuries/sample), determine the number of counts per minute that this standard yields on the particular counting apparatus used in the class. Then calculate the efficiency of the counting apparatus by using the formula

$$\text{Counting efficiency} = \frac{\text{counts/min actually determined}}{\text{counts/min expected from known activity}}$$

Using the results of this determination and the known amounts of $C^{14}$ injected into the rat (number of millimoles acetate and number of μcuries per millimole of $C^{14}$ acetate), calculate the number of "countable counts" of $C^{14}$ acetate injected into the rat.

## Report of Results

Using the planchet tares and the weights of the plated planchets, calculate the appropriate $J$ factors for counting each of the plated samples at infinite thinness (using the radioactive wax curve for organic compounds; see Appendix III). Then summarize the results of the experiment by completing a table such as that shown below. Determine the values for the last column (percentage of injected counts isolated in each fraction) using the value of countable counts injected obtained in the previous section. In all cases subtract the background reading from the observed counts before starting the calculations.

After completing the table, rationalize each result in light of the physiological condition of the rat and the known pathways of acetate metabolism. If both methyl- and carboxyl-labeled açetate have been used, contrast the results of these two isotopes. Trace the path of acetate into each of the compounds in question.

## Discussion

Usually the operations performed in Counting-efficiency Correction are replaced by counting a diluted sample of the initial substrate. In this particular experiment, the alternate procedure of counting a known standard is necessary, due to the unfavorable nature of sodium acetate.

### Exercises

1. Why is it necessary to work in the cold and in the hood while homogenizing the liver in 10% TCA?
2. Suppose you wished to learn the specific sites of the labeling in the isolated glycogen. What steps would you perform to determine the labeling in the various atoms of the glucose residues in glycogen?
3. What labeling pattern in the glucose residues of liver glycogen would you expect from 1-$C^{14}$ acetate?
4. What labeling pattern would you expect in cholesterol following the injection of 1-$C^{14}$ or 2-$C^{14}$ acetate?
5. Trace the acetate label into the Krebs cycle and into $CO_2$.
6. Look up the "dicarboxylic acid shuttle," and predict how this will influence the labeling patterns.
7. If the reaction catalyzed by pyruvate kinase is not appreciably reversed in liver (which seems to be the case), how does this fact influence the expected glucose labeling patterns (see question 4)? Is this fact related to question 6?

### References

Tipson, N. et al., *J. Biol. Chem.*, **176,** 1263 (1948).

Fruton, J. S., and S. Simmonds, *General Biochemistry* (2nd ed.), Chapters 18–20, 25, 26, Wiley, New York, 1958.

Summary of Experiment 34

| Material | Observed Counts/Min/Planchet | Milligrams of Sample/$cm^2$ on Planchet | J Factor | Actual Counts/Min/Planchet | Aliquot Size on Planchet | Total Sample Size | Total Actual Counts/Minute | % of Injected Counts |
|---|---|---|---|---|---|---|---|---|
| Respired $CO_2$ | | | | | 0.3 ml | 300 ml | | |
| Glycogen | | | | | 0.2 ml | 10 ml | | |
| TCA supernate | | | | | 0.5 ml | | | |
| Cholesterol | | | | | 0.5 ml | 5 ml | | |
| Nonsaponifiable lipids | | | | | 1.0 ml | 30 ml | | |
| Fatty acids | | | | | 0.5 ml | 10 ml | | |
| Proteins and nucleic acids | | | | | 0.5 ml | 10 ml | | |

# APPENDIX I

# Photometry

## Principles

If white light is passed through a solution containing a colored compound, certain wavelengths of light are selectively absorbed. The resultant color observed is due to the transmitted light.

The absorption spectrum of riboflavin, in which one plots the amount of light absorbed at given wavelengths, illustrates this point (Fig. 28). Riboflavin appears yellow to the eye. This is because the only absorption within the visual range is in the blue region, where the maximum absorption is 450 mμ. The ultraviolet absorptions having maxima at 260 and 370 mμ are not visually recorded, but can be recorded with special instruments.

Measuring light absorption aids us in both the identification and quantification of substances. For example, the absorption spectrum above is characteristic of riboflavin, and the amount of absorption by riboflavin at a given wavelength is a function of the riboflavin concentration.

Quantitative photometric measurement is based on two formalized laws. The first is *Lambert's law,* which states that the *proportion* of incident light absorbed by a medium is independent of its intensity, and that each successive unit layer of the medium absorbs an equal fraction of the light passing through it. For example, if the intensity of light incident upon the

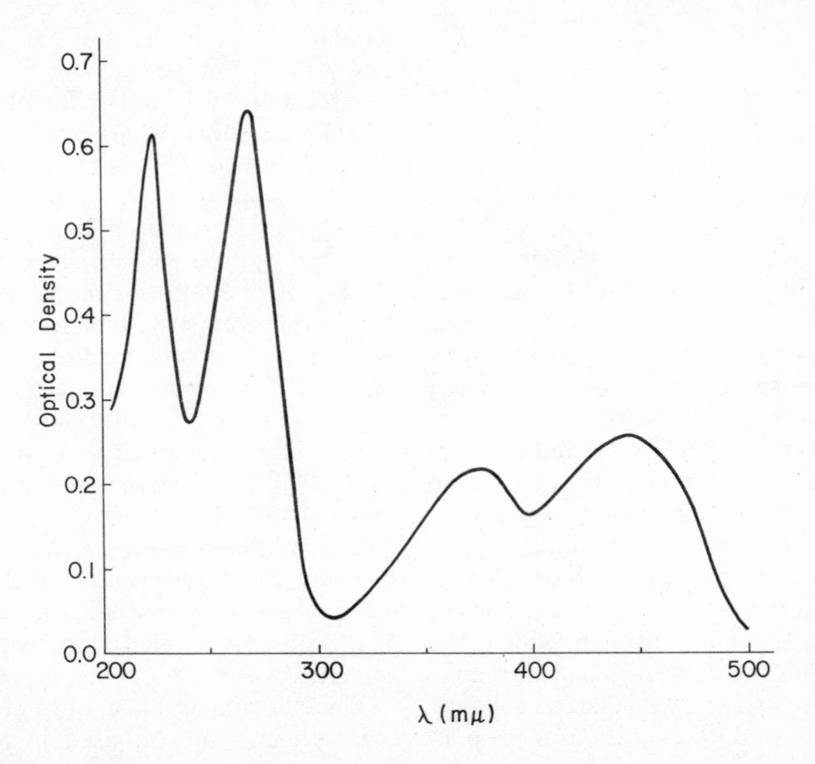


FIG. 28. *Absorption spectrum of riboflavin* ($2.2 \times 10^{-5}M$ *in 0.1M sodium phosphate, pH 7.06*).

medium is unity, and the absorption of each unit thickness of the absorbing medium is equal to one-tenth of the incident light, then the light intensity will be diminished successively on passing through each unit layer of medium as follows: 1.0, 0.9, 0.81, 0.73, 0.67,···. This statement leads to the following mathematical expression:

$$I = I_0 \cdot e^{-\alpha l} \tag{1}$$

where $I_0$ = intensity of the incident light; $I$ = intensity of the transmitted light; $l$ = thickness of the layer (in cm); and $\alpha$ = absorption coefficient of the medium.

Converting equation (*1*) to the logarithmic form, we have

$$\ln \frac{I_0}{I} = \alpha l \tag{2}$$

Utilizing logarithms to the base 10 the absorption coefficient $\alpha$ is converted into the *extinction coefficient K*.

$$\alpha = 2.303K \tag{3}$$

Thus

$$\log \frac{I_0}{I} = Kl \tag{4}$$

Log $(I_0/I)$ is termed the *optical density* (O.D.), or the *absorbance* (*A*).

It is apparent that the coefficients $\alpha$ and $K$ include a concentration factor. The other important law of photometry, *Beer's law*, deals with this variable. Beer demonstrated that the intensity obtained when light passes through a solution of concentration $c$ and length $l$ is equal to that obtained when light passes through a solution of the same substance at concentration $c/2$ and length $2l$. Generalized, Beer's law states that the light absorption is proportional to the number of molecules of absorbing substance through which the light passes. Thus if the absorbing substance is dissolved in transparent solvent, the absorption of the solution is proportional to its molar concentration.

It is apparent that Beer's law and Lambert's law can easily be combined by recognizing that $\alpha$ and $K$ include a concentration factor. Hence, we write

$$Ec = K \tag{5}$$

where $E$ is the molar extinction coefficient, and $c$ is the concentration in moles per liter.

This equation leads to equation *6*.

$$I = I_0 \cdot 10^{-Ecl} \tag{6}$$

$$\text{or } \log_{10} \frac{I_0}{I} = Ecl = \text{optical density (or absorbance)}$$

The molar extinction coefficient, $E$ (often written as $\epsilon$), is numerically equivalent to the optical density ($\log_{10} I_0/I$) of a $1M$ solution of the substance, with a light path of 1 cm. Since optical density is dimensionless, $E$ is in terms of liters/(conc. × length). Then $E^{1M}_{1\,\text{cm}}$ is in units of liters moles$^{-1}$ cm$^{-1}$; that is, cm$^2$/millimole. The millimolar extinction coefficient $E^{1\,M}_{1\,\text{cm}}$ would then be in units of cm$^2$/micromole, and would be numerically equivalent to the optical density of a 1-millimolar solution of the substance where the light path is 1 cm. The term $E^{1\%}_{1\,\text{cm}}$ is often used to refer to a 1% solution (1 g/100 ml).

In practice, log $I_0/I$ is not determined directly by the measurement of $I_0$ and then $I$ on the solution alone. Rather, the solution is compared to a suitable blank—that is, the solvent alone in an identical cuvette. The term $I_0$ is obtained from the blank; $I$, from the solution. The ratio of $I$ and $I_0$ is referred to as the transmittance, $T$, the fraction of incident light transmitted: $T = I/I_0$.

It should be remembered that a constant is given for a *particular* wavelength and solvent, and that another constant applies at a different wavelength or with another solvent.

The following is a summary of the terms (with their symbols) used in absorption spectrophotometry:

| *Term* | *Symbol* |
|---|---|
| Optical density (absorbance) | OD (or $A$) |
| Extinction | $E$ |
| Transmittance | $T$ |
| Molar extinction coefficient | $E_{1\ \mathrm{cm}}^{1\ M}$ (cm²/millimole) |
| $E$ value | $E_{1\ \mathrm{cm}}^{1\ M}$ |
| Intensity of transmitted light after passage through cell | $I$ |
| Length of cell in centimeters | $l$ |
| Concentration in moles per liter | $c$ |

Unlike Lambert's law, which holds in all cases examined to date, Beer's law has some exceptions; usually, however, the anomalies can be attributed to a change in the nature of the solute, with a concomitant change in concentration. Some acids, bases, and salts in solution do not obey Beer's law because they are more completely ionized with increasing dilution and because the light absorption of the ions differs from that of the nonionized molecules. Thus concentrated nitric acid shows the absorption of $HNO_3$ molecules, whereas the diluted acid shows the characteristic absorption band of the nitrate ion. For a similar reason, concentrated cupric chloride solution is green (color of $CuCl_2$), although the diluted solution is blue (color of $Cu^{+2}$ ions). Some organic compounds have varying tendencies to aggregate at various concentrations. For a substance which follows Beer's law, a plot of the optical density versus concentration yields a straight line.

The absorption spectrum obtained from quantitative photometric measurement is characteristic for a particular compound or class of compounds. The shape of the curve—that is, the location of maxima and minima—is therefore commonly used in identifying classes of compounds, and the intensities of absorption are used for quantitative estimation.

## Photometer: Construction and Properties

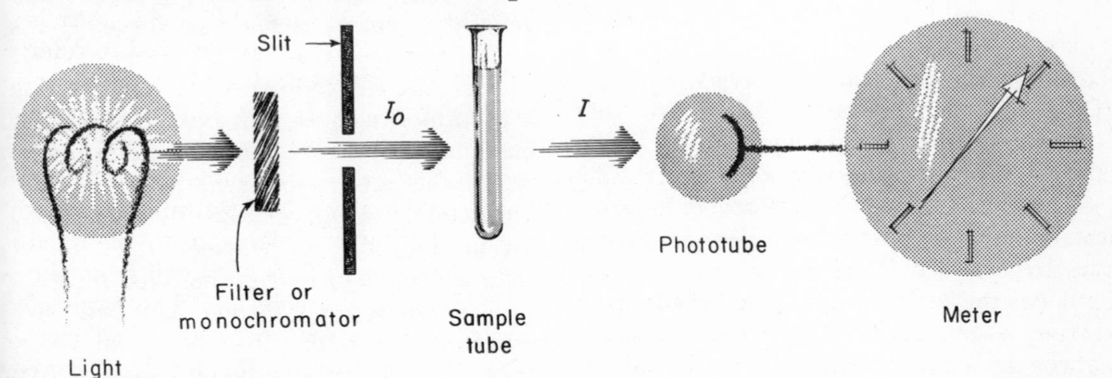


FIG. 29. *Typical photometer.*

### *Light Source*

The light source must have sufficient energy in the range of wavelengths required for analysis of the sample. The hydrogen-discharge lamp is a useful source in the ultraviolet range and in part of the visible range, and the tungsten lamp can be used at wavelengths greater than about 340 mμ.

### *Wavelength Selector*

Photometry is performed with monochromatic light selected from a suitable light source. The term *monochromatic* may be somewhat misleading, since it is usually not possible to have a light source with only a single wavelength represented. When one speaks of a monochromatic light of, say, 500 millimicrons, one usually means a source which has its *maximum* emission at this wavelength and has progressively less energy at longer and shorter wavelengths. Thus a correct description of this light must, in addition, specify the range of wavelengths represented; for example, 95% of the energy from a 500 m$\mu$ source falls between 495 and 505 m$\mu$.

For a given light source it is clear that one obtains less intensity with light of greater purity. On the other hand, the greater the spectral purity of monochromatic light, the greater is the sensitivity and resolution of measurements.

Absorption filters provide a simple and inexpensive means for obtaining light of different wavelengths, and many photometers employ them. Filters generally have a rather broad transmission, and the resolution in taking an absorption spectrum is low (peaks are "smeared" or obscured). Moreover, one may make measurements only at wavelengths for which filters are available.

Monochromatic light can also be obtained by means of a prism or a diffraction grating, and many photometers employ one of these. Light of relatively high purity may generally be obtained, and the wavelength selection is continuously variable.

In choosing the wavelength to use when the concentration of a colored compound is being measured, one generally selects the region of maximal absorption. This is not absolutely necessary, since Beer's law may be expected to hold at all wavelengths where there is appreciable absorption. Thus when an interfering substance absorbs at the wavelength of maximal absorption of the substance being measured, another wavelength may be readily chosen.

### *Slit*

The intensity of light emitted through a given filter or monochromator may be too intense or too weak for a given light-sensing device to record. It is therefore necessary to be able to adjust the intensity of the incident light ($I_0$). This is performed by opening or closing a pair of baffles placed in the light path to form a slit.

### *Sample Tubes*

From Lambert's law it follows that if optical density is to be used as a measure of concentration, the length of the path traversed by light in the sample tube must be the same as that in the blank; that is, the thickness of the sample tube must be the same as that of the blank. One cannot assume that all tubes are identical in this respect; rather, the tubes or cuvettes used in colorimetry must be checked for uniformity (see p. 187 for this procedure).

### *Light Detection by Phototubes*

There are two different types of photoelectric devices used in the usual colorimeter or spectrophotometer: the photovoltaic cell and the vacuum phototube. The photovoltaic cell (Fig. 30) consists of a photosensitive semiconductor (selenium) sandwiched between a transparent metal film and an iron plate. Light falling on the surface of the cell causes a flow of electrons from the selenium to the iron, thereby generating an electromotive force. If the circuit is closed through a microammeter, the current induced is proportional, within a certain range, to the intensity of light incident upon the selenium. The limitations of the cell are its low sensitivity (it does not detect light of very low intensity) and its insensitivity to wavelengths shorter than 270 m$\mu$ and longer than 700 m$\mu$.

The vacuum phototube has two electrodes with a maintained potential difference (Fig. 31). The cathode (negative electrode) consists of a plate coated with a photosensitive metal. Radiation incident upon the photosensitive cathode causes an emission of electrons (the photoelectric effect), which are collected at the anode. The resultant photocurrent can be readily amplified and measured.

A variation of the phototube is the photomultiplier tube, which has, in addition to a

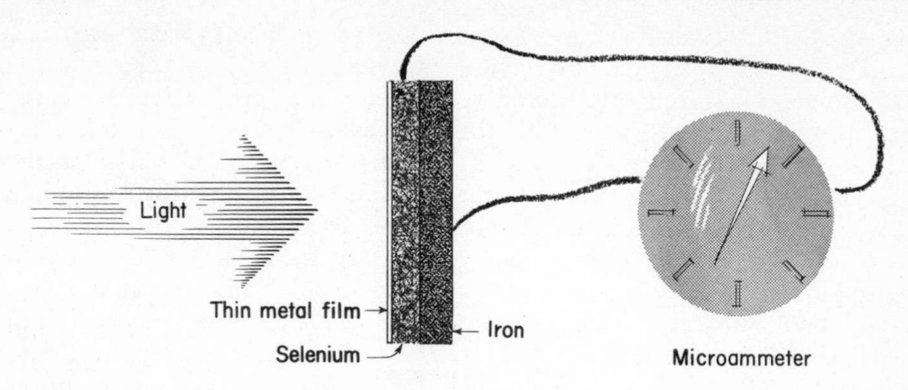


FIG. 30. *Photovoltaic cell.*

photocathode and an anode, several intermediate electrodes known as dynodes. Electrons emitted from the cathode strike the first of these dynodes, whereupon by secondary emission several secondary electrons are emitted. Each of these in turn strikes the second dynode, and the process is then repeated until the original photocurrent is amplified as much as 100,000 times. Photomultipliers are very sensitive light detectors, and the spectral sensitivity may be varied by use of different photosensitive coatings.

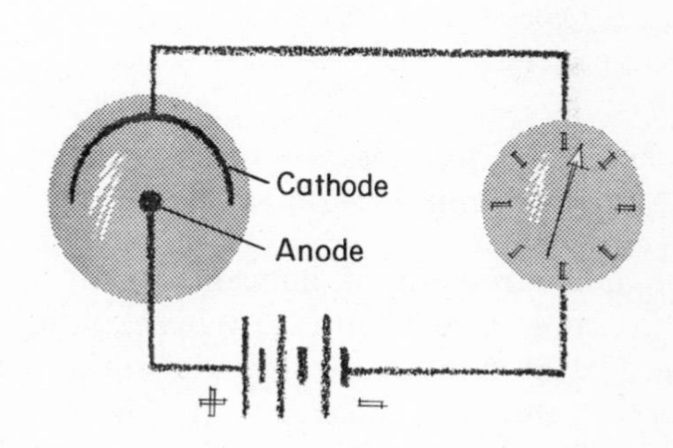


FIG. 31. *Phototube circuit.*

## Procedures

### *Selection of Uniform Tubes*

Select and carefully clean eight photometer tubes which are free of scratches. (Ordinary Pyrex tubes are suitable for many photometers.) Fill one of the tubes half-full with $H_2O$ (use this tube as the blank), and fill the remaining seven half-full with a colored solution (for example, dilute $CuSO_4$ or dilute $K_2Cr_2O_7$). Place the blank in the light path, and, using any filter or wavelength, adjust the colorimeter to 100% transmission (zero optical density). If Pyrex tubes are used, be sure to turn each tube so that the word Pyrex or some other identifying mark faces you, for Pyrex tubes do not have uniform curvature. Then determine the optical density of the seven remaining tubes, using the same filter or wavelength of light. Any tube whose optical density varies more than 4% from the average should be eliminated. Select and check other tubes until you have seven uniform tubes. Then empty one of the sample tubes, wash it, and half-fill it with $H_2O$. Empty the blank, half-fill it with the dilute colored solution used previously, and check the last tube (old blank) against the new blank to see if the optical density of the colored solution viewed through the new tube corresponds with that of the other tubes containing colored solution. When you have selected eight uniform tubes, mark them permanently, and save them for all future colorimetric assays.

### *Procedure for Measuring the Absorption Spectrum*

Using a suitable concentration of a colored substance ($5 \times 10^{-5} M$ riboflavin is a good choice), measure the optical density at suitably spaced intervals over the range of the instrument which you have available. If possible, measurements should be made at 20- or 25-mμ intervals. Plot the OD versus wavelength.

Riboflavin has absorption peaks at about 375 mμ and 450 mμ. It is of interest that photometers which employ filters usually give a spectrum showing only a single peak in this region. A substance such as riboflavin may be used to illustrate that Beer's law is obeyed at all wavelengths at which the compound absorbs. Select three different wavelengths, and illustrate this fact by the procedure outlined below for measuring concentration.

### *Procedure for Measuring Concentration Photometrically*

The measurement of concentration of a given substance requires a blank and a series of standards. A series of unknown samples may be measured at the same time. The blank contains all the constituents used in the analysis *except* the substance to be measured. We can assume, then, that the difference in color between the blank and sample tubes is due only to the sample. The standards contain a range of known concentrations of the standard substance. Measurement of the unknown samples should also be made at several different concentrations to be sure of linear response.

Select the appropriate filter, or set the wavelength selector so as to obtain the desired monochromatic light, usually the wavelength of one absorption maximum. Then adjust the colorimeter so that the optical density of the blank is zero (100% transmission). The optical density of an opaque substance (for example a piece of black cardboard or the opaque portion of the cuvette holder) should be infinite (zero transmission). Adjust this if necessary.

Now measure and record in tabular form both the $\% T$ and the optical density values for each of the samples and each of the standards. Plot both OD and $\% T$ against concentration for both standards and samples.

A precise colorimetric analysis is most easily obtained when the optical density is a linear function of the concentration of the substance—that is, when Beer's law is obeyed. Whether Beer's law obtains can be readily ascertained by drawing a standard curve (optical density versus concentration) or by determining the extinction coefficient, $E$.

From the combined laws, $OD = Ecl$, where $l$ is constant because the tubes are uniform. From this equation it is possible to calculate a relative $E$, since the relationship $E = OD/c$ holds. The extinction coefficient should be constant for a range of concentrations of the standards and the unknowns. For example, given the following data concerning the colorimetric determination of riboflavin standard (all optical densities are, of course, relative to the blank, which has zero optical density), the following results would be obtained:

Riboflavin Standard Curve at 450 mμ

| c, or Concentration (μmoles riboflavin/tube) | $OD_{450}$ | $E = OD/c$ |
|---|---|---|
| 0.1 | 0.122 | $\frac{0.122}{0.1} = 1.22$ |
| 0.2 | 0.244 | $\frac{0.244}{0.2} = 1.22$ |
| 0.3 | 0.366 | $\frac{0.366}{0.3} = 1.22$ |
| 0.5 | 0.610 | $\frac{0.610}{0.5} = 1.22$ |

If the above determination is run with 10-ml samples in tubes with a 1-cm light path, then the molar extinction coefficient $E_{1\,\text{cm}}^{1\,M}$ of riboflavin is $12.2 \times 10^3$ liters mole$^{-1}$ cm$^{-1}$. Once an $E$ value is obtained, the concentration of the compound in question is given by the equation

$$c = \frac{\text{OD}}{E}$$

Thus with an $E$ value of 1.22, optical density values of two unknowns of 0.305 and 0.500 represent, respectively, concentrations of 0.25 and 0.41 $\mu$moles of riboflavin per tube.

The determination of the concentration of a compound in an unknown can also be derived directly from the standard curve (Fig. 32) by graphical means. Both the constancy in the extinction coefficients and the linearity of the standard curve indicate that Beer's law is valid for this system over the concentration range studied.

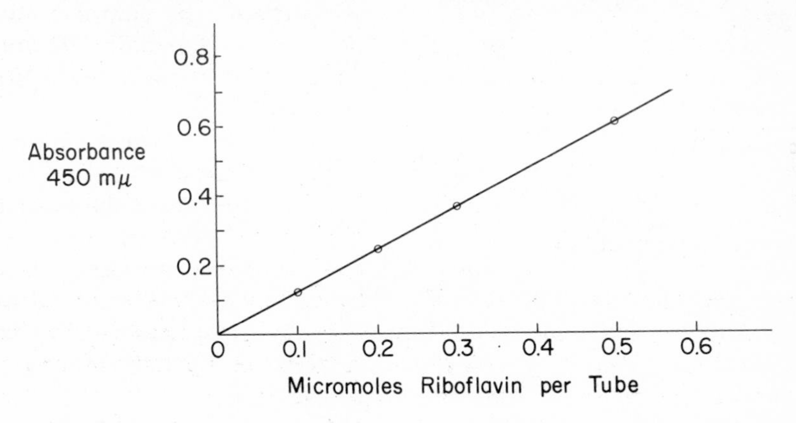


FIG. 32. *Standard curve for riboflavin.*

## References

Gillam, A., and E. S. Stern, *Introduction to Electronic Absorption: Spectroscopy in Organic Chemistry* (2nd ed.), St. Martin's, New York, 1958.

Weissberger, Arnold (Editor), *Physical Methods in Organic Chemistry* (2nd ed.). (Vol. 1, Part II), Interscience, New York, 1949.

Brode, W. R., *Chemical Spectroscopy* (2nd ed.), Wiley, New York, 1952.

Koller, L. R., *Ultraviolet Radiation*, Wiley, New York, 1952.

# APPENDIX II

# Chromatography

Chromatographic techniques represent one of the most powerful tools available for the resolution of complex mixtures. Chromatography may be defined as "an analytical technique for resolution of solutes, in which separation is made by differential migration in a porous medium and migration is caused by flow of solvent."* Thus the separation of mixtures is affected by the differential affinity of the components for a stationary phase (a solid or liquid) and for a mobile phase (a gas or liquid). The chief physicochemical phenomena responsible for this affinity for the nonmobile phase are adsorption, ion exchange, or solution in a stationary solvent. The three types of chromatography: adsorption, ion exchange, and partition chromatography derive their names from these phenomena. The distinction among the three types is more or less arbitrary since it is often uncertain which one or whether more than one of the three phenomena operate in a given procedure.

## Adsorption Chromatography

In its simplest, classical form, adsorption chromatography consists of separation of substances by filtration of a solution through a column of finely powdered adsorbent in a glass tube. The substances are adsorbed at the top of the column, and then slowly move down the column as a suitable solvent or solvent mixture flows through the column. This process is termed "development." If the substances are pigments, the separation is readily visualized, and colored bands are seen to move down the column during development. The substances which have more affinity for the solvent (that is, those which have a lesser tendency to be adsorbed) move more quickly down the column. At this stage in the procedure, solvents can be removed, the column extruded from the glass tube, and the colored zones cut out or separated mechanically (with a spatula). If the substances are not colored, the contents of the column can be arbitrarily divided and the fractions examined by suitable analytical methods. Alternately, the development can be allowed to proceed until the separated compounds emerge from the column and appear in the filtrate. A series of fractions of the filtrate from the column is then collected and the separated solutes detected by suitable means.

The relative affinity of the solutes for the adsorbent is a function of the chemical constitution of the substances being separated, the nature of the solvent, and the nature of the adsorbent (see Lederer and Lederer in References). Solid adsorbents commonly used are aluminum oxide, magnesium silicate, calcium hydroxide, florisil, charcoal, and sucrose. Among the solvents used are hydrocarbons such as hexane, benzene, ether, chloroform, various alcohols, ketones and others. These solvents are used in adsorption chromatography involving the separation and purification of substances which are soluble in organic solvents—phospholipids, steroids, carotenoids, and chlorophyll.

More recently (1950), aqueous solutions have been used in adsorption chromatography, with columns containing adsorbents such as charcoal, cellulose, and starch. For example, the enzyme $\alpha$-amylase has been purified in starch columns—a particularly interesting use of an enzyme's affinity for its substrate. (Consult the references for other examples; for example, Lederer and Lederer, pp. 23, 33–37.)

*Strain, H. H., *Chem. Eng. News*, **30**, 1372 (1952).

## Ion-exchange Chromatography

Ion-exchange chromatography, as the name implies, utilizes the differential affinity of charged molecules in solution for inert, immobile (insoluble), charged substances.

The first ion-exchange materials (still used today for water softening) were natural silicates, such as the zeolites. An example of these zeolites is chabazite ($Na_2Al_2Si_4O_{12} \cdot 6H_2O$), which possesses an insoluble, highly porous, charged, chainlike, crystalline silicate structure, in which the sodium ions are readily accessible to calcium ions or any other ions small enough to penetrate "holes" in the lattice. Small ions such as calcium ions undergo exchange with the sodium ions. The extent of the exchange depends on the calcium ion concentration and on the relative affinity of the silicate for sodium and calcium ions. This is an example of a cation exchanger. Substances which permit exchange of crystal-lattice anions are anion exchangers. For example, the hydroxide ions of the silicate clay minerals, kaolinite and montmorillonite, can undergo exchange with chloride, sulfate, and phosphate ions.

General descriptions of cation and anion exchangers are given below ($Z$ indicates the remainder of the exchanger).

Cation exchanger:

$$Ca^{+2} + Na_2^{+}Z^{-2} \rightleftharpoons 2Na^{+} + Ca^{+2}Z^{-2}$$

Anion exchanger:

$$Cl^{-} + Z^{+}OH^{-} \rightleftharpoons OH^{-} + Z^{+}Cl^{-}$$

The ion exchangers commonly used in the laboratory are synthetic resins, which can be tailor-made for a particular job. These ion-exchange resin particles can be described as elastic, three-dimensional hydrocarbon particles. The hydrocarbon portion of many of the commonly used resins is formed by the copolymerization of styrene and divinylbenzene.

Divinylbenzene thus adds the third dimension—that is, cross-linkage, to the polymer. The degree of cross-linkage can be varied merely by changing the ratio of divinylbenzene to styrene. The desired functional groups can be added to this matrix by varying the chemical composition of the monomers. For example, a resin possessing strong acid groups (for example, $—SO_3H$) can be made by substituting a vinylbenzenesulfonic acid for styrene; similarly, a resin with strong basic groups ($—N^{+}R_3$), weak acid groups ($—COOH$), or weak basic groups ($—NH_3^{+}$) can be made. The acidic resins are cation-exchange resins; the basic resins are anion-exchange resins.

The affinity of a given ion-exchange resin for ionized solutes depends on the nature of the charged group on the resin and on the nature of the polar group on the solute. Thus carboxylic acid and amino resins form only very unstable salts with other weak bases and acids. In contrast, resins which contain "strong" ionizable groups can form stable salts with "weak" as well

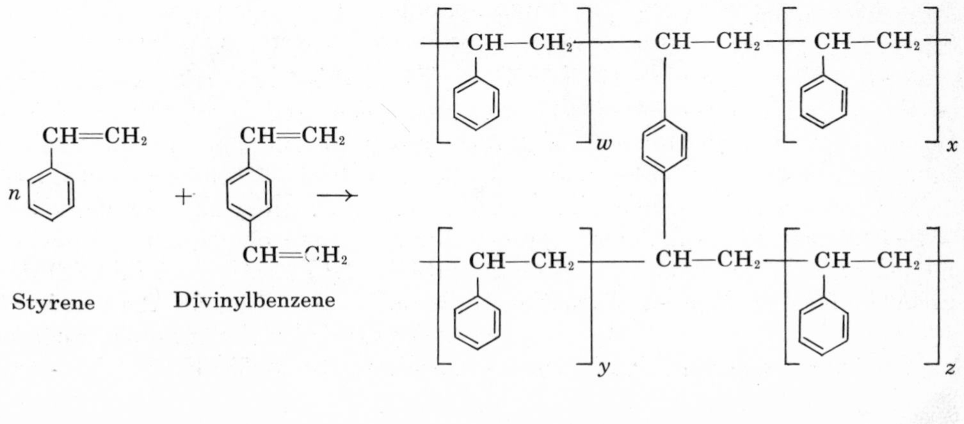

$n$ CH=CH₂ (benzene ring) + CH=CH₂ (benzene ring) CH=CH₂ →

Styrene Divinylbenzene

$$-\left[CH-CH_2\right]_w - CH-CH_2 - \left[CH-CH_2\right]_x -$$

$$-\left[CH-CH_2\right]_y - CH-CH_2 - \left[CH-CH_2\right]_z -$$

as "strong" anions and cations. Thus, sufonic acid groups form stable salts with weak bases, and quarternary amino groups form stable salts with weak acids.

In addition, a given ion-exchange resin exhibits a preference for certain ions. In general, the greater the valency of an ion, the more tightly it is bound by the resin—that is, trivalent > divalent > monovalent ions. Calcium ions displace sodium ions on such a resin. At each ionic valency there also exists a preferential order of exchange. In general, the higher the atomic number, the more tightly is the ion bound by the resin.

For Dowex 50: $Ba^{+2} > Sr^{+2} > Ca^{+2} > Mg^{+2} > Be^{+2}$ (divalent cations)
$Ag^{+} > Tl^{+} > Cs^{+} > Rb^{+} > NH_4^{+} > K^{+} > Na^{+} > H^{+} > Li^{+}$ (monovalent cations)

For Dowex 1: $CNS^{-} > I^{-} > NO_3^{-} > Br^{-} > CN^{-} > HSO_4^{-}$
$> HSO_3^{-} > NO_2^{-} > Cl^{-} > HCO_3^{-} > CH_3COO^{-} > OH^{-}$

For Dowex 3: $I^{-} > NO_3^{-} > Br^{-} > CN^{-} > HSO_3^{-} > BrO_3^{-} > Cl^{-} > OH^{-} > HCO_3^{-} > F^{-}$

These series serve only as a guide. Increasing the concentration of an ion of lower preference permits the removal of ions located several steps earlier in the sequence.

Finally, the affinity of a given ion-exchange resin for ionized solutes depends on accessibility of the ionized groups on the resin to the ions in solution. If the solutes are too large to penetrate the resin, ion exchange will not occur. The size of the "holes" in the resin is a function of the degree of cross linkage: the lower the cross-linkage, the larger the holes and the greater the accessibility. Thus separation of protein molecules whose molecular weight is 100,000 or more can be accomplished only with ion-exchange resins having a low degree of cross-linkage, such as XE-64 or other similar resins. In the Dowex resins the degree of cross-linkage is a function of the relative amounts of divinylbenzene monomer used; hence "4% cross-linkage" means 4% divinylbenzene, 96% styrene and other monovinyl monomers. Furthermore, the lower the cross-linkage, the faster the internal diffusion rates (and therefore the rate of ion exchange), the lower the selectivity for specific ions, and the lower the mechanical strength of the resin. Low cross-linked resins take up relatively large amounts of water, and swell extensively. In contrast, high cross-linked resins have low diffusion rates, are rather brittle, and take up little water.

Many different ion-exchange materials have been developed by modification of natural products. Ion-exchange materials derived from cellulose have been used with great success for the separation of polyelectrolytes of high molecular weight, such as proteins (see Sober and Peterson in References). These materials are di- and triethylaminoethylcellulose (anion-exchange resins) and carboxymethyl-, phospho-, and sulfomethylcellulose (cation- exchange resins):

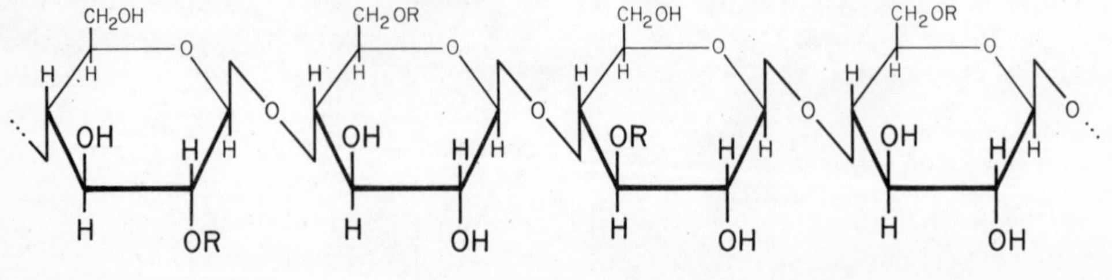

where R is one of the following:

—$PO_3H_2$ (phosphocellulose)
—$CH_2SO_3H$ (sulfomethylcellulose)
—$CH_2$–$CH_2$–N=$(CH_2$–$CH_3)_2$ (diethylaminoethylcellulose, DEAE)
—$CH_2$–$CH_2$—$N^+$≡$(CH_2$–$CH_3)_3$ (triethylaminoethylcellulose, TEAE)
—$CH_2$–COOH (carboxymethylcellulose, CMC)
—$CH_2$–$CH_2$–$CH_2$–$N^+$≡$(CH_2$–$CH_2OH)_3$ and similar forms (ECTEOLA)

Attention should also be drawn to the process of "molecular sieving" carried out on various cross-linked dextran polymers (Sephadexes, Pharmacia, Uppsala, Sweden). These useful "cage-like" polymers are classified according to the apparent dimensions (molecular weight) of solutes which will be held. In addition to the pore-size differences, which are related to the extent of cross linking, the Sephadexes have also been modified chemically to yield such derivatives as DEAE-Sephadex. These materials have been widely used recently for desalting and fractionating peptides and proteins.

For separation of mixtures of ionic substances, a glass column is loaded with the buffer-equilibrated form of the desired ion-exchange resin. Then the mixture of substances in the same buffer solution is applied to the top of the column in a quantity far less than the total capacity of the resin. The compounds exchange for the ions at the top of the resin column. The column is then "developed" by passing buffer solutions through it. Substances are released from the resin as a result of changes in variables affecting the affinity of the resin for various components of the mixture (for example, by increasing the concentration of the buffer solution with respect to one or more ions or by changing the temperature or *p*H of the buffer). In an ideal separation each substance moves down the column at a different rate, and emerges pure from the column. An example of such a procedure is the separation of amino acids by means of Dowex 50 resin, (see Moore and Stein in References).

## Partition Chromatography

In partition chromatography, separation is achieved by differential migration of solutes resulting from differences in distribution between two immiscible solvents. One solvent, *the stationary phase*, is successively washed with a second phase, *the mobile phase*, in such a manner that the solutes partition, or separate, themselves into various areas in the solvent series. The extent of these separations can be roughly predicted from the partition coefficients, $\alpha$, of the substances.

$$\alpha = \frac{\text{conc of compound in solvent A at equilibrium}}{\text{conc of compound in solvent B at equilibrium}}$$

The partition coefficient can be readily determined experimentally from the distribution of the substance between the two solvents at equilibrium in a static situation (for example, in a test tube or separatory funnel). The partition coefficient is a constant at a given temperature. Under ideal conditions the value of $\alpha$ is unaffected by the presence of other substances or by the concentration of the solute in question.

Paper chromatography, countercurrent distribution, and partition-column chromatography all make use of the principles of partition chromatography. These principles are best illustrated by the method of countercurrent distribution which serves as a model for all these forms.

Countercurrent distribution is usually accomplished with a special apparatus perfected by Craig and King (see References). This apparatus consists of a series of tubes interconnected such that a stationary liquid phase is equilibrated with an immiscible liquid phase which can be moved from tube to tube (see the schematic representations on this page).

To illustrate the application of this technique, consider two substances A and B. The partition coefficient for A in two immiscible solvents in the same tube may be defined as

$$\alpha = \frac{[\text{A}]\text{ upper phase}}{[\text{A}]\text{ lower phase}}$$

where the brackets denote concentration. If the volumes of the two phases are equal, then

$$\alpha = \frac{\text{amount of A in upper phase}}{\text{amount of A in lower phase}}$$

If $x_A$ is that *fraction* A which is in the upper layer and $y_A$ is that fraction of A in the lower layer, then

$$\alpha = \frac{x_A}{y_A}$$

Similarly for B,

$$\alpha = \frac{x_B}{y_B}$$

Consider a series of tubes containing equal volumes of the lower, or stationary, phase:

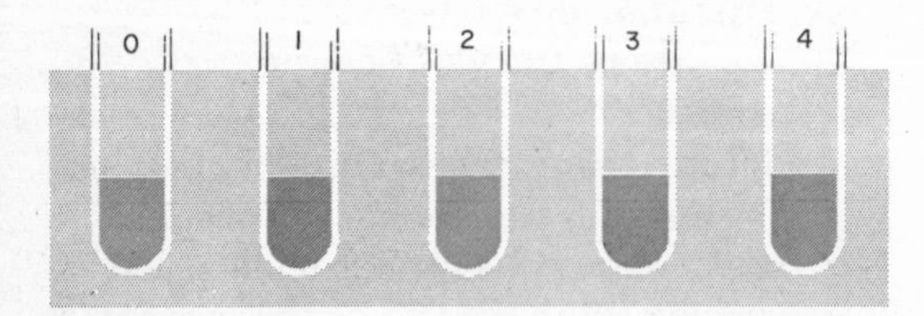


A known amount of solute A is added to tube 0. Then an equal volume of upper-phase solvent is mixed with the lower phase of tube 0 until equilibrated:

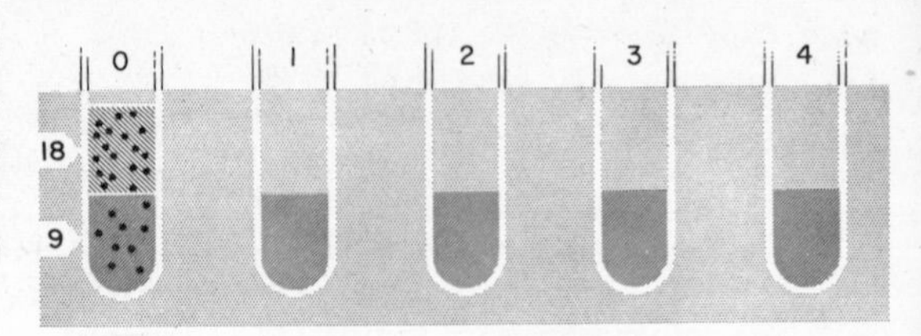


The quantity of A in the upper and lower phases is defined by $x_A$ and $y_A$, respectively (solute is denoted by dot pattern).

The upper phase is moved to tube 1, and fresh upper phase is added to tube 0:

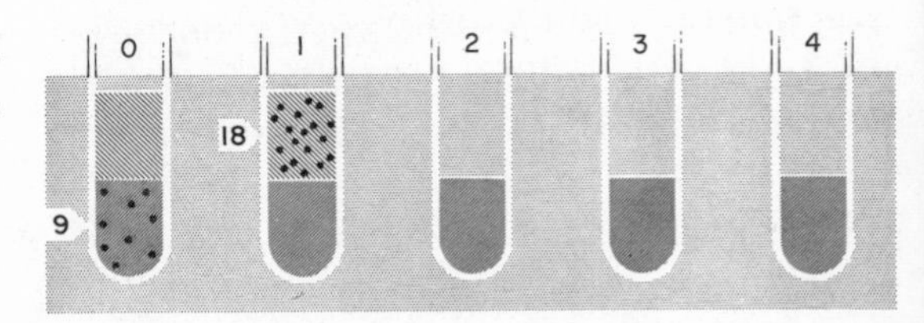


Tubes 0 and 1 are mixed; this constitutes one transfer, with the following results:

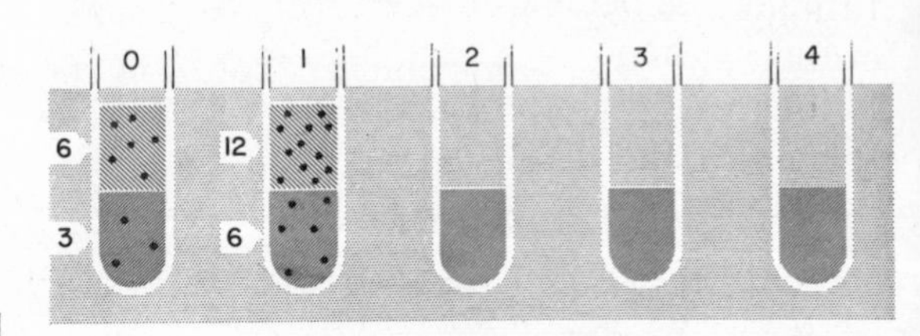


The relative quantities of A in each tube are fixed by $\alpha$. Thus in tube 0 (dropping the subscript A), the upper phase must contain $xy$ fraction of total A and the lower phase $y^2$ fraction, since

$$\alpha = \frac{xy}{y^2} = \frac{x}{y}$$

for tube 0. Similarly for tube 1,

$$\alpha = \frac{x^2}{xy} = \frac{x}{y}$$

The results of several transfers, tabulated on p. 195, list the fraction of A (the sum of the fractions of the two phases) in *each tube* after each transfer. The sum of fractions in all the tubes is given in the final column.

| Transfer Number | Tube Number 0 | 1 | 2 | 3 | 4 | Sum of Fractions in All Tubes |
|---|---|---|---|---|---|---|
| 0 | 1 | | | | | $(x + y)^0 = 1$ |
| 1 | $y$ | $x$ | | | | $(x + y)^1 = 1$ |
| 2 | $y^2$ | $2yx$ | $x^2$ | | | $(x + y)^2 = 1$ |
| 3 | $y^3$ | $3y^2x$ | $3yx^2$ | $x^3$ | | $(x + y)^3 = 1$ |
| 4 | $y^4$ | $4y^3x$ | $6y^2x^2$ | $4yx^3$ | $x^4$ | $(x + y)^4 = 1$ |

It is seen that the distribution in the fractions is merely a binominal expansion of $(x + y)^n$, where $n$ is the number of transfers. Since

$$(x + y)^n = x^n + nx^{n-1}y + \frac{n(n-1)}{2!}x^{n-2}y^2 + \frac{n(n-1)(n-2)}{3!}x^{n-3}y^3 + \cdots$$

the distribution in the series may be calculated for any $n$.

These principles are illustrated by the tabulations below. For example, when substance A has $\alpha = \frac{1}{3}$ and B has $\alpha = 1$, the following data are calculated, after four transfers and after eight transfers:

| Number of Transfers | Substance | Total Fraction in Tube Number 0 | 1 | 2 | 3 | 4 | 5 | 6 | 7 | 8 |
|---|---|---|---|---|---|---|---|---|---|---|
| 4 | A | 0.32 | 0.41 | 0.21 | 0.05 | <0.01 | | | | |
| | B | 0.06 | 0.25 | 0.38 | 0.25 | 0.06 | | | | |
| 8 | A | 0.10 | 0.26 | 0.32 | 0.21 | 0.09 | 0.03 | <0.01 | <0.01 | <0.01 |
| | B | 0.01 | 0.03 | 0.11 | 0.22 | 0.27 | 0.22 | 0.11 | 0.03 | <0.01 |

These data are shown graphically in Fig. 33. As can be seen, an increasing degree of separation is obtained with a greater number of transfers.

Further, if the original $\alpha$ values of the solutes in question are more dissimilar, a greater separation is obtained (Fig. 34).

Paper chromatography and column-partition chromatography operate in a manner and principle similar to countercurrent distribution. In these techniques the stationary phase is absorbed on a relatively inert supporting material such as cellulose and is successively washed by a passing mobile phase. Thus we may envision these techniques as operating with hundreds of very small vessels (stationary phase bound to inert support) instead of the fixed number of vessels of the countercurrent distribution system. This allows a greater resolution but limits the quantities of solutes. It is therefore possible to predict the expected separation with these techniques from the partition coefficients of the solutes. These calculated values sometimes vary from the experimental values due to adsorption of the solutes on the supporting material and to the frequent lack of complete equilibration of solvents before transfer.

In paper chromatography the substance, or a mixture of substances, is applied in solution along a base line on a sheet of filter paper. The sheet is then air-dried. The appropriate solvent mixture is subsequently allowed to migrate past the spots or bands, either descending the vertically oriented paper by gravity and capillarity or ascending the paper by capillarity, "developing" the chromatogram. The former

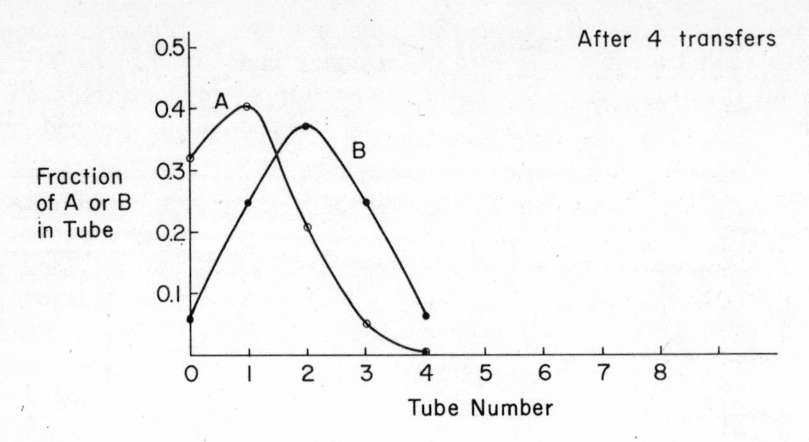


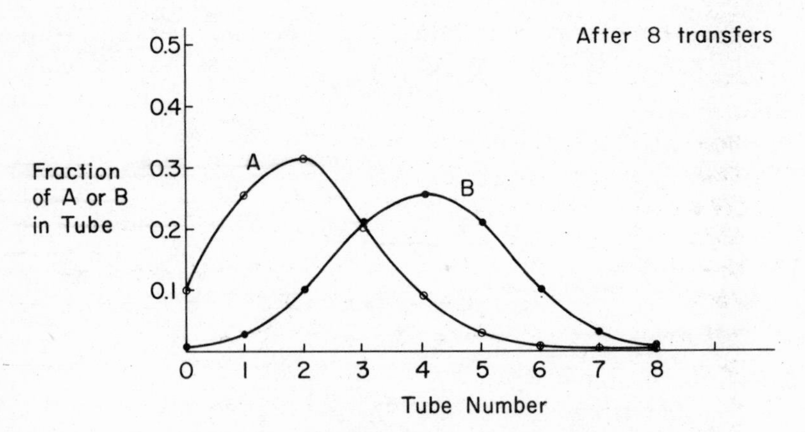


FIG. 33. *Fraction of A and B after 4 and 8 transfers, where* $\alpha_A = \frac{1}{3}$, $\alpha_B = 1$.

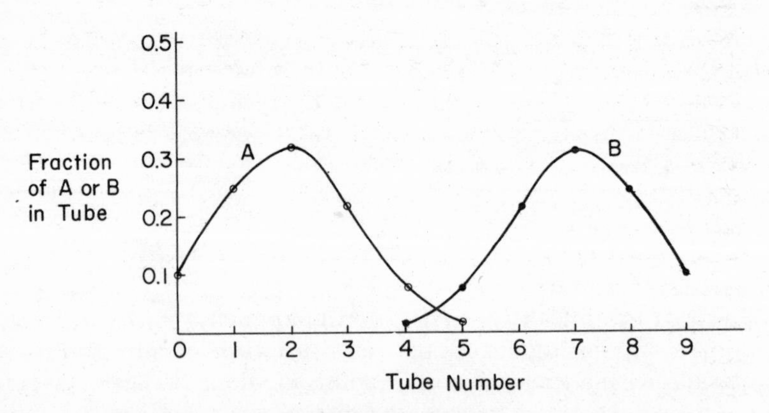


FIG. 34. *Fraction of A and B after 8 transfers, where* $\alpha_A = \frac{1}{3}$, $\alpha_B = 3$.

method is called descending chromatography; the latter, ascending chromatography (Fig. 35).

Although ascending chromatography is often preferred because of the simplicity of the setup, the flow of solvent is faster in the descending technique, and, if necessary, the solvent may be allowed to run off the end of the descending paper to elute the solute or enhance resolution. The dis-

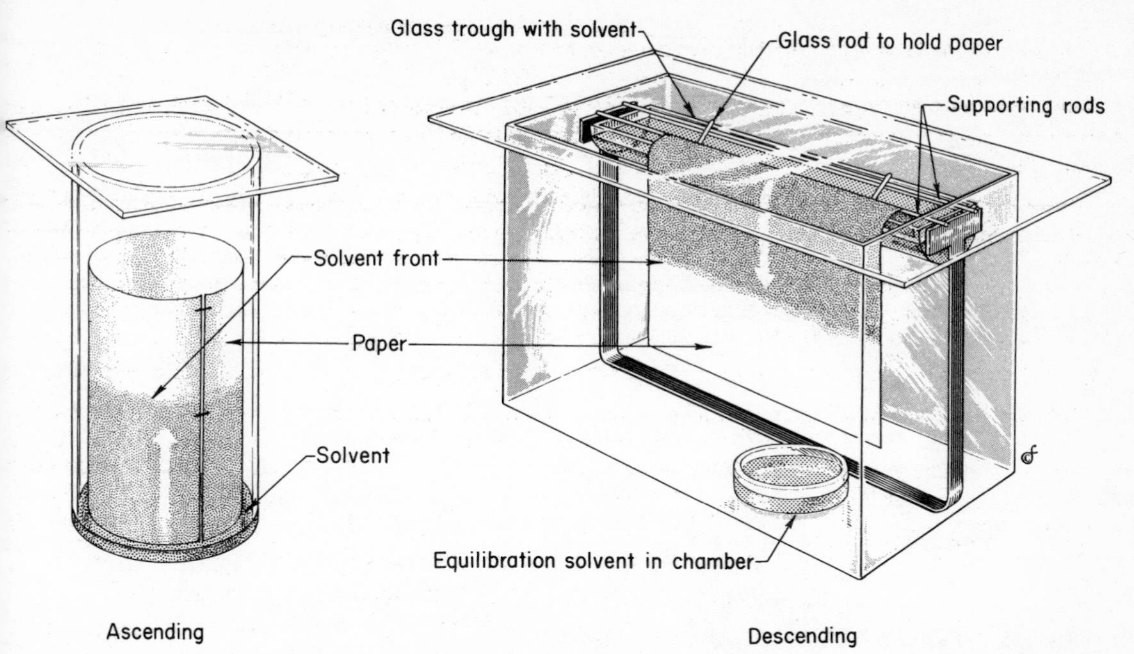


FIG. 35. *Paper chromatography.*

tance traveled by each compound from the origin, or base line, relative to the solvent front is defined as the $R_f$.

$$R_f = \frac{\text{distance from base line traveled by compound}}{\text{distance from base line traveled by solvent}}$$

The $R_f$ value is a characteristic of the particular compound measured under specified conditions—solvent system, temperature, manner of development (whether ascending or descending), the type of paper, and the grain or "machine direction" of the paper. Since it is affected by so many variables, the $R_f$ of a substance in a known system is only a rough indication of identity. Therefore, it is common practice to chromatograph a known sample of a known material—the material presumed to be identical to the unknown—along with the unknown. It should also be noted that two different compounds may have the same $R_f$ in one solvent system. Therefore, results related to the identity of a component obtained by paper chromatography (even though confirmed in several different solvent systems) should be verified by other means.

For increased resolution, the technique of two-dimensional paper chromatography is used (Fig. 36). The material to be chromatographed is applied *near one corner of the paper*. The sheets are developed in one direction as above, allowed to dry, and then developed with another solvent system in a direction at right angles to the first development.

If the materials being separated on paper chromatograms are ultraviolet-absorbing, or if they fluoresce, the spots may be detected readily by examining the chromatogram under a strong ultraviolet light (for example, Mineralite). Other techniques for locating the compounds on paper include color tests, frequently carried out by spraying the chromatogram with appropriate reagents.

Quantitative analysis of a compound separated by paper chromatography can be accomplished in one of two ways: (1) the intensity of colored, or U.V.-absorbing spots may be measured directly on the paper, or (2) the compound may be eluted from the paper and analyzed by any suitable method.

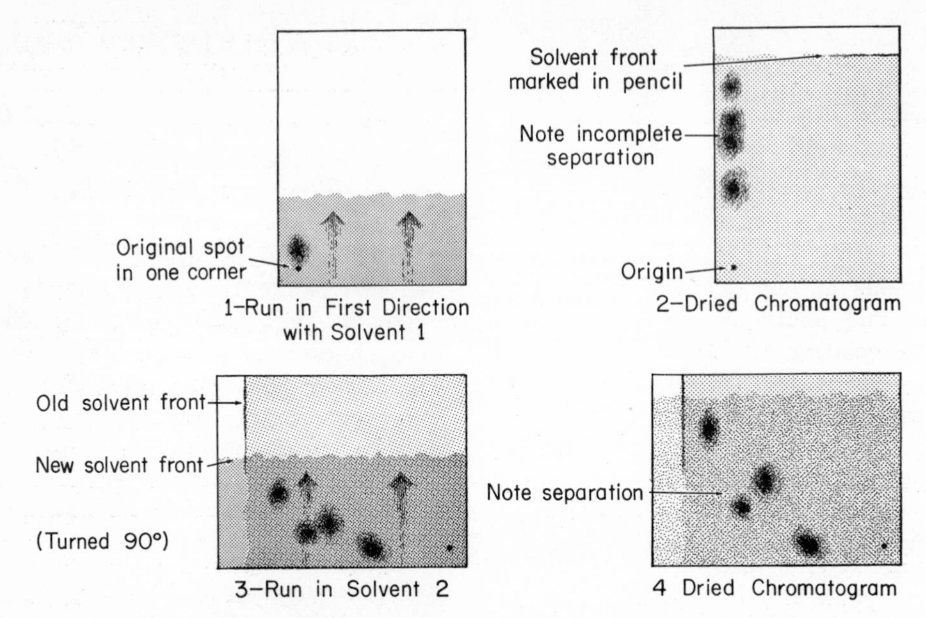


FIG. 36. *Two-dimensional paper chromatography.*

Partition-column chromatography is a macromodification of the paper chromatographic technique. Columns packed with cellulose, starch, or some other material to support the stationary phase, receive a sample of a solute mixture at the top, and are developed by the passage of solvent mixtures through the column. Compounds with different $R_f$ values migrate at different rates and emerge from the column at different times. Again the results are roughly predictable from the partition coefficients of the solutes. This technique allows resolution of larger quantities of solutes than does paper chromatography.

**References**

Block, R. J., E. L. Durrum, and G. Zweig, *A Manual of Paper Chromatography and Paper Electrophoresis* (2nd ed.), Academic, New York, 1958.

Craig, L. C., and T. P. King, *Federation Proceedings*, **17,** 1126 (1958).

Fruton, J. S., and S. Simmonds, *General Biochemistry* (2nd ed.), Wiley, New York, 1958.

Kunin, R., *Ion Exchange Resins* (2nd ed.), Wiley, New York, 1958.

Lederer, E. and M. Lederer, *Chromatography* (2nd ed.), Van Nostrand, Princeton, 1957.

Peterson, E. A., and H. A. Sober, *J. Am. Chem. Soc.*, **78,** 751 (1956).

Sober, H. A., and E. A. Peterson, *Federation Proceedings*, **17,** 1116 (1958).

——— et al., *J. Am. Chem. Soc.*, **78,** 756 (1956).

Moore, S., and W. H. Stein, *J. Biol. Chem.* **211,** 893 (1954).

# APPENDIX III

# Radioactive Isotopes

## Nuclear Transformations

According to present-day nuclear physics, the nucleus of any atom is composed of protons and neutrons. The mass of any atom is relative to that of oxygen 16, which is exactly 16.000. Although the electrons of any atom are relatively insignificant in weight, both the proton and the neutron have a mass number of 1. Protons bear a unit positive charge; neutrons are uncharged. The chemical properties of an atom are largely determined by the nuclear charge (atomic number). The addition of neutrons to the nucleus changes the mass without changing the nuclear charge; thus there are elements whose nuclei vary in nuclear mass, but most of whose chemical properties are essentially the same. These are called isotopes. The description of a nucleus usually includes the mass number, symbolized by a superscript $A$ to the right of the element symbol, and the atomic number—the number of unit positive charges carried by the nucleus—symbolized by a subscript $Z$ to the left of the element symbol; for example, element $X$ is written $_ZX^A$. Thus $_1H^1$, $_{11}Na^{23}$, and $_{15}P^{31}$ describe the nuclei of the elements hydrogen, sodium, and phosphorus, respectively. On this basis the mass number $A$ is the total number of protons plus neutrons in the nucleus. The difference $A - Z$ is the number of neutrons.

Some elements have only one stable isotope (for example, $_{11}Na^{23}$ and $_{15}P^{31}$); others are mixtures of two or more stable isotopes. For example, sulfur has four isotopes: $_{16}S^{32}$, $_{16}S^{33}$, $_{16}S^{34}$, $_{16}S^{36}$. Their nuclei contain 16 protons each, and 16, 17, 18, or 20 neutrons, respectively.

Only certain combinations of neutrons and protons are stable. In stable nuclei the ratio of neutrons to protons varies from about 1 for elements of low atomic number to 1.5 for elements of high atomic number. An excess of one or the other component leads to a spontaneous redistribution of particles while the neutron:proton ratio is brought to a more stable state. The nuclear reactions usually involve the following physical entities:

1. neutron, $n$;
2. proton, $p$;
3. deuteron, $d$, which is the heavy hydrogen nucleus $_1H^2$;
4. alpha particle, $\alpha$, which is the helium nucleus $_2He^4$;
5. negative electron, or negative beta particle, $\beta^-$, and the positive electron, or positron, $\beta^+$, which apparently do not exist in the nucleus but are produced in certain nuclear transformations; and
6. gamma radiations, $\gamma$, high-energy photons — electromagnetic radiation quanta—which do not exist in the nucleus but are produced in certain nuclear transformations.

Several kinds of nuclear transformation occur: (1) transformation of a neutron into a proton by $\beta^-$ particle emission; (2) transformation of a proton into a neutron and a positron—$\beta^+$ emission, or capture of an orbital electron from the nearest electron shell ($K$ capture); and (3) emission of an $\alpha$ particle. In all of these types of nuclear transformation, there *may* also be emission of electromagnetic radiation; for example, gamma rays, X-rays.

As an example of $\beta^-$ particle emission, consider a simple nuclear transformation of carbon. For stable combinations, no more than six or seven neutrons can be held in the nucleus with the six protons.

These combinations correspond to the two stable carbon nuclei found in nature, ${}_6C^{12}$ and ${}_6C^{13}$. Eight neutrons form ${}_6C^{14}$, an unstable radioactive configuration. By transformation of a neutron into a proton, ${}_6C^{14}$ is transformed to ${}_7N^{14}$. This process requires emission of a $\beta^-$ particle:

$$n \longrightarrow p + \beta^-$$

Einstein's concept of equivalence of mass, $m$, and energy, $E$, expressed by the relation $E = mc^2$, where $c$ is the velocity of light, is fundamental in nuclear reactions. The magnitude of the energies involved in nuclear transformations can be derived from the mass-energy relation if the changes in either mass or energy which occur in the reaction are accurately known. A mass unit of 1 ($\frac{1}{16}$ of $O^1$) corresponds to 931 million electron volts (mev).

Carbon 14 ($C^{14}$) is known to emit negative electrons ($\beta^-$ particles) with a maximum kinetic energy of 0.15 mev. This process forms the residual, stable nucleus, $N^{14}$.

$$C^{14} \longrightarrow N^{14} + \beta^- + 0.15 \text{ mev}$$

If five neutrons are associated with six protons to form ${}_6C^{11}$, this nucleus represents an unstable configuration of eleven particles, the stable configuration being the naturally occurring nonradioactive isotope of boron, ${}_5B^{11}$. The process

$$p \longrightarrow n + \beta^+$$

occurs with the concurrent production of 0.981 mev.

The emission of $\alpha$ particles is confined almost entirely to the heavy elements ($Z > 82$). Radioactivity produced in elements with $Z < 82$ is associated almost entirely with the emission of negative or positive electrons.

All radioactive disintegrations proceed spontaneously at rates which cannot be altered by ordinary chemical or physical means. For any given element the rate of decay is proportional to the number, $N$, of nuclei of the element present:

$$\frac{-dN}{dt} = \lambda N$$

Where $t$ is time, and $\lambda$ is the disintegration constant characteristic of the isotope.

At $t = 0$ (zero time), $N_0$ atoms are present. At any time, $t$, thereafter, the number present, $N$, can be found by integration of the above equation within the limits 0 to $t$ and $N_0$ to $N$.

$$\int_{N_0}^{N} \frac{dN}{N} = -\lambda \int_0^t dt$$

which is

$$N = N_0 e^{-\lambda t}$$

or when expressed in natural logarithms:

$$\ln \frac{N}{N_0} = -\lambda t$$

It is convenient to express the rate of radioactive decay by the term half-life, $T_{1/2}$. When the radioactive intensity drops to half its initial value, the number of atoms of the unstable isotope must also have dropped to half the number initially present. Thus, substituting into the above equation, we have

$$\ln \tfrac{1}{2} = -\lambda T_{1/2}$$

$$T_{1/2} = \frac{0.693}{\lambda}$$

For different radioactive species, $T_{1/2}$ may vary from $10^{-11}$ sec to $1\text{–}3 \times 10^{10}$ yr. For example, $C^{11}$ has a half-life of 20.33 min, whereas $C^{14}$ has a half-life of 5500 yr.

The three different types of radiation, $\alpha$, $\beta$, and $\gamma$ rays, possess radically different physical characteristics. The $\alpha$ rays are streams of doubly positive-charged helium nuclei moving with relatively slow velocities ($1\text{–}2 \times 10^9$ cm/sec). The $\alpha$ particle, heavy and sluggish, rarely deviates from its straight path; only direct collision with other nuclei will cause deflection. In addition, because electrostatic interaction with other electrons is so intense, the penetrating power of $\alpha$ particles in an absorbing material is very small. The $\beta$ rays are composed of positively or negatively charged electrons moving at various velocities, depending upon the mass-energy differences in the particular nuclear transformation. At low energies (soft $\beta$ rays) velocities may be only slightly faster than those of $\alpha$ rays, but at higher energies (hard $\beta$ rays) velocities may approach the speed of light ($3 \times 10^{10}$ cm/sec).

The $\beta$ particles generally move more erratically than do the $\alpha$ particles. Many interactions or collisions occur in which $\beta$ particles are deflected at a large angle from their initial direction.

The $\gamma$ rays are, of course, uncharged photons which move at the speed of light. For a given energy, $\gamma$ rays possess the greatest penetrating power of all three types of radiations. Photons lose energy only by collision.

A summary of the decay properties of several radioactive isotopes of biochemical interest is presented in the following table.

Decay Properties of Isotopes of Major Biochemical Interest

| Symbol | Reaction | Energy (mev) of Emitted Particle | Half-life |
|---|---|---|---|
| $_1H^3$ | $_1H^3 \longrightarrow {_2He^3} + \beta^-$ | 0.015 | 12.3 yr |
| $_6C^{14}$ | $_6C^{14} \longrightarrow {_7N^{14}} + \beta^-$ | 0.15 | 5500 yr |
| $_{15}P^{32}$ | $_{15}P^{32} \longrightarrow {_{16}S^{32}} + \beta^-$ | 1.72 | 14.3 days |
| $_{16}S^{35}$ | $_{16}S^{35} \longrightarrow {_{17}Cl^{35}} + \beta^-$ | 0.107 | 87.1 days |

## Measurement

Radioactivity is commonly measured by the ionization produced when radiation particles interact with the planetary electrons of the atoms in the space traversed by the particles. The ion pair produced by the radiation consists of the ejected electron and the residual, positively charged atom. The electron can be ejected from any of the orbitals, not necessarily only the outer-valence shell. The ions, of course, have a great tendency to recombine. Quantitative measurement then depends on separating and counting the ions before appreciable recombination occurs. Separation is accomplished by allowing the ionization to occur in an electric field so that the electrons are accelerated very rapidly toward the anode, and so that the heavy, positive ions move slowly toward the cathode.

The electric field can be controlled so that (1) only the electrons liberated by the passage of the ionizing particle are measured, or (2) the electrons produced by radiation particles are accelerated and cause ejection of other electrons from other atoms of gas, resulting in further multiplication of the original ionization. In the first method, single electrons released as a result of collision of an ionizing particle with a gas atom are measured in an *ionization chamber* connected to a sensitive electrometer. This method measures the total ionization in the volume of the chamber. In the second method, showers of electrons produced by multiplication of a single ionization are measured in a Geiger-Muller counter or a proportional counter attached to a scaling circuit and a mechanical register. The general relationship between ionization and potential of the electric field is presented in Fig. 37. Ionization chambers are, in general, less sensitive than Geiger-Muller or proportional counters, but the ionization chambers have a lower background correction.

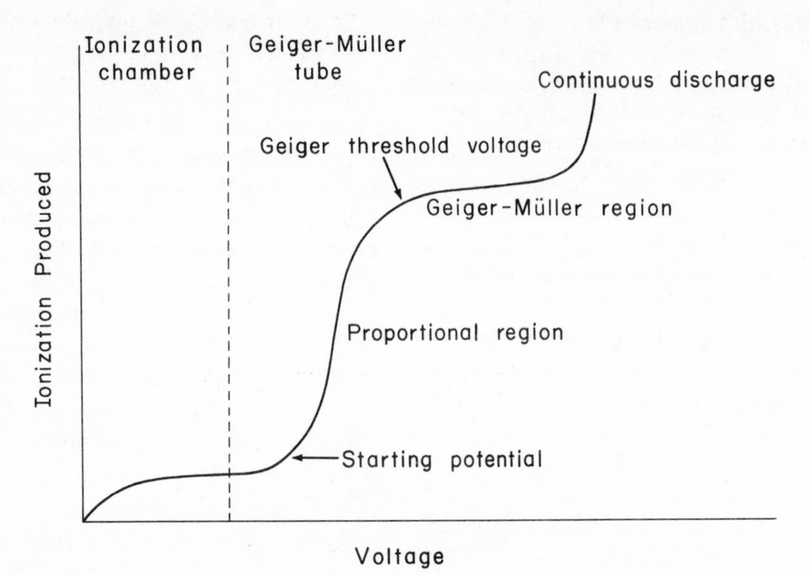


FIG. 37. *Ionization produced by radioactive particles in an electric field.*

Other methods for the detection and measurement of radioactivity depend on interaction of radiation particles with other atoms. For example, the scintillation counter measures light emission as a result of collisions of particles with a suspended phosphor. Another method, radioautography, is based on the effect of radiation on photographic emulsions. This method is more valuable for locating radiation than for measuring it.

## *Ionization Chamber*

The ionization chamber (Fig. 38) consists primarily of two electrodes within a closed gas space, with either a thin window through which the radiations from an external source can pass or some other means of introducing the radioactive material (for example, as a gas) into the chamber. In addition, there are devices for applying the potential differences across the electrodes and for measuring total ionization.

As the potential difference applied to the electrodes is increased, there is less probability of recombination of ions, so that the amount of ionization detected increases. There is, however, a plateau where increased potential has no additional effect on the amount of ionization detected. The amount of ionization can be measured either from the rate at which the potential difference is reduced by the collection of electrons at the anode or from the current produced by passing the collected charge through a known resistance. Because the charge is very small, a sensitive electrometer, (for example, a vibrating-reed electrometer) is required.

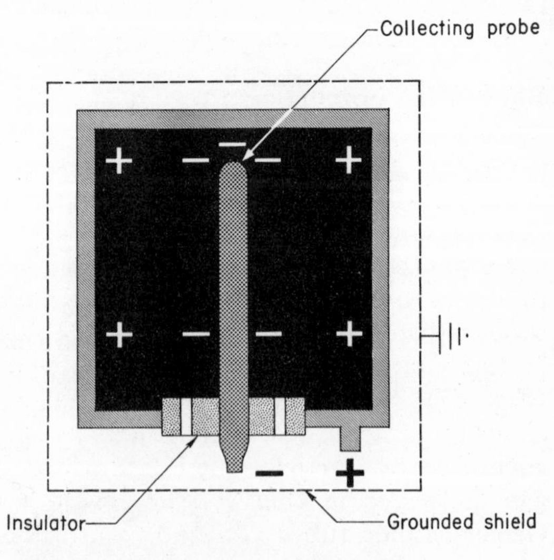


FIG. 38. *An ionization chamber.*

### *Proportional Counters*

As the potential of the electric field is increased above the region at which ionization is constant, the electrons produced by the ionizing current are accelerated at such high velocities that they themselves cause ionizations. Consequently, the number of electrons reaching the anode represents multiplication of the effect of the ionizing particle. In this region the number of electrons is still proportional to the number of ion pairs produced. The proportionality factor depends on whether a small number of ion pairs are produced ($\beta$ particle) or several thousand ($\alpha$ particle). Originally, proportional counters were used to count $\alpha$ particles at relatively low voltages with a consequently low proportionality factor. With considerably increased voltage, proportional counters can be used with great efficiency to count $\beta$ particles.

Both proportional counters and Geiger-Muller counters require the presence of a specific polyatomic "counting gas" (for example, methane or 1% butane in helium) in order to provide the proper atoms for multiple ionizations. The efficiency of these counters at any given voltage is a function of the nature of the counting gas used.

### *Geiger-Muller Counters*

Beyond the proportional region, there is a transition region at which proportionality decreases, and, further beyond, there is a zone at which the number of electrons reaching the anode is approximately the same whether the original ionizing particle produces only one ion pair or several thousand within the sensitive volume. This is the Geiger-Muller region, or plateau (Fig. 37). The counters used to detect showers of ions (Townsend avalanches) generated by particles hitting polyatomic gas molecules in this voltage region are Geiger-Muller tubes. These tubes do not operate at low voltages.

The usual Geiger-Muller tube consists of a cylindrical cathode at ground potential and a fine, high-voltage central wire, which is the anode. The tube is filled with a gas mixture. The Geiger tubes most commonly used in biochemical research today are the end-window and the flow-counter tubes. In the end-window tubes, the counting gas mixture is held in the Geiger tube by a window of mica (1–3 mg/cm$^2$); therefore, radioactive particles must pass through the window in order to be detected. High-energy $\beta$ particles (for example, $P^{32}$) usually penetrate such end-window counters, but many low-energy particles (for example, $C^{14}$) are screened out by the window and are therefore not detected. This inefficiency is overcome in the gas-flow counter (Fig. 39C), in which the sample is placed directly in the tube. The construction of a flow counter includes an opening through which the counting gas can escape. Flow counters, therefore, require a pre-flushing before counting may be started, and a continual flow of counting gas during operation, hence the name gas-flow counter.

Since 1955 the thin-window, or flow-window, tube—a modification of the two types of Geiger tubes shown above—has come into common use, owing to its efficiency and ease of automation. This tube has an extremely thin window of plastic ($\frac{1}{4000}$ in. thick). This thin window, which passes a large percentage of even the low energy $\beta$ particles, is more efficient than the mica end window. Yet the extreme thinness of the window allows a slow leakage of counting gas. Hence, the thin-window tubes are flushed with counting gas during operation.

The efficiency of Geiger-Muller tubes in detecting $\beta$ particles which enter the sensitive volume is practically 100%. For $\gamma$ rays, the situation is different; most Geiger-Muller counters have less than 1% efficiency for $\gamma$ rays.

The passage of the ionizing particle into the sensitive volume, the production of ion pairs, and the resulting Townsend ava-

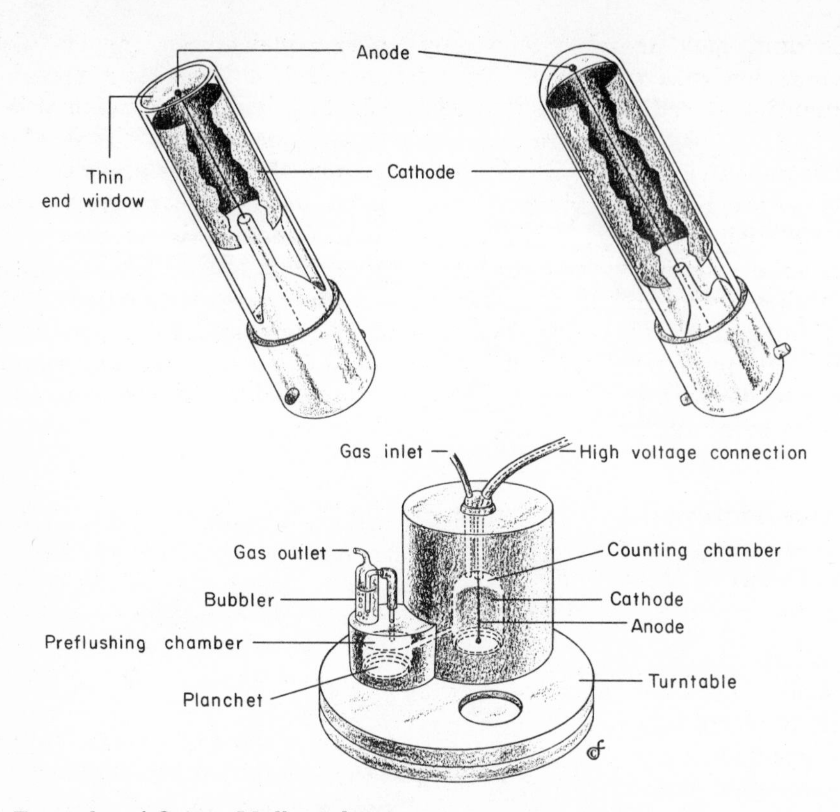


FIG. 39. *Examples of Geiger-Muller tubes.*

lanche occur in less than a microsecond. The collection of electrons at the anode lowers the anode's potential markedly, and the positive ions formed around the anode wire move slowly toward the cathode because of their large mass. The time required to reach the cathode is about 200 μsec, during which time the counter cannot respond to the passage of a second ionizing particle. This interval is known as "dead time." The radioactivity of the sample should therefore be adjusted ($<3000$ cpm) so that there is no appreciable probability that a second particle will enter the sensitive volume within the dead time. The discharges from Geiger-Muller tubes or proportional counters are usually registered by some mechanical counting device. These devices, called scaling circuits (scalers), consist of vacuum tubes wired so as to yield a binary (2, 4, 8, 16, $\cdots$) or digital (1, 10, 100, $\cdots$) counting system connected through a system of lights for easy assay.

### *Scintillation Counter*

Since certain organic phosphors emit photons when hit by charged particles, the number of radioactive disintegrations of an added sample can be counted by measuring the number of light flashes produced. This extremely sensitive method can be used accurately for samples of low specific activity and for low-energy $\beta$ particles (for example, from tritium) which are difficult to detect efficiently by other methods.

## Units of Radioactivity

The basic unit of radioactivity, the curie, *c*, was originally defined as the number of disintegrations per second emitted from 1 g of radium (approximately $3.7 \times 10^{10}$/

sec). This unit, now divorced from any formal connection with radium, is defined as that quantity of radioactive substance which emits $3.7 \times 10^{10}$ disintegrations per second. The millicurie, *mc* ($3.7 \times 10^{7}$ disintegrations per second) and the microcurie, $\mu c$ ($3.7 \times 10^{4}$ disintegrations per second) are useful subdivisions of the curie. The units of radioactivity define the *rate* of disintegration, and not the amount of material. For any particular isotope, however, the rate of disintegration is determined by the total number of radioactive atoms present and by the disintegration constant of the isotope; therefore, the rate can be directly related to the amount of labeled material. This relationship is termed the specific radioactivity. The particular units of specific radioactivity depend on the units of measurement available; the most commonly used expressions are millicuries/millimole (mc/mmole), millicuries/gram (mc/g), disintegrations per minute/gram (dpm/g), and counts per minute/milligram (cpm/mg).

## Radioactive Isotope Determination in the Geiger Range

### *Determination of Operating Voltage, or Plateau*

After becoming thoroughly familiar with the operation of the instrument, allow the instrument to warm up for 10 min. If the unit is a gas-flow counter, preflush the counting chamber with the appropriate gas.

The proper operating voltage for the Geiger counter depends on the type of ionizing gas in, or flushed through, the tube. This voltage must be empirically determined for a particular counting setup. Place a radioactive source that is active enough to give about 2000 cpm within the Geiger-Muller region in the counter. With the count switch "on," increase the operating voltage cautiously until the machine begins to count. Then record the counts accumulated over a 1-min interval versus the voltage of the system. Repeat this process at 50-volt intervals over a 500-volt span, or until the system reaches a voltage of continuous discharge (see Fig. 37, p. 202).

Your plot of counts per minute versus operating voltage will resemble the curve of Fig. 37, except that the initial plateau in the ion-chamber region will be much lower than the starting potential of the instrument, and therefore unobservable.

The Geiger-Muller, or plateau, portion of the curve should extend for 150–200 volts, with a slope of less than 10% per 100 volts. The optimal operating voltage should be about one-third of the way up the plateau from the Geiger threshold voltage.

The operating voltage, once determined for a particular end-window tube or flow counter with a particular gas and flow rate, remains relatively constant and need not be checked for the remainder of the experiment.

### *Preparation and Counting of Sample*

#### BACKGROUND

Background radiation is due to (1) indigenous radiation, (2) radioactive contaminants on planchets or walls of the counter, and (3) "noise" in the detecting circuits. Determine the background counting rate by placing a clean, nonradioactive sample holder (planchet) into the counter and recording the counting rate over a 5-min period. *Substract this background level from the observed counting rate of all samples.* For samples with low counting rates, the background is a highly significant correction, and must be known accurately (measured for a longer time).

#### SAMPLE

Carefully place the sample on a tared planchet. Often the sample is added in solution or as a slurry and the solvent subsequently evaporated, leaving a uniform solid deposit on the surface of the planchet. *Under no circumstances should a sample containing a volatile radioactive substance be placed in a gas-flow counter!* Dry the planchet using a heat lamp, and weigh the sample. Record the total sample weight and the planchet number.

#### COUNTING TIME

The length of time required for counting the sample and the background depends on the amount of radioactivity of the sample relative to the background and on the degree of accuracy desired. Since radioactive decay is a random process, the observed rate approaches an average count per minute with an increase in the total number of disintegrations observed (time of counting). If possible, count the sample and the background long enough to bring the deviation of the observed counts per minute from the actual counts per minute down to $\pm 3\%$ (about 1000 counts for each sample).

### *Factors Affecting Counting Rate*

Many factors affect the counting rate of a radioactive sample. The variation of the count as a function of the thickness of an end window, and relative to the applied voltage, has already been discussed. In addition, since particles are emitted in all directions, the counting efficiency will be governed by the solid angle subtended by the measuring element. Each system therefore has an efficiency factor related to geometry, which is fixed as long as the physical setup is not changed.

Further, there are losses caused by absorption within the sample. The magnitude of self-absorption depends on the energy of the emitted particles and on the thickness of the layer through which the radiation must pass (relatable to the weight per unit area of the sample). An allowance for self-absorption must be made in each radioactive analysis. This can be accomplished by one of several methods explained below:

#### COUNTING AT INFINITE THINNESS

In certain situations the weight of sample can be restricted so that self-absorption is negligible—that is, so that the sample counts as if it were infinitely thin. This method can be used for samples of high specific activity and/or with isotopes of high-energy radiation.

#### COUNTING AT INFINITE THICKNESS

If a sample is beyond a certain thickness (weight in milligrams per square centimeter), radiation from the bottom layers of the sample is completely absorbed and does not contribute to the observed counting rate. The minimum sample thickness for infinitely thick counting can be estimated from the penetrating power of the emitted particle. For example, the limit of penetration of $C^{14}$ particles is a sample containing about 28.5 mg of organic material/cm$^2$. Therefore, samples of over 30 mg of organic material/cm$^2$ can be considered infinitely thick. In practice, the thickness of a particular sample is increased until the counting rate is uniform and independent of the quantity of sample counted.

#### SELF-ABSORPTION CORRECTION CURVES

In practice, it is not always possible to prepare samples at infinite thinness or thickness. Accordingly, an empirical correction can be made which permits conversion of the counting rate of any sample to either infinite thickness or infinite thinness. This correction relates the observed specific activity (for example, cpm/mg) to the actual specific activity as a function of the thickness of the sample. An example of such a self-absorption curve for $C^{14}$ is given in Fig. 40. For the most reliable measure-

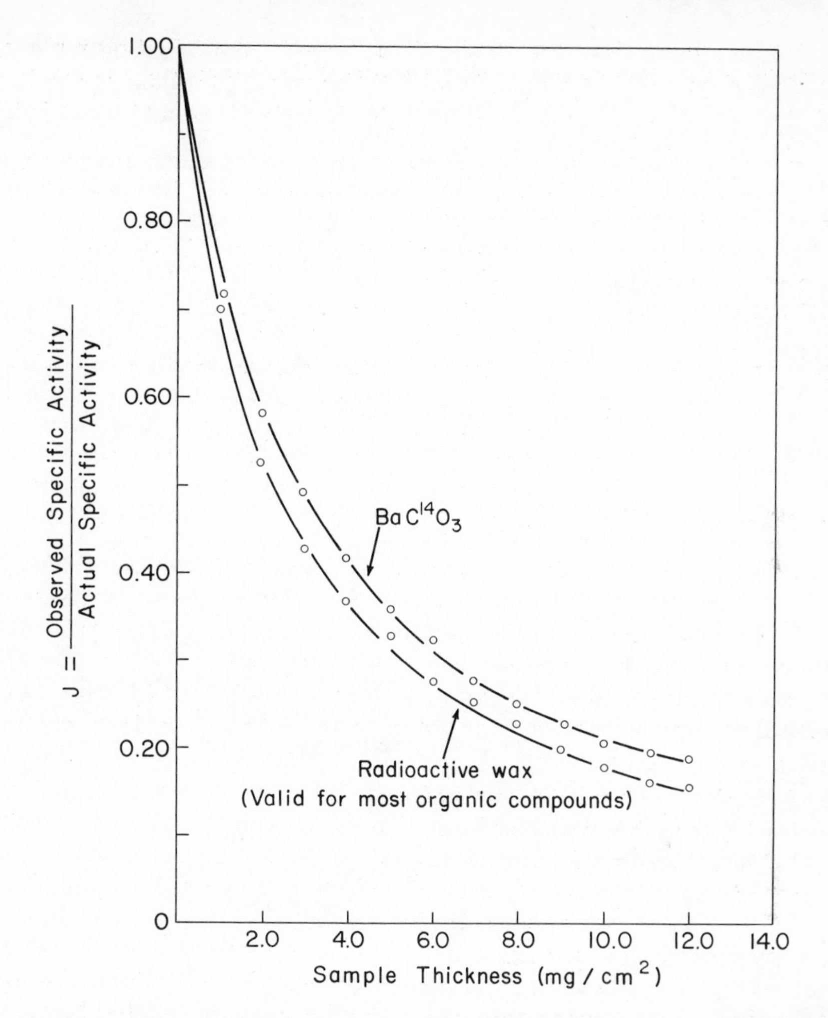


FIG. 40. *Carbon-14 self-absorption curve.* [*Adapted from Calvin et al.*, Isotopic Carbon, *New York, John Wiley & Sons, Inc., 1959.*]

ments, a self-absorption curve should be prepared for the particular material to be studied, using the counter and sample holders which are to be used in further work.

EXPRESSION OF RESULTS

Express the results of counting both as total counts per minute (at infinite thickness or thinness), corrected for background, and as specific radioactivity. In addition, give some indication of the statistical error (standard deviation) of the counts.

An experiment utilizing radioactive isotopes should include the description of the total amount and specific activity of the substances used. A discussion of the results should include the same information concerning the products. *Whenever possible, all of the original isotopic material should be accounted for.*

**References**

Kamen, M. D., *Isotopic Tracers in Biology* (3rd rev. ed.), Academic, New York, 1957.

Steinberg, D., and S. Udenfriend, "The Measurement of Radioisotopes," *In* S. P. Colowick, and N. O. Kaplan (Editors), *Methods in Enzymology* (Vol. IV), Academic, New York, p. 425, 1957.

# APPENDIX IV

# Use of Common Logarithms

## *General Theory*

The basic principle of logarithms may be demonstrated by considering numbers that differ by powers of 10. In the expression $10^2 = 100$, the exponent 2 is the common logarithm, or logarithm to the base 10 ($\log_{10}$), of 100. Thus:

$$10^2 = 100, \quad \log_{10} 100 = 2$$

$$10^1 = 10, \quad \log_{10} 10 = 1$$

$$10^0 = 1, \quad \log_{10} 1 = 0$$

$$10^{-1} = 0.1, \quad \log_{10} 0.1 = -1 \text{ or } 9 - 10$$

Therefore, the common logarithm expresses the power to which 10 must be raised to give the number desired.

For numbers whose logarithms are not as obvious as those in the above examples, tables of logarithms must be used. In general, common logarithms are made up of a *characteristic* (an integer part) and a *mantissa* (a decimal part). The characteristic is determined by inspection; the mantissa is obtained from a log table. To obtain the $\log_{10}$ of 30, for example, first determine the characteristic. It is clear that the $\log_{10}$ of 30 will be a number between 1 and 2, since $10^1 = 10$ and $10^2 = 100$. Hence, for the $\log_{10}$ 30, the characteristic must be 1. In the log table (see inside back cover), find, opposite the number 30, the mantissa: 4771. Thus, 1.4771 is the power to which 10 must be raised to give 30, or $\log_{10} 30 = 1.4771$. Check the log table to confirm the following facts: $\log_{10} 224 = 2.3502$, $\log_{10} 0.15 = 9.1761 - 10$.

Practical use of logarithms requires both determination of the logs of numbers and the determination of numbers from logarithms—that is, finding the *antilogarithms*. Thus, if you are given that the logarithm of a number is 1.6085, a log table will show 407 opposite the mantissa 6085. Since the mantissa 6085 is preceded by a 1, the number whose logarithm is 1.6085 must be greater than 10 ($10^1$) and less than 100 ($10^2$), therefore 40.7 is the antilogarithm of 1.6085 (that is, $\log_{10} 40.7 = 1.6085$). Check your knowledge of these points by confirming the following antilogarithms: 881 is the antilogarithm of 2.9450, 0.074 is the antilogarithm of 8.8692 − 10.

## *Application of Logarithms to the Henderson-Hasselbalch Equation*

The Henderson-Hasselbalch equation relating *p*H, *p*Ka, and the concentration of ionizable species is a frequently used biochemical expression involving logarithms. The following examples illustrate its application.

(1) Find the ratio of $COO^-$ to COOH that exists in a glycine solution at *p*H 3.6 (glycine $pKa_1 = 2.2$, $pKa_2 = 9.7$).

$$pH = pKa_1 + \log_{10} \frac{[COO^-]}{[COOH]}$$

$$3.6 = 2.2 + \log_{10} \frac{[COO^-]}{[COOH]}$$

$$1.4 = \log_{10} \frac{[COO^-]}{[COOH]}$$

Since the antilogarithm of 1.4 is 25.12, the ratio is

$$\frac{[COO^-]}{[COOH]} = 25.12$$

(2) What would be the average charge on the amino groups of glycine in solution at *p*H 8.6?

$$pH = pKa_2 + \log_{10} \frac{[NH_2]}{[NH_3^+]}$$

$$8.6 = 9.7 + \log_{10} \frac{[NH_2]}{[NH_3^+]}$$

$$-1.1 = \log_{10} \frac{[NH_2]}{[NH_3^+]}$$

Converting the negative logarithm to a positive value by adding 10 to −1.1 and then subtracting 10, we find

$$\frac{\begin{array}{l}10.0 - 10\\ -1.1\end{array}}{8.9 - 10} = \log_{10} \frac{[NH_2]}{[NH_3^+]}$$

The antilogarithm of the mantissa, 9, is 7944, and the characteristic is −2. Therefore,

$$0.07944 = \frac{[NH_2]}{[NH_3^+]}$$

or

$$0.07944\ [NH_3^+] - [NH_2] = 0$$

but

$$[NH_3^+] + [NH_2] = 1$$

or represents the total amine forms. Therefore, by simultaneous equations, we obtain

$$1.07944\ [NH_3^+] = 1$$

or

$$[NH_3^+] = \frac{1}{1.07944} = 0.926$$

Thus, the average charge on the amine groups of glycine at $p$H 8.6 is $0.926^+$.

# APPENDIX V

# Suggestions to Instructors

*Teaching Staff*

A detailed consideration of objectives, procedures, and possible pitfalls, and a necessity for frequent consultation with students during laboratory work have shown in our laboratories the need for one senior staff member per 30–35 students and one experienced teaching assistant per 16–18 students. In addition, it is desirable that a qualified individual be responsible for the preparation of reagents. It is also helpful if a graduate assistant can perform certain of the experiments ahead of the class so as to check materials.

*Establishment of Effective Research Procedures*

Perhaps the most important single problem lies in helping the student to develop satisfactory research procedures and care and precision in recording and treating his data. To accomplish these ends requires extensive criticism of individual research books, together with personal consultation, especially in the beginning of the course.

*Group Discussions and Experiments*

We have found it highly useful to consider the results of experiments on a classwide basis. After individual data have been tabulated, we then discuss the purpose of the experiment and criticize the data and conclusions.

*Cooperative Experiments*

Cooperative experiments require careful and specific delineation of responsibility among students. Summary reports and a public posting of all data are most helpful.

*Lab Schedule*

The material presented in this manual is designed to cover about 53 laboratory periods—clearly more than can be done by any student or group of students in an average semester. This allows for flexibility of planning and emphasis on quality of performance rather than on quantity. The following table shows (1) the amount of time required for satisfactory completion of individual experiments by a class of thirty students and (2) the way in which students participate in experiments—that is, whether they work as individuals or in pairs or groups. From these experiments a schedule may be chosen to fit various class sizes, availability of time, and availability of starting materials and equipment.

For example, in a recent semester one of our classes completed the following schedule: Experiments 19, 20, 21, 24, 1–10, 29–33, 25–28, 15, 16, 34.

| Experiment Number | No. of Laboratory Periods | Student Participation |
|---|---|---|
| 1 | 1 | individuals |
| 2 | 1 | individuals |
| 3 | 1 | individuals |
| 4 | $\frac{1}{2}$–1 | individuals |
| 5 | $\frac{1}{2}$–1 | individuals |
| 6 | 2 | individuals |
| 7 | 1 | individuals or pairs |
| 8 | 1 | individuals or pairs |
| 9 | 1 | individuals or pairs |
| 10 | 3–4 | pairs |
| 11 | 1 | pairs |
| 12 | 1 | individuals or pairs |
| 13 | 1 | individuals or pairs |
| 14 | 1 | pairs |
| 15 | 1 | pairs |
| 16 | 2 | pairs |
| 17 | 1–2 | pairs |
| 18 | 1 | groups |
| 19 | 1 | pairs |
| 20 | 1 | pairs |
| 21 | 2 | individuals |
| 22 | 3–4 | pairs or groups |
| 23 | 3–4 | pairs or groups |
| 24 | 3–4 | pairs or groups |
| 25 | 1 | pairs |
| 26 | 1 | pairs |
| 27 | 1 | pairs or groups |
| 28 | 1 | pairs or groups |
| 29 | 1 | individuals |
| 30 | 2 | groups |
| 31 | 3 | groups |
| 32 | 2 | groups |
| 33 | 1 | pairs or groups |
| 34 | 3 | groups |

# APPENDIX VI

# Reagents and Materials

The following quantities of reagents are sufficient (they allow for waste) for 10 students. Larger classes will need some multiple of these quantities.

## EXPERIMENT 1

10 Sugar unknowns: 5-g samples of monosaccharide or disaccharide (for example, glucose, galactose, mannose, fructose, xylose, arabinose, maltose, lactose, sucrose).

50 ml 1% Glucose: Dissolve $\frac{1}{2}$ g of glucose in 50 ml water. Store at 0–5°C.

20 ml 5% Alcoholic $\alpha$-naphthol: Dissolve $\frac{1}{2}$ g of $\alpha$-naphthol in 20 ml of 95% ethanol. Put in dropper bottle.

60 ml Conc. $H_2SO_4$.

500 ml Nelson's A reagent: Dissolve 12.5 g $Na_2CO_3$ (anhydrous), 12.5 g potassium sodium tartrate, 10 g $NaHCO_3$, and 100 g $Na_2SO_4$ (anhydrous) in 350 ml $H_2O$, and dilute to 500 ml. Save for future experiments.

50 ml Nelson's B reagent: Dissolve 7.5 g $CuSO_4 \cdot 5H_2O$ in 50 ml $H_2O$, and add 1 drop of conc. $H_2SO_4$. Save for future experiments.

500 ml Arsenomolybdate reagent: Dissolve 25 g $(NH_4)_6 \cdot Mo_7O_{24} \cdot 4H_2O$ in 450 ml $H_2O$, and add 21 ml of conc. $H_2SO_4$. Dissolve 3.0 g $Na_2HAsO_4 \cdot 7H_2O$ in 25 ml $H_2O$, and add to above acid molybdate. Store in brown bottle for 24 hr at 37°C. Reagent should be yellow with no green tint. Save for future experiments.

100 ml Glucose standard (100 $\mu$g/ml): Dissolve 100 mg glucose (dextrose) in 1000 ml $H_2O$; store in cold until until ready for use. Save for Experiment 10 also.

## EXPERIMENT 2

500 ml 0.5*M* $NaIO_4$: Dissolve 53.5 g $NaIO_4$ in 500 ml $H_2O$. Discard after use.

1000 ml 1*N* $H_2SO_4$: Mix 28 ml conc. $H_2SO_4$ with 972 ml $H_2O$.

200 ml 20% KI: Dissolve 40 g KI in 200 ml $H_2O$.

2000 ml 0.1*N* $Na_2S_2O_3$: Dissolve 50 g $Na_2S_2O_3 \cdot 5H_2O$ in 2 liters of *previously boiled water*. Add 2 g sodium tetraborate (borax) as a preservative. Standardize by titration with standard $I_2$ using a starch end point (0.1*N* $I_2$: 12 g KI in 10 ml $H_2O$, add 6.5 g $I_2$; dissolve $I_2$, and dilute to 500 ml).

100 ml 1% Soluble starch: Boil approximately 100 ml $H_2O$; add 2 g soluble starch, and stir until dissolved; add 100 ml $H_2O$, and cool.

30 ml Ethylene glycol.

20 ml 1% Phenolphthalein: Dissolve 200 mg phenolphthalein in 10 ml ethanol; add 10 ml $H_2O$, place in dropper bottle; save for future experiments.

500 ml 0.05*N* NaOH: Dissolve 1 g NaOH in 500 ml $H_2O$; titrate against standard acid.

## EXPERIMENT 3

10 ml 1% Glucose: From Experiment 1.

10 ml 1% Solutions of fructose, xylose, arabinose, ribose, galactose, mannose, or others used as unknowns: Dissolve 100 mg of each sugar in 10 ml $H_2O$.

2000 ml Isopropanol:acetic acid:$H_2O$ (3:1:1 v/v). Mix 1200 ml isopropanol, 400 ml glacial acetic acid, 400 ml $H_2O$ (stable for one week).

200 ml Aniline-acid-oxalate reagent: Add 0.9 g oxalic acid to 200 ml $H_2O$; when dissolved, add 1.8 ml aniline. Place part of the reagent in a spray bottle, and store remainder in a brown bottle for future experiments.

400 ml 0.5% $NaIO_4$: Dissolve 2 g $NaIO_4$ in 400 ml $H_2O$. Place part in a spray bottle, store remainder in a brown bottle for future experiments.

400 ml 0.5% Benzidine in ethanol: glacial acetic acid (4:1 v/v). Add 320 ml 95% EtOH to 80 ml glacial HAc, and disperse 2 g of powdered benzidine in solution. Place in spray bottle, and store remainder for future experiments.

1 Measuring stick.

4 Large sheets of Whatman No. 1 paper.

## EXPERIMENT 4

30 ml Dilute HCl: Mix 1 ml conc. HCl with 30 ml $H_2O$.

## EXPERIMENT 5

500 ml Seliwanoff's reagent: Dissolve 0.25 g resorcinol in 500 ml 6*N* HCl (conc. HCl: $H_2O$; 1:1).

200 ml Bial's reagent: Dissolve 0.6 g orcinol in 200 ml conc. HCl; add 8–10 drops 10% $FeCl_3$.

1500 ml Dry methanol.

300 ml *n*-Butyl alcohol.

100 ml Conc. $HNO_3$.

30 g Dry Dowex 50 ($H^+$ form): Wash 30 g Dowex 50 in 6*N* HCl (conc. HCl: $H_2O$; 1:1), and then wash several times with water until elute yields no strong $Cl^-$ test (1% $AgNO_3$); then dry at approximately 100°C in an oven.

1000 ml Ethyl acetate: *n*-propanol: water (5:3:2 v/v): Mix 500 ml ethyl acetate with 300 ml *n*-propanol and 200 ml $H_2O$.

Periodate spray of Experiment 3.

Benzidine spray of Experiment 3.

50 g Each of several monosaccharides (same as unknowns of Experiment 2).

4 Large sheets Whatman No. 1 paper.

1 Microscope.

## EXPERIMENT 6

3 Sets dissecting tools.

5 Rats.

200 ml 10% Trichloroacetic acid: Dissolve 20 g TCA in 200 ml $H_2O$. Store at 0–5°C.

200 ml 5% Trichloroacetic acid: Dissolve 10 g TCA in 200 ml $H_2O$. Store at 0–5°C.

10 g Washed sand.

1000 ml 95% Ethanol.

200 ml Ether.

Ice.

## EXPERIMENT 7

200 ml 2*N* HCl: Mix 33 ml conc. HCl into 167 ml $H_2O$.

200 ml 1.2*N* NaOH: Dissolve 9.6 g NaOH in 200 ml $H_2O$.

500 ml 3,5-Dinitrosalicylate reagent: Dissolve, with warming, 5 g 3,5-dinitrosalicylic acid in 100 ml 2*N* NaOH (8 g NaOH/100 ml). Add 150 g sodium potassium tartrate to 250 ml $H_2O$, and warm to dissolve. Mix the two solutions, and make up to 500 ml with water.

100 ml 0.02*M* sodium phosphate buffer, 0.005*M* NaCl (*p*H 6.9): Dissolve 170 mg $Na_2HPO_4$ (anhydrous), 140 mg $NaH_2PO_4 \cdot H_2O$, and 30 mg NaCl in 100 ml $H_2O$.

## EXPERIMENT 8

300 ml $H_2O$, boiled, $CO_2$-free (protect with soda lime tube).

200 ml 0.35*M* $NaIO_4$: Dissolve 15 g $NaIO_4$ in 200 ml $H_2O$.

50 ml Ethylene glycol.

2000 ml 0.010*N* NaOH: Dissolve approximately 800 mg of NaOH in 2 liters of previously boiled $H_2O$, then titrate with standard acid to 3 decimal places.

Phenolphthalein from Experiment 2.

## EXPERIMENT 9

1 g Phenylmercuric nitrate slurry: Suspend 1 g phenylmercuric nitrate in about 10 ml $H_2O$. Record concentration in mg/ml.

2000 ml 0.8*M* Potassium phosphate buffer: Dissolve 190 g $K_2HPO_4 \cdot 3H_2O$ and 109 g $KH_2PO_4$ in approximately 1800 ml $H_2O$, and adjust to *p*H 6.7 with HCl or KOH before making up to 2 liters.

25 g Celite 535 or similar material.

200 ml 14% $NH_4OH$: Mix 100 ml conc. $NH_4OH$ with 100 ml $H_2O$.

200 ml 2*N* HCl: Mix 32 ml conc. HCl with 170 ml $H_2O$.

200 ml 2*N* NaOH: Dissolve 16 g NaOH in 200 ml $H_2O$.

6 potatoes: Stored 24 hr at 1–5°C and tested for enzymatic activity.

300 g $Mg(Ac)_2 \cdot 4H_2O$.

2000 g Dowex 50 ($H^+$ form): Mix 2000 g of Dowex 50, X-8, with 4000 ml of dilute HCl (300 ml conc. HCl + 3700 ml $H_2O$); stir for 5 min; fill with additional $H_2O$, and decant. Wash by decantation or suspension and vacuum filtration until the eluate no longer yields any titratable acidity (*p*H 3–4); drain until moist.

400 ml IR-45 ($OH^-$ form): Add 1000 ml wet IR-45 to a battery jar. Wash with 2 liters 4% $Na_2CO_3$, and decant. Wash with 2 liters 4% $NH_4OH$, and decant. Wash with 2 liters 4% NaOH, and decant. Rinse with deionized water with stirring and decanting until *p*H of wash water is 9 or below. Store under water.

500 ml Reducing agent: Dissolve 15 g $NaHSO_3$ and 5 g *p*-methylaminophenol (Elon) in 500 ml $H_2O$. Use brown bottle; stable 10 days.

1500 ml Absolute methanol: Use reagent grade stock that has not been opened.

1 Waring blendor.

Cheesecloth.

Glass wool.

5 Glass tubes, approximately 1 inch diam. and 2 feet long.

20 g Charcoal.

1000 ml Acid molybdate reagent: Add 136 ml conc. $H_2SO_4$ to 360 ml $H_2O$ (**Caution**); cool. Dissolve 25 g $(NH_4)_6Mo_7O_{24} \cdot 4H_2O$ in 500 ml $H_2O$, and add to above sulfuric acid solution; make up to 1 liter.

100 g Soluble starch.

1000 ml Phosphate standard: 1 μmole/ml; dissolve 136 mg $KH_2PO_4$ in 1 liter $H_2O$.

2000 ml 5% KOH: Dissolve 100 g KOH in 2000 ml $H_2O$.

50 ml 0.01*N* KI, 0.01*N* $I_2$: Dissolve 90 mg KI in 50 ml $H_2O$. Add 65 mg $I_2$, and mix until dissolved.

## EXPERIMENT 10

Nelson's A reagent from Experiment 1.

Nelson's B reagent from Experiment 1.

Arsenomolybdate reagent from Experiment 1.

100 ml 1*N* HCl: Mix 8.3 ml conc. HCl with 92 ml $H_2O$.

100 ml 1*N* NaOH: Dissolve 4 g NaOH in 100 ml $H_2O$.

40 ml Glucose standard (100 μg/ml): Dissolve 4 mg glucose (dextrose) in 40 ml $H_2O$. Always prepare fresh.

Phosphate standard from Experiment 9.

Reducing reagent from Experiment 9.

200 ml 2.5% Ammonium molybdate: Dissolve 5.0 g $(NH_4)_6Mo_7O_{24} \cdot 4H_2O$ in 200 ml $H_2O$; save for Experiment 16.

100 ml 10*N* $H_2SO_4$; mix 27.8 ml conc. $H_2SO_4$ with 72.2 ml $H_2O$.

50 ml 30% $H_2O_2$, purchased as such.

4 Large sheets Whatman No. 1 paper.

Periodate spray from Experiment 3.

Benzidine spray from Experiment 3.

10 ml 1% Glucose-1-phosphate dipotassium dihydrate: Dissolve 100 mg glucose-1-phosphate potassium salt in 10 ml $H_2O$; store cold.

10 ml 1% Glucose-6-phosphate: Dissolve 100 mg glucose-6-phosphate Na or K salt in 10 ml $H_2O$; store cold.

10 ml 1% Glucose in 1*N* HCl: Dissolve 100 mg glucose in 10 ml 1*N* HCl (conc. HCl:$H_2O$; 1:12); *refrigerate.*

1000 ml Methanol:$NH_4OH$:$H_2O$ (6:1:3 v/v). Mix 600 ml methanol, 100 ml formic acid, and 300 ml $H_2O$.

200 ml Phosphate spray: Mix 9 ml of 70%

(w/w) $HClO_4$, 1.7 ml conc. HCl, and 2 g ammonium molybdate; then add 0.2 g $Na_2$ EDTA, and dilute to 200 ml.
1 U.V. Lamp.
1 Oil or sand bath.

## EXPERIMENT 11

400 g Active dried Baker's yeast.
4 liters 95% ethanol.
300 ml 20% $ZnSO_4$: Dissolve 60 g $ZnSO_4$ in 300 ml $H_2O$.
1000 ml Saturated $Ba(OH)_2$: Saturate 1000 ml of previously boiled $H_2O$ with $Ba(OH)_2$; store in $CO_2$-free system.
50 g Charcoal.
Phenolphthalein from Experiment 2.
100 g Filter aid (such as Celite 535).
100 ml 0.1*M* HCl: Add 0.83 ml of conc. HCl to 99 ml of $H_2O$.

## EXPERIMENT 12

200 g Crab or lobster shells from a fish market; or frozen cracked crab.
2000 ml 10% HCl: Mix 700 ml conc. HCl with 1300 ml $H_2O$.
4 liters Conc. HCl.
100 g Charcoal.
1 g Glucosamine hydrochloride.
500 ml 95% Ethanol.

## EXPERIMENT 13

400 ml Standard HCl: Approximately 5.0*N*; add 166 ml conc. HCl to 234 ml $H_2O$, then standardize against standard alkali.
2000 ml 5.6% Alcoholic KOH: Dissolve 112 g KOH in 2000 ml 95% ethanol (don't store in glass-stoppered bottle).
Phenolphthalein solution from Experiment 2.
2000 ml 95% EtOH.
2000 ml 0.10*N* NaOH: Dissolve 8 g NaOH in 2 liters $H_2O$; standardize with acid (don't store in glass-stoppered bottle). Save for Experiment 16, using Ascarite tube to protect from $CO_2$.
1000 ml Ether.
15 ml Glycerol.
100 ml 2% NaHClO (use Clorox).
10 g Potassium bisulfate.
20 ml 5% $\alpha$-Naphthol in 95% EtOH: Dissolve 1 g $\alpha$-naphthol in 20 ml 95% EtOH; store in brown bottle.
10 10-ml Samples of unknown oil (olive oil, corn oil, cottonseed oil, linseed oil, others). Number the vials, and record the contents for the instructor.
200 ml Conc. HCl.
200 ml Conc. $H_2SO_4$.

## EXPERIMENT 14

75 ml Artificial bile solution: Mix 5 g sodium taurocholate, 25 ml $H_2O$, and 50 ml glycerol; warm on steam bath until clear.
50 ml 2% Detergent: Dissolve 1 g Dreft in 50 ml $H_2O$.
200 ml 0.05*M* $NH_4Cl$ [$NH_4OH$ buffer (*p*H 8)]: Prepare 0.05*M* $NH_4Cl$ (2.7 g $NH_4Cl$/liter) and 0.05M $NH_4OH$ (3.4 ml conc. $NH_4OH$/liter). To 200 ml of $NH_4Cl$ solution slowly add $NH_4OH$ solution, checking with *p*H meter until solution is *p*H 8 (should take approximately 20 ml).
200 ml 0.1*M* $CaCl_2$: Dissolve 2.2 g $CaCl_2$ or 2.9 g $CaCl_2 \cdot 2H_2O$ in 200 ml $H_2O$.
Phenolphthalein solution from Experiment 2.
4000 ml Ethanol:ether (9:1 v/v): Mix 3600 ml 95% ethanol and 400 ml ether. Store in vessel with siphon or spigot.
1000 ml 0.02*N* KOH: Dissolve 1.1 g KOH in 1 liter of $H_2O$; standardize with acid.
100 ml 5*N* $NH_4OH$: Mix 33.8 ml conc. $NH_4OH$ with 66 ml $H_2O$.
10 g "Pancreatin" (Difco Co.).
50 Glass beads.
Oils of Experiment 13.

## EXPERIMENT 15

300 g Calf brain (may be previously frozen).
4000 ml Diethyl ether.
4000 ml Acetone.

300 ml Chloroform: methanol (1:1 v/v): Mix 150 ml chloroform and 150 ml methanol; store in dark or brown bottle.
1 Waring blendor.
500 ml 95% Ethanol.

## EXPERIMENT 16

4 g Cholesterol.
1000 ml Chloroform: Store in dark bottle.
100 ml Conc. $H_2SO_4$.
100 ml Acetic anhydride.
400 ml 0.5% Digitonin in 50% ethanol: Mix 200 ml $H_2O$ with 200 ml ethanol, and dissolve 2 g digitonin therein.
500 ml Acetone: absolute ethanol (1:1 v/v): Mix 250 ml acetone with 250 ml absolute ethanol.
200 ml 95% Ethanol.
200 ml 2*N* HCl: Mix 33.2 ml conc. HCl with 167 ml $H_2O$.
Phenolphthalein from Experiment 2.
0.1*N* NaOH from Experiment 13.
20 ml 10% Glycerol: Mix 2 ml glycerol with 18 ml $H_2O$.
20 ml 30% $H_2O_2$.
100 ml 2.5% Ammonium molybdate from Experiment 11.
100 ml 10*N* $H_2SO_4$: Mix 27.8 ml conc. $H_2SO_4$ with 72 ml $H_2O$.
Phosphate standard from Experiment 9.
Acid molybdate reagent from Experiment 9.
Reducing reagent (1% Elon, 3% $NaHSO_3$): Prepare as in Experiment 9.
100 ml 3% NaHClO (use Clorox).
5% α-Naphthol as in Experiment 13.
1800 ml *n*-Butanol: diethylene glycol: $H_2O$ (4:1:1 v/v). Mix 1200 ml *n*-butanol, 300 ml diethylene glycol, and 300 ml $H_2O$.
200 ml 0.2% Ninhydrin in ethanol: Dissolve 0.2 g ninhydrin in 100 ml 95% ethanol.
50 g Stannous chloride (solid): Label on bottle should read "dissolve 0.4 g in 100 ml 3*N* HCl just before use."
1000 ml 3*N* HCl: Mix 250 ml conc. HCl with 750 ml $H_2O$.
1000 ml 2% Aqueous phosphomolybdic acid: Dissolve 20 g phosphomolybdic acid in 1 liter $H_2O$.
1 Glass tray or pie plate.
20 ml Serine solution: Dissolve 40 mg D,L-serine in 20 ml $H_2O$.
20 ml Ethanolamine solution: Mix 0.05 ml ethanolamine with 20 ml $H_2O$.
20 ml Choline chloride solution: Dissolve 0.2 g choline chloride in 20 ml $H_2O$.
20 ml Sphingosine solution: Dissolve 200 mg sphingosine in 20 ml $H_2O$.
1000 ml Carbohydrate chromatographic solvent: Mix 200 ml pyridine, 300 ml water, and 600 ml *n*-butanol together; separate the upper phase, and mix it with an equal volume of pyridine. Store in a closed vessel in the hood.
Aniline-acid-oxalate spray from Experiment 3.
0.5% $NaIO_4$ spray from Experiment 3.
0.5% Benzidine spray from Experiment 3.
4 Large sheets Whatman No. 1 paper.
10 ml 1% Solutions of glucose, galactose, glucosamine · HCl, and inositol; dissolve 100 mg of each in 10 ml $H_2O$.
1000 ml Methanol.
10 mg TLC standards, cerebroside, sphingomyelin, stearic acid, phosphatidyl ethanolamine, lecithin.
100 ml 0.2% Dichlorofluorescein in ethanol: Dissolve 0.2 g dichlorofluorescein in 100 ml 95% ethanol.
1000 ml Hexane: ethyl ether: acetic acid (90:10:1). Mix 900 ml of hexane with 100 ml of ether and 10 ml of glacial acetic acid.
100 ml Conc. $NH_4OH$.

## EXPERIMENT 17

100 ml D,L-Ethionine in isotonic saline: Dissolve 900 mg NaCl and 1.67 g of D,L-ethionine in 100 ml $H_2O$.
100 ml Isotonic saline: Dissolve 900 mg NaCl in 100 ml $H_2O$.
100 ml Saturated KOH.
200 ml 95% Ethanol.
400 ml Chloroform.
1 pack Cotton.
1 roll Aluminum foil.

20 ml Bromcresol green 0.1%: Dissolve 20 mg bromcresol green in 20 ml $H_2O$.
200 ml 6*N* HCl: Mix 100 ml conc. HCl with 100 ml $H_2O$.
4 Female rats.

## EXPERIMENT 18

Students make the reagents for this experiment; supply the quantities listed in the text of the experiment.

## EXPERIMENT 19

10 g Glycine, leucine, alanine, and other amino acids.
3 *p*H Meters.
500 ml 2*N* $H_2SO_4$: Slowly add 27.8 ml conc. $H_2SO_4$ to 472 ml $H_2O$.
500 ml 2*N* NaOH: Dissolve 40 g NaOH in 500 ml $H_2O$, and dilute to 100 ml.

## EXPERIMENT 20

*Note:* Much class time can be saved in this experiment if the proteins, casein and gluten, are hydrolyzed before class.

500 ml $H_2O$-Saturated phenol: Dissolve 40 ml of $H_2O$ in 460 ml of chromatographic grade liquid phenol. **Caution:** Phenol causes painful burns on the skin. Label bottle with a warning.
5 g Protein (casein and/or gluten).
10 ml Amino acid standards: 0.1% Solutions of glycine, valine, leucine, methionine, threonine, serine, phenylalanine, tryptophan, lysine, arginine, histidine, aspartic acid, alanine, and glutamic acid. Mix 10 mg of each with 10 ml $H_2O$.
200 ml 0.3% $NH_4OH$: Add 2 ml conc. $NH_4OH$ to 200 ml $H_2O$.
50 ml 8*N* $H_2SO_4$. **Caution:** Slowly add 11.1 ml conc. $H_2SO_4$, with stirring, to 38.9 ml water.
200 g $Ba(OH)_2 \cdot 8H_2O$.
100 ml Saturated $Ba(OH)_2$: Saturate 100 ml of previously boiled water with $Ba(OH)_2$.
1000 ml Biuret reagent: Dissolve 1.50 g $CuSO_4 \cdot 5H_2O$ and 6.0 g sodium potassium tartrate (Rochelle salt, $NaKC_4H_4O_6 \cdot 4H_2O$) in about 500 m water in a 1-liter volumetric flask. To this add (with constant swirling) 300 ml of freshly prepared, carbonate-free 10% sodium hydroxide. Make up to 1 liter with distilled water and store the mixture in a polyethylene bottle. Discard if a black or red precipitate appears.
500 ml Ninhydrin solution: Dissolve 400 mg of reagent-grade $SnCl_2 \cdot 2H_2O$ in 250 ml citrate buffer, 0.2*M*, *p*H 5 (4.3 g citric acid + 8.7 g $Na_3$ citrate·$2H_2O$ in 250 ml—adjusted to *p*H 5.0 with NaOH or HCl, *not* $NH_4OH$). Add this solution to 250 ml methyl cellosolve containing 10 g of dissolved ninhydrin. Store in cold.
500 ml Ninhydrin spray: Dissolve 1 g of Ninhydrin in 500 ml *n*-butanol; add 0.5 ml collidine just before use.
1000 ml 50% Aqueous *n*-propanol (v/v): Mix 500 ml *n*-propanol with 500 ml $H_2O$.
100 ml 0.001*M* Glycine standard: Dissolve 7.5 mg glycine in 100 ml $H_2O$.
1 Stapler.
50 ml 16*N* $H_2SO_4$: Slowly stir 25 ml conc. $H_2SO_4$ into 25 ml $H_2O$.
20 ml 1*N* KOH: Dissolve 1.1 g KOH in 20 ml $H_2O$.
500 ml Diethyl ether.
1 Glass tray or pie plate.
1000 ml *n*-Butanol.
500 ml Formic acid.
4 Large sheets of Whatman No. 1 paper.

## EXPERIMENT 21

5 ml Fluorodinitrobenzene solution (FDNB): Add 0.25 ml FDNB (**Caution**) to 4.75 ml of absolute ethanol. Use Pro-pipette for this transfer. *Do not spill!* If an accident occurs, flush skin with ethanol immediately. Do *not* use mouth pipette for FDNB!
1000 ml 0.05M Potassium acid phthalate buffer (*p*H 4.5): Mix 500 ml of 0.1*M* KH phthalate, 455 ml of 0.1*M* NaOH and 45 ml $H_2O$.
6 Large sheets phthalate-buffered paper: Dip Whatman No. 1 or No. 4 paper in 0.05*M* phthalate buffer

(*p*H 4.5), then air- or oven-dry.

1000 ml Tertiary amyl alcohol saturated with phthalate buffer (*p*H 6): Mix 1500 ml of tertiary amyl alcohol with 500 ml of 0.1*M* KH phthalate buffer (*p*H 6) (10 g KH phthalate into 500 ml $H_2O$, then KOH until *p*H 6). After thorough mixing remove aqueous (lower) phase, and discard.

300 ml Ninhydrin spray reagent (see Experiment 20).

500 ml 0.3% $NH_4OH$: Add 5 ml of conc. $NH_4OH$ to 500 ml of $H_2O$.

100 ml 6*N* HCl (50 ml conc. HCl + 50 ml $H_2O$).

100 ml 4.2% $NaHCO_3$ (4.2 g of $NaHCO_3$ in 100 ml $H_2O$).

0.2 ml DNP-Amino acid standards (add 1 mg of DNP-amino acid to 1.5 ml of acetone; shake to dissolve; store in cold): DNP-alanine, DNP-glycine, DNP-valine, DNP-leucine, DNP-serine, DNP-phenylalanine, and so on, depending upon dipeptides chosen. An alternate choice is for the students to make DNP derivatives from amino acids.

10 ml 0.1% Amino acid standards of Experiment 20.

500 ml Peroxide-free diethyl ether.

2 Heat lamps.

24 Plastic planchets or 1 porcelain spot plate.

24 Pasteur disposable pipettes.

500 ml $H_2O$-Saturated phenol (see Experiment 20).

6 Large sheets Whatman No. 1 or No. 4 filter papers.

10 Hydrolysis vials.

500 ml Formic acid.

1000 ml *n*-Butanol.

10 Dipeptide unknowns (10-mg samples).

## EXPERIMENT 22

300 g Standard Brands 20:40 dried yeast.

2000 ml 0.1*M* $NaHCO_3$: Dissolve 16.8 g in $H_2O$, and make up to 2 liters.

2000 ml 0.3*M* Sucrose: Dissolve 205 g sucrose in 1500 ml $H_2O$, and make up to 2 liters.

2000 ml 0.05*M* Sodium acetate buffer (*p*H 4.7): Mix one liter of 0.05*M* acetic acid (2.88 ml glacial acetic acid per liter) with one liter of 0.05*M* sodium acetate (6.8 g $NaOAc \cdot 3H_2O$ per liter), and check *p*H. Adjust if necessary with glacial acetic acid or NaOH.

4000 ml Reducing sugar reagent: 3,5-dinitrosalicylate reagent of Experiment 7.

2000 ml 0.005*M* Glucose, 0.005*M* fructose standard: Dissolve 1.80 g glucose and 1.80 g fructose in 2 liters $H_2O$.

500 ml Picrate solution: Dissolve 15 g picric acid in 65 ml 1*M* NaOH; add 20 ml 0.05*M* acetate buffer (*p*H 4.7), and dilute to 500 ml. Check *p*H and adjust to 4.7 if necessary with HCl or NaOH.

1 Constant-temperature water bath.

1 High-speed centrifuge (Servall SS-1 or RC-2).

1 Package of cotton.

1 Dialysis setup with tubing.

1000 ml 0.1% $Na_2$-EDTA: Dissolve 1 g $Na_2$-EDTA in 1 liter $H_2O$.

4000 ml Acetone (reagent grade), cooled to −15°C.

2 lbs NaCl (for cooling baths).

1000 ml Biuret reagent: Dissolve 1.50 g $CuSO_4 \cdot 5H_2O$ and 6.0 g sodium potassium tartrate ($NaKC_4H_4O_6 \cdot 4H_2O$) in about 500 ml $H_2O$. Add 300 ml of carbonate-free, freshly prepared 10% NaOH. Make up to one liter, and store in a polyethylene bottle. Discard if black or red precipitate forms.

1000 ml Protein standard (10 mg/ml): Use Armour crystallized bovine plasma albumin.

100 ml $1.5 \times 10^{-4}M$ *p*-Hydroxy mercuribenzoate (PMB): Dissolve 5.4 mg PMB in 100 ml of $2 \times 10^{-5}M$ glycylglycine (*p*H 8.0).

500 ml 0.2*M* Sodium citrate (0.01*M* $Na_2$-EDTA buffers of *p*H values 2.5, 3.5, 4.5, 5.5, 6.5, and 7.5): Prepare by appropriate mixing of 0.1*M* citric acid and 0.1*M* sodium citrate, each solution 0.01*M* with respect to $Na_2$-EDTA.

## EXPERIMENT 23

|  |  |
|---|---|
| | Components of "AC" and lactose growth media (reasonable quantities). See *Experiment*. |
| Culture | *E. coli*, strain $K_{12}$. |
| 4000 ml | 0.08*M* Sodium phosphate buffer (*p*H 7.7). Dissolve 51.0 g $Na_2HPO_4$ (anhydrous) and 4.8 g $NaH_2PO_4 \cdot H_2O$ in 4 liters $H_2O$ before adjusting *p*H with HCl or NaOH. |
| 1000 ml | 0.0025*M* *o*-Nitrophenyl-$\beta$-galactoside (ONPG): Dissolve 0.80 g ONPG in 1 liter $H_2O$. |
| 1000 ml | 1*M* $Na_2CO_3$: Dissolve 106 g $Na_2CO_3$ in 1 liter $H_2O$. |
| 500 ml | Phenol reagent-2N (Folin-Ciocalteau). Fisher Scientific Co. |
| 1000 ml | Protein standard (1 mg/ml): Use Armour crystallized bovine plasma albumin. |
| 1000 ml | 2% $Na_2CO_3$ in 0.1*M* NaOH. |
| 200 ml | 0.5% $CuSO_4 \cdot 5H_2O$ in 1% sodium tartrate: Dissolve 1.0 g $CuSO_4 \cdot 5H_2O$ in 100 ml $H_2O$. *Separately* dissolve 2.0 g sodium tartrate in 100 ml $H_2O$. Mix just before using. Prepare fresh every 2 days. |
| 1 | Autoclave. |
| 1 | Shaker. |
| 1 | Fermentation tank and carboys. |
| 1 | Sonic oscillator or 500 g alumina (1557 AB levigated; Buehler Ltd., Evanston, Ill.). |
| 1 | Constant-temperature bath. |
| 1 | Continuous centrifuge. |
| 1 | High-speed centrifuge. |
| 100 g | $P_2O_5$. |

## EXPERIMENT 24

|  |  |
|---|---|
| 50 ml | 0.003*M* $DPN^+$: Dissolve 110 mg $DPN^+ \cdot 4H_2O$ in 40 ml $H_2O$. Titrate to *p*H 7–8 with 0.1*M* NaOH. Make up to 50 ml. |
| 1000 ml | 0.03*M* Sodium pyrophosphate (*p*H 8.4): Dissolve 8 g $Na_2P_2O_7$ in 800 ml $H_2O$. Adjust to *p*H 8.4 with 1*M* HCl. Make up to 1 liter. |
| 100 ml | 0.4*M* $Na_2HAsO_4$: Dissolve 12.48 g $Na_2HAsO_4 \cdot 7H_2O$ in 100 ml $H_2O$. |
| 100 ml | 0.4*M* Sodium phosphate (*p*H 8.3): Dissolve 1.9 g $NaH_2PO_4 \cdot H_2O$ and 7.2 g $Na_2HPO_4 \cdot 7H_2O$ in 90 ml $H_2O$. Adjust *p*H to 8.3 with 6*M* NaOH, and make up to 100 ml. |
| 100 ml | 0.003*M* Iodoacetic acid: Dissolve 55 mg in 100 ml $H_2O$. *Not stable!* Renew every two days. |
| 2000 ml | 0.001*M* $Na_2$-EDTA (*p*H 7.5): Dissolve 750 mg $Na_2$-EDTA (Versene-$Na_2$) in 1900 ml $H_2O$. Adjust *p*H to 7.5 by careful addition of 1*M* NaOH, and make up to 2 liters. |
| 100 ml | 0.06*M* Fructose-1,6-diphosphate (disodium) in 0.03*M* pyrophosphate buffer (*p*H 8.4). Dissolve 2.30 g $Na_2$-HDP in 90 ml 0.03*M* sodium pyrophosphate buffer (*p*H 8.4). Titrate to *p*H 8–9 with a few drops of 1*M* NaOH. Make up to 100 ml with buffer. |
| 500 ml | Phenol reagent-2N (Folin-Ciocalteau): See Experiment 24. |
| 1000 ml | Protein standard (1 mg/ml): See Experiment 23. |
| 1000 ml | 2% $Na_2CO_3$ in 0.1*M* NaOH: See Experiment 23. |
| 200 ml | 0.5% $CuSO_4 \cdot 5H_2O$ in 1% sodium tartrate: See Experiment 23. |
| 100 ml | 7.5*M* $NH_4OH$: Mix 50 ml conc. $NH_4OH$ with 50 ml $H_2O$. |
| 3000 ml | 0.03*M* KOH: Dissolve 5.6 g KOH in 3 liters $H_2O$. |
| 4000 ml | Saturated $(NH_4)_2SO_4$ (*p*H 7.5). This requires about 800 g/liter. Allow time for solution. Make sure that solid $(NH_4)_2SO_4$ is present. Adjust to *p*H 7.6 with conc. $NH_4OH$ (dilute 1:5 for measuring *p*H with a meter). |
| 1300 ml | Saturated $(NH_4)_2SO_4$ in 0.001*M* $Na_2$-EDTA (*p*H 8.4). Dissolve 750 g $(NH_4)_2SO_4$ in 1 liter 0.001*M* $Na_2$-EDTA. Titrate with conc. $NH_4OH$ to *p*H 8.4 (measured with a 1:5 dilution). |
| 50 ml | 5% Nembutal and syringes. |
| 25 ml | Aldolase (1 mg/ml). |
| 1 | Meat grinder. |
| 4 sets | Dissecting tools. |
| 4 | Rabbits. |
| 20 | Whatman No. 1 filter papers (24-cm diam.). |
| 1 g | Cysteine (free base). |
| 4 $yd^2$ | Cheesecloth. |

## EXPERIMENT 25

- 300 ml 10% NaCl: Dissolve 30 g NaCl in approximately 250 ml $H_2O$, and dilute to 300 ml.
- 300 ml 30% Trichloroacetic acid: Dissolve 90 g TCA in 250 ml $H_2O$, and dilute to 300 ml. Store in cold.
- 500 ml Acetone.
- 500 ml Ether.
- 1000 ml Ethanol (absolute).
- 1 Large rat.
- 500 ml 20% Potassium acetate (*p*H 5): Dissolve 100 g potassium acetate in about 400 ml $H_2O$; adjust to *p*H 5 with glacial acetic acid; and dilute to 500 ml.
- 500 ml 90% Phenol ($H_2O$). *Use care!*
- 300 g Frozen liver (beef, pork, or 24-hr fasted rat).
- 4 $yd^2$ Cheesecloth.
- 1 Clinical centrifuge.
- 3 sets Dissecting tools.
- 1 Blendor (Waring type).
- 1 roll Aluminum foil.

## EXPERIMENT 26

- 100 ml 0.5*N* KOH: Dissolve 2.8 g KOH in 100 ml $H_2O$.
- 100 ml 1.0*N* KOH: Dissolve 5.6 g KOH in 100 ml $H_2O$.
- 200 ml 20% Perchloric acid: Dilute 67 ml of 60% $HClO_4$ with 133 ml $H_2O$.
- 200 ml 2*N* HCl: Dilute 33 ml conc. HCl to 200 ml with $H_2O$.
- 2000 ml 0.05*M* Ammonium formate (*p*H 3.5): Mix 3.75 ml of pure formic acid in 1900 ml $H_2O$; adjust *p*H to 3.5 with conc. $NH_4OH$; dilute to 2 liters.
- 1 Paper electrophoresis apparatus.
- 1 Ultraviolet lamp (such as Mineralite).
- 100 ml 1% Starch solution: Dissolve 1 g soluble starch in 50 ml boiling water before cooling with an additional 50 ml $H_2O$.
- 100 ml 0.01*N* $I_2$, 0.01*M* KI: Dissolve 170 mg KI in 100 ml $H_2O$, then add 250 mg $I_2$, and stir until dissolved.
- 5 ml 0.02*M* nucleotides: Dissolve 40 mg of each of the four RNA nucleotides (adenosine-2′,3′-monophosphate, uridine 2′,3′-monophosphate, and so on) in separate 5-ml samples.
- 400 ml Orcinol acid reagent: Add 2 ml of 10% $FeCl_3 \cdot 6H_2O$ to 400 ml of conc. HCl.
- 100 ml $1 \times 10^{-4}$ Nucleotide standard: Dissolve 1 mg of each RNA nucleotide in 100 ml $H_2O$.
- 100 ml 6% Orcinol in 95% ethanol.

## EXPERIMENT 27

- 50 g Frozen *E. coli* cells (General Biochemical Industries, Chagrin Falls, Ohio; or Corn Products, Muscatine, Iowa).
- 200 ml 0.01*M* Sodium citrate: Dissolve 588 mg $Na_3$ citrate$\cdot 2H_2O$ in 200 ml $H_2O$.
- 600 ml 15% Duponol C, 0.14*M* NaCl, 0.01*M* $Na_3$ citrate: Dissolve 1.76 g $Na_3$ citrate$\cdot 2H_2O$, 90 g sodium lauryl sulfate, and 4.92 g NaCl in 600 ml $H_2O$.
- 1000 ml Absolute ethanol.
- 300 ml 1.4*M* NaCl: Dissolve 24.6 g NaCl in 300 ml $H_2O$.
- 300 ml 0.14*M* NaCl-0.015*M* Sodium citrate (*p*H 7.1): Add 2.457 g NaCl and 864 mg citric acid to approximately 200 ml $H_2O$, and titrate to *p*H 7.1 with 1*N* NaOH (4 g NaOH/100 ml, approximately 10 ml) before making up to 300 ml.
- 150 ml 5% Duponol C in 45% aqueous ethanol: Mix 82.5 ml absolute ethanol with 67.5 ml $H_2O$. Then dissolve 7.5 g Duponol C (sodium lauryl sulfate) therein.
- 100 g NaCl (solid).
- 250 ml 75% ethanol: Dilute 200 ml of 95% ethanol to 250 ml with $H_2O$.
- 1 Potter-Elvehjem homogenizer, teflon tipped.
- 1 Stirring motor for homogenizer.
- 1 High-speed centrifuge (e.g., Servall SS-1).

## EXPERIMENT 28

- 500 ml Diphenylamine reagent (freshly prepared): Mix 8 g diphenylamine, 800 ml glacial acetic acid, and 22 ml conc. $H_2SO_4$. Store at 2°C, and warm to room temperature just before use.
- 20 ml Deoxyribonucleic acid (DNA)

standard: Suspend 20 mg DNA in 20 ml 10% trichloroacetic acid, heat at 90°C for 15 min; cool, and place in bottle marked 1 mg DNA/ml.

100 ml  $5 \times 10^{-4}M$ Methylene blue in $0.1M$ $MgSO_4$: Dissolve 2.5 g $MgSO_4 \cdot 7H_2O$ in 100 ml $H_2O$ before adding 20 mg methylene blue.

200 ml  $1.0M$ Sodium acetate buffer: Mix 11.7 ml glacial acetic acid and 50 ml $H_2O$; cool, titrate to *p*H 5.5 with $4N$ NaOH (16 g NaOH/100 ml), and dilute to 200 ml.

5 mg  Deoxyribonuclease (DNAse), crystalline (Worthington or similar).

100 ml  $0.01M$ $MgSO_4$ in $0.5M$ sodium acetate (*p*H 5.5): Dissolve 0.25 g $MgSO_4 \cdot 7H_2O$ in 100 ml $0.5M$ sodium acetate (*p*H 5.5).

200 ml  10% Trichloroacetic acid: Dissolve 20 g TCA in 200 ml $H_2O$.

## EXPERIMENT 29

1  Warburg bath.

10  Warburg flasks and manometers: Assign one student to each flask, and allow him to keep flask in desk.

100 ml  Ferricyanide reagent, *freshly prepared:* Dissolve 823 mg reagent-grade potassium ferricyanide in 100 ml of glass-distilled $H_2O$; store in a brown bottle with a loosely fitting glass stopper.

100 ml  Hydrazine solution: Dissolve 0.25 ml 95% hydrazine sulfate in 50 ml of glass-distilled water.

100 ml  Manometer fluid: Dissolve 4.4 g anhydrous NaBr, 30 mg Triton X-100 (Rohm and Hass Co.), and 30 mg Evan's blue in 100 ml $H_2O$.

5 g  Anhydrous lanolin.

1 tube  Stopcock grease.

1 box  Strong detergent (Alconox).

200 ml  $CHCl_3$.

1 box  Pipe cleaners

## EXPERIMENT 30

200 ml  5.0% Zinc sulfate: Dissolve 10 g $ZnSO_4$ in 200 ml $H_2O$.

500 ml  Nelson's A reagent from Experiment 1.

50 ml  Nelson's B reagent from Experiment 1.

500 ml  Arsenomolybdate reagent from Experiment 1.

100 ml  Glucose standard (100 $\mu$g/ml): Dissolve 10 mg glucose in 100 ml $H_2O$; refrigerate until used. Renew each day.

1000 ml  $0.1M$ $KHCO_3$: Dissolve 10 g $KHCO_3$ in 1 liter $H_2O$.

200 g  Fleischmann's 20:40 dried yeast.

100 ml  $0.04M$ Potassium phosphate buffer (*p*H 6.0): Dissolve 544 mg $KH_2PO_4$ in 90 ml $H_2O$. Titrate to *p*H 6.0 by dropwise addition of $1M$ KOH. Dilute to 100 ml. Alternatively, dilute extra $0.08M$ buffer (below) with an equal volume of $H_2O$.

200 ml  10% Trichloroacetic acid: Dissolve 20 g TCA in 200 ml $H_2O$.

100 ml  Acid molybdate reagent from Experiment 9.

100 ml  Reducing reagent: 3% $NaHSO_3$:1% Elon (freshly prepared): Dissolve 3 g $NaHSO_3$ and 1 g *p*-methylaminophenol in 100 ml $H_2O$.

100 ml  Phosphate standard from Experiment 9 (1 $\mu$mole/ml).

100 ml  Ether.

100 ml  $0.08M$ Potassium phosphate buffer (*p*H 6.0): Dissolve 1.09 g $KH_2PO_4$ in 90 ml $H_2O$, and titrate to *p*H 6.0 by dropwise addition of $1M$ KOH. Dilute to 100 ml.

20 ml  $0.2M$ Fructose-1,6-diphosphate, potassium salt-$0.2M$ $MgCl_2$: Dissolve 2.18 g $K_2$-HDP salt and 0.82 g $MgCl_2 \cdot 6H_2O$ in 20 ml $H_2O$. Adjust *p*H to 5–7 with $1M$ HCl or $1M$ KOH. Alternatively, make up 20 ml of HDP, Mg salt, *p*H 5–7.

100 ml  $1M$ Glucose: Dissolve 18 g glucose in 100 ml $H_2O$.

10 ml  $0.05M$ Iodoacetate: Suspend 93 mg iodoacetic acid in 8 ml $H_2O$. Titrate to *p*H 6.0 with $1M$ KOH, and dilute to 10 ml. Renew every other day.

100 ml  $0.2M$ Glucose: Dissolve 3.6 g glucose in 100 ml $H_2O$. Renew every 3 days.

5 ml  $0.1M$ ADP: Dissolve 0.258 g $K_2$-ADP $\cdot$ $2H_2O$ in 4 ml $H_2O$. Adjust *p*H to 6.0 with $2M$ KOH, and dilute to 5 ml.

10 ml  $0.2M$ KF: Dissolve 188 mg KF$\cdot$ $2H_2O$ in 10 ml $H_2O$.

5 ml 0.2$M$ 3-phosphoglycerate K salt: Dissolve or suspend 186 mg 3-phosphoglyceric acid in 4 ml $H_2O$. Titrate to $p$H 6.0 with 1$M$ KOH, and dilute to 5 ml.

10 ml 0.1$M$ $KH_2AsO_4$: Dissolve 180 mg $KH_2AsO_4$ in 10 ml $H_2O$.

200 ml 0.3$M$ $Ba(OH)_2$: Dissolve 5.1 g $Ba(OH)_2$ in 200 ml of previously boiled $H_2O$. Protect from $CO_2$ with a soda lime tube.

1 Clinical centrifuge.

1 Blendor (Waring type).

1 High-speed centrifuge (e.g., Servall SS-1).

1 Warburg apparatus with manometer and flasks.

## EXPERIMENT 31

1 Meat grinder.

1 *Fresh* pig heart.

4000 ml 0.001$M$ Ethylenediaminetetracetate (EDTA) ($p$H 7.4): Dissolve 1.488 g disodium EDTA in 4000 ml $H_2O$, and titrate to $p$H 7.4 with NaOH (use 1 to 2 liters of this for other reagents).

1000 ml 0.02$M$ Potassium phosphate, 0.001$M$ EDTA ($p$H 7.4): Dissolve 2.79 g $K_2HPO_4$ and 0.54 g $KH_2PO_4$ in 1 liter of the above 0.001$M$ EDTA.

100 ml 1$M$ Acetic acid: Dilute 5.9 ml glacial acetic acid with 94 ml $H_2O$.

200 ml 0.25$M$ Potassium phosphate, 0.001$M$ EDTA ($p$H 7.4): Dissolve 7.0 g $K_2HPO_4$ and 1.34 g $KH_2PO_4$ in 200 ml of the above 0.001$M$ EDTA, and adjust to $p$H 7.4 if necessary.

5 g Sodium hydrosulfite ($Na_2S_2O_4$).

6 ml 0.001$M$ $DPN^+$: Dilute 1 ml of 0.003$M$ $DPN^+$ from Experiment 24 with 2 ml $H_2O$.

1000 ml 0.3$M$ Potassium phosphate ($p$H 7.4): Dissolve 42 g $K_2HPO_4$ and 8.1 g $KH_2PO_4$ in 1000 ml $H_2O$; adjust $p$H if necessary with M KOH or HCl.

10 ml 0.25$M$ Malate (K or Na salt): Dissolve 335 mg malic acid and 100 mg NaOH in approximately 8 ml $H_2O$; adjust $p$H to 6–8 with 1$N$ NaOH, and dilute to 10 ml with $H_2O$.

50 ml 0.5$M$ Sodium succinate: Dissolve 6.78 g sodium succinate in 50 ml $H_2O$.

10 ml 0.05$M$ Sodium succinate: Dilute 0.5$M$ succinate (1 ml + 9 ml $H_2O$).

10 ml 0.25$M$ β-Hydroxybutyrate: Dissolve 260 mg β-Hydroxybutyric acid in 5 ml $H_2O$ by gradual addition of 2.5 ml of 1$N$ NaOH (final $p$H should be in range 6–8); make up to 10 ml. Or use sodium salt (315 mg), and adjust $p$H with HCl.

10 ml $6 \times 10^{-4}M$ Cytochrome $c$: Quantities used depend upon purity of starting material. Molecular weight is 13,000.

5 g Ascorbic acid.

10 ml 1$N$ NaOH: Dissolve 400 mg NaOH in 10 ml $H_2O$.

10 ml 0.5$M$ Malonate: Suspend 520 mg malonic acid in 6 ml $H_2O$; adjust to $p$H 6–8 with 4$N$ NaOH (16 g NaOH/100 ml), and dilute to 10 ml.

10 ml 0.1$M$ KCN: Dissolve 65 mg KCN in 10 ml $H_2O$. Keep in tightly stoppered, clean bottle marked *poison*. Provide a propittete for dispersing.

50 ml 0.01$M$ Methylene blue: Dissolve 187 mg methylene blue in 50 ml $H_2O$.

10 ml Antimycin A (180 μg/ml 95% EtOH): Dissolve 1.8 mg Antimycin A in 10 ml 95% EtOH.

50 ml DPIP solution: Dissolve 13.5 mg 2,6-dichlorophenolindophenol (DPIP) solution in 50 ml $H_2O$.

10 ml 6$M$ KOH: Dissolve 3.4 g KOH in 10 ml $H_2O$.

1 Warburg apparatus with manometers and flasks.

1 Hand spectroscope.

1 Blendor (Waring type).

1 High-speed centrifuge (e.g., Servall SS-1).

10 ml 95% Ethanol.

## EXPERIMENT 32

*Reagents Common to Both Parts*

100 ml 6$N$ KOH: Dissolve 34 g KOH in 100 ml of boiled water. Keep solution away from air.

2 ml Hexokinose (Crude, type II, Sigma Chemical Corp.): Dissolve 20 mg enzyme in 2 ml 0.1% glucose (100

mg glucose/100 ml $H_2O$). Store in refrigerator. Prepare fresh solution each day!

200 ml 5% Trichloroacetic acid: Dissolve 10 g TCA in 200 ml $H_2O$.

200 ml 1% Elon-3% $NaHSO_3$: Dissolve 2 g Elon (*p*-methylaminophenol) and 6 g $NaHSO_3$ in 200 ml $H_2O$. *Must be fresh.*

200 ml Acid molybdate reagent from Experiment 9.

100 ml Phosphate standard (1 μmole/ml) from Experiment 9.

10 ml 0.5*M* Pyruvate (Na or K salt): Dissolve 550 mg sodium pyruvate or 715 mg K pyruvate in 10 ml $H_2O$.

1 Warburg apparatus with manometers and flasks.

1 High-speed centrifuge (e.g., Servall SS-1).

*Additional Reagents for Part A*

1000 ml 0.25*M* Sucrose in 0.001*M* EDTA (*p*H 7.4): Dissolve 85.5 g sucrose in approximately 800 ml of 0.001*M* $Na_2$-EDTA (340 mg $Na_2$-EDTA/liter): titrate to *p*H 7.4 with 0.1*N* NaOH, and make up to 1000 ml.

10 ml 0.2*M* Sodium hydrogen glutamate: Dissolve 370 mg monosodium glutamate in 10 ml $H_2O$.

100 ml 0.005*M* 2,4-Dinitrophenol (DNP): Dissolve 92 mg DNP in 100 ml $H_2O$.

10 ml Mixture containing, in μmoles per ml, 3 ATP; 15 $MgCl_2$, 150 glucose, 0.3 $DPN^+$; 3 EDTA: Dissolve 18.7 mg ATP-$Na_2$-$2H_2O$, 30.5 mg $MgCl_2 \cdot 6H_2O$, 270 mg glucose; 2 mg $DPN^+$, and 10 mg $Na_2$-EDTA in approximately 8.0 ml $H_2O$; titrate to *p*H 6–8 with 0.1*M* NaOH (400 mg NaOH/100 ml), and make up to 10 ml with $H_2O$.

200 ml 2.5*N* NaOH: Dissolve 20 g NaOH in 150 ml $H_2O$ before making up to 200 ml with $H_2O$.

100 ml 0.1*M* Potassium phosphate buffer (*p*H 7.4): Mix 20 ml 0.1*M* $KH_2PO_4$ (1.36 g $KH_2PO_4$/100 ml) and 80 ml 0.1*M* $K_2HPO_4$ (1.74 g $K_2HPO_4$/100 ml); check *p*H, and adjust to *p*H 7.4 with additional $K_2HPO_4$ or $KH_2PO_4$ solution.

100 ml Pyruvate standard (0.5 μmoles/ml): Dissolve 5.5 mg sodium pyruvate in 100 ml $H_2O$.

1 Rat (80–150 g).

1 Potter-Elvehjem homogenizer, teflon tipped, mounted on a drive motor.

1 set Dissecting tools.

*Additional Reagents for Part B*

2000 ml 0.25*M* Sucrose containing 1.85 g $K_2HPO_4$/liter: Dissolve 3.7 g $K_2HPO_4$ and 171 g sucrose in 1500 ml $H_2O$ before making up to 2 liters.

2000 ml 0.25*M* Sucrose: Dissolve 171 g sucrose in 1500 ml before making up to 2 liters.

100 ml 1*M* Sucrose: Dissolve 34.2 g sucrose in 70 ml $H_2O$ before making up to 100 ml.

100 ml 1*M* Potassium phosphate buffer (*p*H 7.2): Mix 70 ml 1*M* $K_2HPO_4$ (17.4 g $K_2HPO_4$/100 ml) with 30 ml 1*M* $KH_2PO_4$ (1.36 g $KH_2PO_4$/100 ml), and adjust *p*H to 7.2 with further additions of 1*M* $K_2HPO_4$ or 1*M* $KH_2PO_4$.

5.0 ml 0.05*M* ATP *p*H 7.0: Dissolve 156 mg $Na_2$-$ATP \cdot 2H_2O$ in 4 ml $H_2O$, and adjust to *p*H 6.5–7.5 with 0.1*M* NaOH (400 mg NaOH/100 ml. Dilute to volume.

10 ml 0.1*M* Magnesium sulfate: Dissolve 246 mg $MgSO_4 \cdot 7H_2O$ in 10 ml $H_2O$.

10 ml 1*M* Glucose: Dissolve 1.8 g glucose in 8 ml $H_2O$, then make up to 10 ml. Prepare fresh.

10 ml 0.5*M* Sodium malate: Suspend 660 mg malic acid in 5 ml $H_2O$; add sufficient 2*N* NaOH (8 g/100 ml; about 2.5 ml needed) to raise *p*H to 6–8, filling to 10 ml.

100 ml 0.0001*M* 2,4-Dinitrophenol (DNP): Dissolve 1.84 mg DNP in 100 ml $H_2O$.

10 ml 0.1*M* Potassium cyanide: Dissolve 65 mg KCN in 10 ml $H_2O$. Mark as *poison*.

10 ml 0.5*M* α-Ketoglutarate: Dissolve 950 mg sodium α-ketoglutarate in 5 ml $H_2O$; titrate to *p*H 6–8 with 2*N* NaOH or KOH, and make up to 10 ml.

10 ml 0.25$M$ sodium $\beta$-hydroxybutyrate: See Experiment 31.
10 ml 0.5$M$ succinate: See Experiment 31.
100 ml 50% Trichloroacetic acid: Dissolve 50 g of TCA in a minimum of $H_2O$ before making up to 100 ml.
1000 ml Biuret reagent of Experiment 20.
100 ml Protein standard (10 mg/ml) of Experiment 20.

## EXPERIMENT 33

20 ml 0.1$M$ ATP-K salt: Dissolve 1.32 g $ATP \cdot K_2 \cdot 4H_2O$ in 15 ml $H_2O$. Adjust $p$H to 6.5–7.5 with 1$N$ KOH. Store in refrigerator.
100 ml 0.1$M$ Tris-chloride ($p$H 7.0): Dissolve 1.21 g tris(hydroxymethyl)-aminomethane in 90 ml $H_2O$. Adjust $p$H to 7.0 with 2$M$ HCl, and dilute to 100 ml.
100 ml 2$M$ $FeCl_3$: Dissolve 54 g $FeCl_3 \cdot 6H_2O$ in 100 ml $H_2O$. Protect from light!
100 ml 0.1$M$ HCl: Add 8.3 ml of conc. HCl to 92 ml $H_2O$.
400 ml 0.05$M$ KCl: Dissolve 1.5 g KCl in 400 ml $H_2O$.
200 ml 3.0$M$ $NH_2OH \cdot HCl$ ($p$H 7): Dissolve 42 g $NH_2OH \cdot HCl$ in 120 ml $H_2O$; adjust to $p$H 7 with saturated KOH, and make up to 200 ml with $H_2O$.
30 ml 0.1$M$ $MgCl_2$: Dissolve 600 mg $MgCl_2 \cdot 6H_2O$ in 30 ml $H_2O$.
10 ml 1.0$M$ Potassium acetate: Dissolve 1 g in 10 ml $H_2O$.
200 ml 100% TCA: Dissolve 200 g of trichloroacetic acid in a minimum of $H_2O$, titrate to $p$H 0.9 with saturated NaOH (**Caution:** Heat is evolved), and make up to 200 ml with $H_2O$.
20 ml 0.01$M$ Hydroxamate standard: Dissolve 15.0 mg acetylhydroxamic acid or 40 mg tyrosine hydroxamic acid in 20 ml $H_2O$. Prepare fresh.
10 ml 0.02$M$ Mixture of 15 L-amino acids. Dissolve 0.02 millimole of each of the following amino acids in 10 ml 0.1$M$ tris-Cl buffer ($p$H 7): alanine, serine, threonine, methionine, leucine, isoleucine, valine, tryptophan, tyrosine, phenylalanine, cysteine, glycine, lysine, histidine, and arginine.
1 Rat.
1 Potter-Elvehjem homogenizer, teflon tipped, mounted on a drive motor.
1 set Dissecting tools.

## EXPERIMENT 34

2 Rats: 100–150 g.
100 ml Absolute ethanol.
2.0 ml Radioactive acetate solution: 1.5 millimoles glucose (270 mg) plus 100 $\mu$moles sodium acetate containing 10$\mu$curies sodium-1-$C^{14}$ acetate or sodium-2-$C^{14}$ acetate in 2 ml solution. (This quantity is required for each injected animal.)
2000 ml 1$N$ NaOH: Dissolve 80 g NaOH in 2 liters of water. Store under ascarite tube to limit $CO_2$ absorption.
300 ml 10% Trichloroacetic acid: Dissolve 30 g TCA in 300 ml $H_2O$.
300 ml 95% Ethanol.
300 ml Chloroform:methanol (1:1 v/v). Mix 150 ml chloroform with 150 ml methanol; store in a dark bottle.
200 ml 15% Ethanolic KOH: Dissolve 30 g KOH in 200 ml 95% ethanol.
400 ml Petroleum ether (30–60°C).
40 g Anhydrous $NaSO_4$.
100 ml Acetone:ethanol (1:1 v/v). Mix 50 ml acetone with 50 ml absolute ethanol.
100 ml 0.5% Digitonin in 50% ethanol: Mix 50 ml of absolute ethanol and 50 ml $H_2O$ before dissolving 500 mg digitonin therein.
100 ml Acetone.
100 ml Acetone:ether (1:1 v/v). Mix 50 ml acetone with 50 ml ether.
100 ml Ether.
200 ml 3$N$ $H_2SO_4$: Dilute 16.8 ml conc. $H_2SO_4$ to 200 ml with water. *Use care!*
20 ml 0.1% Bromcresol green: Dissolve 20 mg bromcresol green (Na salt) in 20 ml $H_2O$.
100 ml 1$M$ $BaCl_2$: Dissolve 24.4 g $BaCl_2 \cdot 2H_2O$ in 100 ml of previously boiled $H_2O$. Store under ascarite tube.
2 sets Dissecting tools.
1 Hypodermic syringe.
2 Respiratory apparatus.
1 Clinical centrifuge.
10 ml 10% Triton WR 1339.

# Index